Cellular Metals

Cellular Metals: Fabrication, Properties and Applications

Editors

Isabel Duarte
Matej Vesenjak
Thomas Fiedler
Lovre Krstulović-Opara

MDPI • Basel • Beijing • Wuhan • Barcelona • Belgrade • Manchester • Tokyo • Cluj • Tianjin

Editors
Isabel Duarte
University of Aveiro
Portugal

Matej Vesenjak
University of Maribor
Slovenia

Thomas Fiedler
The University of Newcastle
Australia

Lovre Krstulović-Opara
University of Split
Croatia

Editorial Office
MDPI
St. Alban-Anlage 66
4052 Basel, Switzerland

This is a reprint of articles from the Special Issue published online in the open access journal *Metals* (ISSN 2075-4701) (available at: https://www.mdpi.com/journal/metals/special_issues/cellular_metals).

For citation purposes, cite each article independently as indicated on the article page online and as indicated below:

LastName, A.A.; LastName, B.B.; LastName, C.C. Article Title. *Journal Name* **Year**, *Volume Number*, Page Range.

ISBN 978-3-0365-1038-5 (Hbk)
ISBN 978-3-0365-1039-2 (PDF)

Contents

About the Editors . . . . **vii**

Isabel Duarte, Thomas Fiedler, Lovre Krstulović-Opara and Matej Vesenjak
Cellular Metals: Fabrication, Properties and Applications
Reprinted from: *Metals* **2020**, *10*, 1545, doi:10.3390/met10111545 . . . . **1**

Francisco García-Moreno, Laurenz Alexander Radtke, Tillmann Robert Neu, Paul Hans Kamm, Manuela Klaus, Christian Matthias Schlepütz and John Banhart
The Influence of Alloy Composition and Liquid Phase on Foaming of Al–Si–Mg Alloys
Reprinted from: *Metals* **2020**, *10*, 189, doi:10.3390/met10020189 . . . . **5**

Dirk Lehmhus, Daniela Hünert, Ulrike Mosler, Ulrich Martin and Jörg Weise
Effects of Eutectic Modification and Grain Refinement on Microstructure and Properties of PM AlSi7 Metallic Foams
Reprinted from: *Metals* **2019**, *9*, 1241, doi:10.3390/met9121241 . . . . **23**

Isabel Duarte, Matej Vesenjak and Manuel J. Vide
Automated Continuous Production Line of Parts Made of Metallic Foams
Reprinted from: *Metals* **2019**, *9*, 531, doi:10.3390/met9050531 . . . . **57**

Takashi Kuwahara, Mizuki Saito, Taro Osaka and Shinsuke Suzuki
Effect of Primary Crystals on Pore Morphology during Semi-Solid Foaming of A2024 Alloys
Reprinted from: *Metals* **2019**, *9*, 88, doi:10.3390/met9010088 . . . . **71**

Takashi Kuwahara, Taro Osaka, Mizuki Saito and Shinsuke Suzuki
Compressive Properties of A2024 Alloy Foam Fabricated through a Melt Route and a Semi-Solid Route
Reprinted from: *Metals* **2019**, *9*, 153, doi:10.3390/met9020153 . . . . **83**

Takashi Kuwahara, Akira Kaya, Taro Osaka, Satomi Takamatsu and Shinsuke Suzuki
Stabilization Mechanism of Semi-Solid Film Simulating the Cell Wall during Fabricationof Aluminum Foam
Reprinted from: *Metals* **2020**, *10*, 333, doi:10.3390/met10030333 . . . . **99**

Nallely Montserrat Trejo Rivera, Jesús Torres Torres and Alfredo Flores Valdés
A-242 Aluminium Alloy Foams Manufacture from the Recycling of Beverage Cans
Reprinted from: *Metals* **2019**, *9*, 92, doi:10.3390/met9010092 . . . . **113**

Ferdinandus Sarjanadi Damanik and Günther Lange
Influence of MWCNT Coated Nickel on the Foaming Behavior of MWCNT Coated Nickel Reinforced AlMg4Si8 Foam by Powder Metallurgy Process
Reprinted from: *Metals* **2020**, *10*, 955, doi:10.3390/met10070955 . . . . **129**

Susana C. Pinto, Paula A.A.P. Marques, Matej Vesenjak, Romeu Vicente, Luís Godinho, Lovre Krstulović-Opara and Isabel Duarte
Mechanical, Thermal, and Acoustic Properties of Aluminum Foams Impregnated with Epoxy/Graphene Oxide Nanocomposites
Reprinted from: *Metals* **2019**, *9*, 1214, doi:10.3390/met9111214 . . . . **139**

Susana C. Pinto, Paula A. A. P. Marques, Romeu Vicente, Luís Godinho and Isabel Duarte
Hybrid Structures Made of Polyurethane/Graphene Nanocomposite Foams Embedded within
Aluminum Open-Cell Foam
Reprinted from: *Metals* **2020**, *10*, 768, doi:10.3390/met10060768 . . . . **157**

**Matej Vesenjak, Masatoshi Nishi, Toshiya Nishi, Yasuo Marumo, Lovre Krstulović-Opara,
Zoran Ren and Kazuyuki Hokamoto**
Fabrication and Mechanical Properties of Rolled Aluminium Unidirectional Cellular Structure
Reprinted from: *Metals* **2020**, *10*, 770, doi:10.3390/met10060770 . . . . **173**

Chengxing Yang, Kai Xu and Suchao Xie
Comparative Study on the Uniaxial Behaviour of Topology-Optimised and Crystal-Inspired
Lattice Materials
Reprinted from: *Metals* **2020**, *10*, 491, doi:10.3390/met10040491 . . . . **185**

Zhimin Xu, Kangpei Meng, Chengxing Yang, Weixu Zhang, Xueling Fan and Yongle Sun
Elasto-Plastic Behaviour of Transversely Isotropic Cellular Materials with Inner Gas Pressure
Reprinted from: *Metals* **2019**, *9*, 901, doi:10.3390/met9080901 . . . . **201**

Fei Yang, Xinmin Shen, Panfeng Bai, Xiaonan Zhang, Zhizhong Li and Qin Yin
Optimization and Validation of Sound Absorption Performance of 10-Layer Gradient
Compressed Porous Metal
Reprinted from: *Metals* **2019**, *9*, 588, doi:10.3390/met9050588 . . . . **213**

**Yoshihiko Hangai, Ryusei Kobayashi, Ryosuke Suzuki, Masaaki Matsubara and Nobuhiro
Yoshikawa**
Aluminum Foam-Filled Steel Tube Fabricated from Aluminum Burrs of Die-Castings by
Friction Stir Back Extrusion
Reprinted from: *Metals* **2019**, *9*, 124, doi:10.3390/met9020124 . . . . **231**

Isabel Duarte, Thomas Fiedler, Lovre Krstulović-Opara and Matej Vesenjak
Brief Review on Experimental and Computational Techniques for Characterization of
Cellular Metals
Reprinted from: *Metals* **2020**, *10*, 726, doi:10.3390/met10060726 . . . . **243**

About the Editors

Isabel Duarte Chemical Engineering Degree in University of Coimbra (FCTUC, 1992), Materials Engineering Master's Degree in University of Lisbon (IST/UL, 1998) and PhD in Sciences of Engineering at University of Porto (FEUP, 2005). She is a researcher at the Department of Mechanical Engineering of the University of Aveiro (Portugal). Her research interests focus on the fabrication and characterization of multifunctional porous and cellular materials for engineering applications. She has a strong multi-international cooperation in these thematic fields, resulted several original scientific journal articles. She is also a member of the editorial board of MDPI Metals journal.

Matej Vesenjak is a Full Professor at the Faculty of Mechanical Engineering, University of Maribor, Slovenia. His research interests include design and characterization of advanced cellular structures and metamaterials with strong emphasis on geometrical properties, experimental testing and computational simulations. Prof. Matej Vesenjak has published over 120 original scientific journal articles and over 250 contributed conference papers. He has a strong multi-international cooperation and gave over 70 invited lectures at foreign university. He is also a visiting professor and has received many national and international fellowships and awards for his academic achievements.

Thomas Fiedler is an Associate Professor at the University of Newcastle (Australia). His expertise combines advanced numerical modelling and experimentation with a particular focus on materials science. His group is spearheading the development of low-cost metallic foams and works towards their large-scale utilization in industry. Dr Fiedler has published over 100 peer-reviewed journal publications in this field and is a member of the editorial boards of the MDPI journals Metals and Materials.

Lovre Krstulović-Opara is a full professor on University of Split, Faculty of Electrical, Eng., Mechanical Eng. and Naval Architecture (Croatia). His expertise is in advanced experimental testing, infrared thermography and non-destructive methods. His work is based on deformation evaluation by infrared acquisition of composites and metals, mostly involving cellular structures. Several published papers are product of strong collaboration with University of Maribor, University of Aveiro and University of Newcastle.

Editorial

Cellular Metals: Fabrication, Properties and Applications

Isabel Duarte [1,*], Thomas Fiedler [2], Lovre Krstulović-Opara [3] and Matej Vesenjak [4]

[1] Center for Mechanical Technology and Automation, Department of Mechanical Engineering, University of Aveiro Campus Universitário de Santiago, 3810-193 Aveiro, Portugal

[2] Centre for Mass and Thermal Transport in Engineering Materials, School of Engineering, The University of Newcastle, Callaghan NSW 2308, Australia; thomas.fiedler@newcastle.edu.au

[3] Faculty of Electrical Engineering, Mechanical Engineering and Naval Architecture, University of Split, 21000 Split, Croatia; Lovre.Krstulovic-Opara@fesb.hr

[4] Faculty of Mechanical Engineering, University of Maribor, 2000 Maribor, Slovenia; matej.vesenjak@um.si

* Correspondence: isabel.duarte@ua.pt

Received: 2 November 2020; Accepted: 17 November 2020; Published: 20 November 2020

1. Introduction and Scope

Cellular solids and porous metals have become some of the most promising lightweight multifunctional materials due to their superior combination of advanced properties mainly derived from their base material and cellular structure. They are used in a wide range of commercial, biomedical, industrial, and military applications. In contrast to other cellular materials, cellular metals are non-flammable, recyclable, extremely tough, and chemically stable and are excellent energy absorbers. The manuscripts of this Special Issue provide a representative insight into the recent developments in this field, covering topics related to manufacturing, characterization, properties, specific challenges in transportation, and the description of structural features. For example, a presented strategy for the strengthening of Al-alloy foams is the addition of alloying elements (e.g., magnesium) into the metal bulk matrix to promote the formation of intermetallics (e.g., precipitation hardening). The incorporation of micro-sized and nano-sized reinforcement elements (e.g., carbon nanotubes and graphene oxide) into the metal bulk matrix to enhance the performance of the ductile metal is presented. New bioinspired cellular materials, such as nanocomposite foams, lattice materials, and hybrid foams and structures are also discussed (e.g., filled hollow structures, metal-polymer hybrid cellular structures).

2. Contributions

Sixteen original scientific contributions (fifteen original articles and one review paper) on different topics of cellular metals have been published in this Special Issue. Seven contributions focus on the advances in production and manufacturing technologies for aluminum alloy foams and the understanding of the foaming behavior and the fundamental relationships between their structure and their properties, with some even exploring nano-sized reinforcement elements. García-Moreno et al. [1] investigate the influence of the alloy composition and the role of the liquid phase and its influence on the gas nucleation and foaming of Al–Si–Mg alloys prepared by the powder metallurgy method using in-situ X-ray radioscopy, X-ray tomography, and X-ray tomoscopy. Lehmhus et al. [2] analyze the effects of the solidification rates and addition of Sr, B, and TiB2/TiAl3 on the foaming behavior and microstructural, morphological, and mechanical properties of aluminum foams fabricated by the powder metallurgy method. Duarte et al. [3] present an automated production line composed of a continuous furnace, a cooling sector, and a robotic system to fabricate metal foam parts and/or components using a powder metallurgical method. Several case studies are also provided. Three original contributions from Kuwahara et al. [4–6] related to direct foaming methods, as an alternative to the powder metallurgy method, are also presented. In the first contribution, Kuwahara et al. [4] investigate the pore formation of Al-Cu-Mg foams fabricated by the melt route using magnesium as a thickening agent or by the semi-solid route using primary crystals at a semi-solid temperature with a solid fraction of 20%. They analyze the nano-sized and micro-sized precipitates of the cell wall and the effect of the cell wall structure and pore morphology on the compressive properties. Kuwahara et al. [5] characterize the compressive properties of Al-Cu-Mg foam fabricated by the melt route and the semi-solid route. They also provide a morphological, microstructural, and metallurgical analysis. Kuwahara et al. [6] analyze and numerically evaluate the stabilization mechanism of the cell walls of the aluminum alloy foams prepared by a semi-solid route in which the melt is thickened by primary crystals. Trejo-Rivera et al. [7] present a methodology based on a direct foaming method to fabricate aluminum alloy foams made of A-242 alloy from the recycling of secondary aluminum obtained from beverage cans, using calcium and titanium hydride as a thickening and blowing agent, respectively. They provide microstructural and metallurgical analyses and the mechanical properties. Three contributions discuss the reinforcing role of carbon nanostructures (e.g., multiwalled carbon nanotubes and graphene oxide) in developing new cellular structures with enhanced properties. Damanik and Lange [8] investigate the effect of multiwalled carbon nanotubes coated with nickel on the foaming behavior and morphology of reinforced AlMg4Si8 foam prepared by a powder metallurgy process. Pinto et al. [9,10] present novel aluminum foam–polymer nanocomposite hybrid structures, providing the structural, mechanical, thermal, and acoustic properties. These hybrid structures are prepared by infiltrating an open-cell aluminum foam with a polymer (epoxy [9] and polyurethane foams [10] with or without graphene oxide) to enhance the performance of the resulting hybrid structures. The three following contributions are focused on new cellular structures to be applied for structural applications. Vesenjak et al. [11] present a novel approach to fabricate aluminum unidirectional cellular structures with longitudinal pores by rolling a thin aluminum foil with acrylic spacers and finally compacting by explosion welding. They provide the metallographic, quasi-static, and dynamic compressive properties. The fabrication process and influencing parameters have been further studied using computational simulations. Yang et al. [12] present a comparative numerical study of the uniaxial behavior of three lattice materials—face center cube, edge center cube, and vertex cube—generated using topology optimization and crystal inspiration. They provide a characterization of their deformation modes, mechanical properties, and energy absorption capabilities. Xu et al. [13] analytically and numerically describe the elasto-plastic behavior of transversely isotropic cellular materials with inner gas pressure, focusing on a gas-filled ellipsoidal-cell face center cube foam with a relative density of 0.5. Yang et al. [14] report the findings of an experimental and numerical work developed to study the sound absorption performance of 10-layer gradient compressed porous metal to be applied for noise reduction. Hangai et al. [15] introduce a new manufacturing method to fabricate steel tubes filled with aluminum foam from aluminum burrs of die-castings by friction stir back extrusion. They further determine the compressive behavior and potential applications for impact energy absorption. Finally, Duarte et al. [16] present a brief review of the main experimental and numerical techniques and standards to investigate and quantify the structural, mechanical, thermal, and acoustic properties of cellular metals.

3. Conclusions and Outlook

The contributions of this Special Issue present recent advances in cellular metal research. Cellular metals enable an immediate weight reduction and material saving of components. Simultaneously, they can perform multiple functions due to their three-dimensional cellular structures. Despite these advantages, the commercial use of cellular metals is still limited due to the high production costs required to create regular cellular structures (topology and morphology). Another inhibitor to their application is the random variation in their mechanical, acoustic, and thermal behavior. However, new technical solutions ensure that these materials can be produced in a commercially profitable way with the required quality and reproducibility, such that their integration into engineering structures can fully utilize their unique properties.

Acknowledgments: The guest editors would like to thank all the authors for their contributions and the reviewers for their outstanding efforts to improve the scientific quality of the different contributions. We would also like to thank the staff at the Metals Editorial Office, particularly Betty Jin, who supported and managed the publication process.

Conflicts of Interest: The authors declare no conflict of interest.

References

1. García-Moreno, F.; Radtke, L.A.; Neu, T.R.; Kamm, P.H.; Klaus, M.; Schlepütz, C.M.; Banhart, J. The Influence of Alloy Composition and Liquid Phase on Foaming of Al–Si–Mg Alloys. *Metals* **2020**, *10*, 189. [CrossRef]
2. Lehmhus, D.; Hünert, D.; Mosler, U.; Martin, U.; Weise, J. Effects of Eutectic Modification and Grain Refinement on Microstructure and Properties of PM AlSi7 Metallic Foams. *Metals* **2019**, *9*, 1241. [CrossRef]
3. Duarte, I.; Vesenjak, M.; Vide, M.J. Automated Continuous Production Line of Parts Made of Metallic Foams. *Metals* **2019**, *9*, 531. [CrossRef]
4. Kuwahara, T.; Saito, M.; Osaka, T.; Suzuki, S. Effect of Primary Crystals on Pore Morphology during Semi-Solid Foaming of A2024 Alloys. *Metals* **2019**, *9*, 88. [CrossRef]
5. Kuwahara, T.; Osaka, T.; Saito, M.; Suzuki, S. Compressive Properties of A2024 Alloy Foam Fabricated through a Melt Route and a Semi-Solid Route. *Metals* **2019**, *9*, 153. [CrossRef]
6. Kuwahara, T.; Kaya, A.; Osaka, T.; Takamatsu, S.; Suzuki, S. Stabilization Mechanism of Semi-Solid Film Simulating the Cell Wall during Fabrication of Aluminum Foam. *Metals* **2020**, *10*, 333. [CrossRef]
7. Trejo-Rivera, N.M.; Torres Torres, J.; Flores Valdés, A. A-242 Aluminium Alloy Foams Manufacture from the Recycling of Beverage Cans. *Metals* **2019**, *9*, 92. [CrossRef]
8. Damanik, F.S.; Lange, G. Influence of MWCNT Coated Nickel on the Foaming Behavior of MWCNT Coated Nickel Reinforced AlMg4Si8 Foam by Powder Metallurgy Process. *Metals* **2020**, *10*, 955. [CrossRef]
9. Pinto, S.C.; Marques, P.A.; Vesenjak, M.; Vicente, R.; Godinho, L.; Krstulović-Opara, L.; Duarte, I. Mechanical, Thermal, and Acoustic Properties of Aluminum Foams Impregnated with Epoxy/Graphene Oxide Nanocomposites. *Metals* **2019**, *9*, 1214. [CrossRef]
10. Pinto, S.C.; Marques, P.A.A.P.; Vicente, R.; Godinho, L.; Duarte, I. Hybrid Structures Made of Polyurethane/Graphene Nanocomposite Foams Embedded within Aluminum Open-Cell Foam. *Metals* **2020**, *10*, 768. [CrossRef]
11. Vesenjak, M.; Nishi, M.; Nishi, T.; Marumo, Y.; Krstulović-Opara, L.; Ren, Z.; Hokamoto, K. Fabrication and Mechanical Properties of Rolled Aluminium Unidirectional Cellular Structure. *Metals* **2020**, *10*, 770. [CrossRef]
12. Yang, C.; Xu, K.; Xie, S. Comparative Study on the Uniaxial Behaviour of Topology-Optimised and Crystal-Inspired Lattice Materials. *Metals* **2020**, *10*, 491. [CrossRef]
13. Xu, Z.; Meng, K.; Yang, C.; Zhang, W.; Fan, X.; Sun, Y. Elasto-Plastic Behaviour of Transversely Isotropic Cellular Materials with Inner Gas Pressure. *Metals* **2019**, *9*, 901. [CrossRef]
14. Yang, F.; Shen, X.; Bai, P.; Zhang, X.; Li, Z.; Yin, Q. Optimization and Validation of Sound Absorption Performance of 10-Layer Gradient Compressed Porous Metal. *Metals* **2019**, *9*, 588. [CrossRef]
15. Hangai, Y.; Kobayashi, R.; Suzuki, R.; Matsubara, M.; Yoshikawa, N. Aluminum Foam-Filled Steel Tube Fabricated from Aluminum Burrs of Die-Castings by Friction Stir Back Extrusion. *Metals* **2019**, *9*, 124. [CrossRef]
16. Duarte, I.; Fiedler, T.; Krstulović-Opara, L.; Vesenjak, M. Brief Review on Experimental and Computational Techniques for Characterization of Cellular Metals. *Metals* **2020**, *10*, 726. [CrossRef]

Publisher's Note: MDPI stays neutral with regard to jurisdictional claims in published maps and institutional affiliations.

Article

The Influence of Alloy Composition and Liquid Phase on Foaming of Al–Si–Mg Alloys

Francisco García-Moreno [1,2,*], Laurenz Alexander Radtke [2], Tillmann Robert Neu [1,2], Paul Hans Kamm [1,2], Manuela Klaus [3], Christian Matthias Schlepütz [4] and John Banhart [1,2]

[1] Institute of Applied Materials, Helmholtz-Zentrum Berlin für Materialien und Energie, Hahn-Meitner-Platz 1, 14109 Berlin, Germany; t.neu@tu-berlin.de (T.R.N.); paul.kamm@helmholtz-berlin.de (P.H.K.); banhart@helmholtz-berlin.de (J.B.)

[2] Institute of Materials Science and Technology, Technische Universität Berlin, Hardenbergstr. 36, 10623 Berlin, Germany; laurenz.rdt@googlemail.com

[3] Department of Microstructure and Residual Stress Analysis, Helmholtz-Zentrum Berlin für Materialien und Energie, Albert-Einstein-Str. 15, 12489 Berlin, Germany; klaus@helmholtz-berlin.de

[4] Swiss Light Source, Paul Scherrer Institute, 5232 Villigen, Switzerland; christian.schlepuetz@psi.ch

* Correspondence: garcia-moreno@helmholtz-berlin.de

Received: 31 December 2019; Accepted: 24 January 2020; Published: 28 January 2020

Abstract: The foaming behaviour of aluminium alloys processed by the powder compaction technique depends crucially on the exact alloy composition. The AlSi8Mg4 alloy has been in use for a decade now, and it has been claimed that this composition lies in an "island of good foaming". We investigated the reasons for this by systematically studying alloys around this composition by varying the Mg and Si content by a few percent. We applied in situ X-ray 2D and 3D imaging experiments combined with a quantitative nucleation number and expansion analysis, X-ray tomography of solid foams to assess the pore structure and pore size distribution, and in situ diffraction experiments to quantify the melt fraction at any moment. We found a correlation between melt fraction and expansion height and verified that the "island of good foaming" actually exists, and foams outside a preferred range for the liquid fraction—just above T_S and between 40–60%—show a poorer expansion performance than the reference alloy AlSi8Mg4. A very slight increase of Si and decrease of Mg content might further improve this foam.

Keywords: metal foam; CALPHAD; liquid fraction, X-ray diffraction; X-ray radioscopy; X-ray tomography; X-ray tomoscopy

---

1. Introduction

Although the unusual properties of aluminium foams promise a wide range of applications [1] and various commercial manufacturers are available, the overall production and selling volumes are still quite low compared to, for example, wrought aluminium products or cellular polymers. One reason may be the still too high costs, but it is certainly also necessary to further improve the potential of metal foams by optimising their structure and properties. Two main production routes are commercially available, namely, the melt route and the powder metallurgical route [2]. Foams made by the latter method are stabilized by the action of metal oxide networks hindering film thinning and rupture, as shown by Körner et al. [3], which is crucial for achieving a homogeneous foam. Beside stabilising the liquid foam, other factors, such as the way of creating and distributing gas bubbles during nucleation and further bubble growth, have to be considered. These requirements have led to optimization strategies and related works of various kinds in the past.

The influence of gas nucleation on different alloys was studied by Rack et al. [4] and Kamm [5]. The influence of the different powder compaction methods was described by Neu [6]. In all these cases, the alloy composition played an important role [4,7]. The foamability of different aluminium alloy systems, such as AlSi [7–9], AlMg [10], AlSiCu [11], AlSiMg, [10,12–14], AlSiCuZn [15] or AlSiMgCu [13], has been reported. The alloy AlSi8Mg4 (in wt.%) has been found to be specially suited and was therefore patented [16] due to the fact of its good foaming behaviour and other favourable properties such as good corrosion resistance. It is now used commercially by the company Pohltec Metalfoam for the production of aluminium foam sandwiches (AFS) [17]. Helwig et al. [13] reported that the large amount of liquid available at low temperatures plays a major role in obtaining a good foam structure, enhancing gas nucleation, and preventing crack formation and propagation in early stages. The presence of Mg in the alloy also facilitates powder compaction and metallic bonding among powder particles due to the breaking of the aluminium oxide layers [18]. On the downside, the macroscopic expansion of Mg-containing alloys can be possibly compromised by increased surface oxidation as demonstrated by Simancik et al. [10].

There has been the claim from Helwig et al. [16] of a narrow "island of good foaming" for alloy AlSi8Mg4 (where numbers define the amount of the alloying element in weight percent, wt.%) with a width of the alloy composition of ±1 wt.% for both components, however, without giving a precise definition of foam quality. The aim of this work was to verify, disprove or further specify this claim in terms of measured quantities, to investigate the role of the liquid phase and its influence on gas nucleation and foaming behaviour as well as to find the most suitable alloy composition for a favourable combination of expansion and morphology.

2. Materials and Methods

2.1. Sample Preparation

Different alloy compositions in the vicinity of the reference alloy AlSi8Mg4 were produced by mixing elemental or alloyed powders with the blowing agent TiH_2. The latter was heat treated for 3 h at 480 °C to match to the alloy's solidus temperature [19]. The powders used are listed in Table 1. The amount of blowing agent was kept constant at 0.5 wt.% for all compositions. For each composition, a total of 30 g of powders were mixed in a tumbling mixer for 30 min and then filled into a cylindrical steel mould of 36 mm diameter. The powders were first cold compacted for 10 s at a uniaxial pressure of 300 MPa to displace as much air as possible and then, in a second step, hot compacted at 400 °C and 300 MPa for 900 s as described in more detail by Helwig et al. [20]. From the resulting cylindrical tablet, rectangular samples of 10 mm × 10 mm × 4 mm and 4 mm × 4 mm × 2 mm size were prepared with a CNC mill to be used in laboratory and synchrotron X-ray radioscopy experiments, respectively. The larger sample surfaces were always perpendicular to the uniaxial compression direction.

Table 1. List of powders used.

Powder	Provider	Purity	D_{50} (μm)
Al	Aluminium Powder Company Ltd.	99.7	63.9
Si	Elkem AS	97.5	25.7
AlMg50	Possehl Erzkontor	-	63.8
TiH_2	Chemetall GmbH	98.8	14.4

2.2. Laboratory X-Ray Radioscopy and Tomography

For foaming, samples were placed on top of a resistive heating plate (900 W power) leaving a free path for X-ray observation. A thermocouple inserted into the heating plate on its obverse side combined with a CAL 3300 thermo-controller from CAL Controls, Hertfordshire, UK, allowed for the adjustment of the desired temperature profile. The corresponding temperature in the samples was calibrated by reference measurements with a thermocouple inserted into dummy precursors. To analyse the foaming behaviour of the samples, the temperature of the heating plate was raised to 700 °C (corresponding to 673 °C inside the sample) with an average heating rate of ~16.5 K/s and held there for 180 s, after which the power was switched off and natural cooling led to solidification and conservation of the resulting foam.

The whole foaming process was observed by in situ laboratory X-ray radioscopy. The system consisted of a micro-focus X-ray source from Hamamatsu, Photonics, Hamamatsu, Japan, with a spot size of 5 μm and a power of 10 W (100 kV and 100 μA). The radiographic images were detected by a flat panel detector, also from Hamamatsu, with a 120 mm × 120 mm large field of view and 2240 × 2368 pixels, each one 50 μm × 50 μm in size. Due to the geometrical magnification of 3.5×, an effective pixel size at the sample site of 14.3 μm was achieved. A quantitative analysis of the projected images allows, among others, a detailed calculation of the area expansion evolution of the foams. The equipment has been explained in more detail elsewhere [21].

Post-solidification tomographic images of the foams based on 1000 projections distributed over an angle of 360° were recorded with the same system just by replacing the heating stage by a rotation stage from Huber, Rimsting, Germany.

2.3. Synchrotron X-Ray Radioscopy and Diffraction

Simultaneous energy-dispersive X-ray diffraction (ED-XRD) and radioscopy measurements were performed at the Energy Dispersive Diffraction (EDDI) instrument, Bessy II, Berlin, Germany. A scheme of the setup is shown in Figure 1. A detailed description of the system can be found in the literature [22,23]. Samples of 4 mm × 4 mm × 2 mm size were placed on a steel holder on top of a M-660 rotation stage from Physik Instrumente PI, Karlsruhe, Germany, and heated up to 640 °C at a rate of 3.4 K/s with an infrared (IR) heating lamp of 150 W power. The temperature was controlled by a thermocouple inserted into the steel holder beneath the sample. The samples were rotated at 0.2 Hz during the foaming process for better statistics. The transmitted X-ray image was converted to visible light by a 200 μm thick LuAG:Ce scintillator, and the corresponding image was projected onto a Complementary metal-oxide-semiconductor (CMOS) sensor with an effective pixel size of 2.5 μm (DIMAX, PCO, Kelheim, Germany) by a mirror/lens system. We recorded 200 images per second with an exposure time of 3 ms for each. Full diffraction patterns could be recorded in transmission with a multi-channel Ge-detector, model GL0110, Canberra, Lingolsheim, France, under a fixed angle of $2\theta = 6°$.

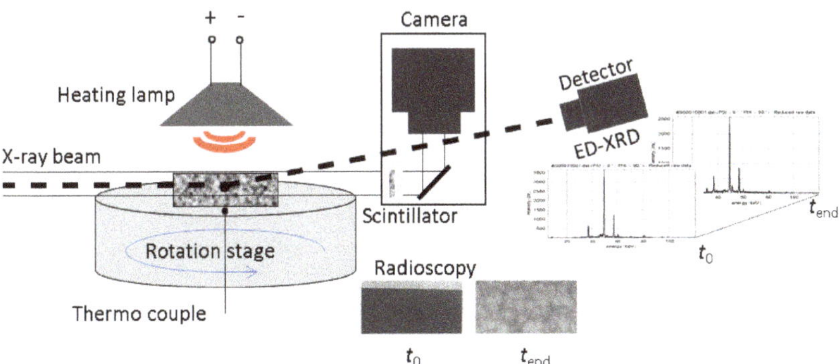

Figure 1. Schematic view of the experimental setup for simultaneous X-ray radioscopy and diffraction measurements during foaming of samples at the Energy Dispersive Diffraction (EDDI) beamline, Bessy II, Berlin, Germany.

The ED-XRD patterns were acquired at a speed of 0.4 diffractograms per second throughout the foaming process including melting, foam aging, and solidification. Using the software available at the beamline [22], the solid fraction of the different phases could be calculated from the diffraction patterns by integrating the area of the peaks, each of which corresponded to the volume fraction of the corresponding phase [24]. The liquid fraction in the evolving foam was calculated comparing the measured peak areas with the areas of the peaks in the solid state. For that purpose, several corrections for the primary beam (wiggler) spectrum, absorption, and detector dead time were applied to the raw data. The individual diffraction lines in the diffraction patterns were fitted by pseudo Voigt functions. Finally, the results were corrected for the changing materials density by taking into account the average X-ray attenuation derived from radioscopy data and Beer–Lambert's attenuation law [25]. As the alloys reduce their density during heating due to the fact of thermal expansion, and this especially during foaming, a density correction was applied. The corresponding density evolution was obtained from the radioscopic images using our own software AXIM [25].

2.4. Synchrotron X-Ray Tomoscopy

To further study the foam structure evolution in the liquid state, the recently developed tomoscopy setup at the TOMCAT beamline, PSI, Villigen, Switzerland, was used [26]. The X-ray tomoscopy allowed us to resolve, in 3D, the real foam structure in situ during the foaming procedure, i.e., in the liquid state. Thus, we were able to resolve the gas nucleation stage and the size and shape of the first evolving bubbles, which are a determining factor for the later structural quality of the foam. Samples of 4 mm × 4 mm × 2 mm size were placed in an X-ray transparent boron nitride crucible of 8 mm diameter which was rotated at 10 Hz while acquiring 10 tomograms per second (each covering a 180° angle with 400 projections of 90 μs exposure time, for a total individual scan duration of 0.05 s, separated by 180° rotation without data acquisition) of over a time span of several minutes. The samples were heated up at a rate of ~2 K/s with two IR lasers as described in detail elsewhere [27]. The transmitted X-ray image was converted to visible light by a 150 μm thick LuAG:Ce scintillator and transferred to the GigaFRoST high-speed camera [28] by an optical system with a high numerical aperture and 4 fold magnification [29]. The tomographic projections images were filtered using the propagation-based phase contrast algorithm by Paganin et al. [30] and reconstructed with the gridrec algorithm [31]. The resulting effective pixel size was 2.75 μm.

2.5. Calculation of the Liquid Fraction

To predict phase evolution throughout the foaming process, including the fraction of liquid available over the solidification range for the different alloys studied, the commercial software Thermo-Calc 2016a, Thermo-Calc Software AB, Solna, Sweden [32] and a Scheil–Gulliver model [33] was applied. Thermodynamic data were taken from the COST507B database. Thermo-Calc calculates the equilibrium of the phases according to the CALculation of PHAse Diagrams (CALPHAD) method.

3. Results

Calculations performed with Thermo-Calc indicate that the amount of liquid fraction should increase with the Si and decrease with the Mg content (see Appendix A and Figure A1). The effects of the variation of the Si and Mg content in the range of ±1% and ±2% around the reference alloy's composition (AlSi8Mg4) on gas nucleation, foam expansion, pore structure, and liquid phase were studied systematically. For all experiments, except the tomoscopy and diffraction analyses, each reported value was the average of three measurements. When showing individual curves or images, a representative sample was selected.

3.1. Gas Nucleation and Bubble Evolution

Gas nucleation and bubble growth in the early foaming stage were studied by X-ray tomoscopy. Figure 2 shows representative tomographic slices extracted from tomograms recorded during the stage of bubble nucleation and early growth at three different foaming times and for four representative alloys. At 474 °C, the first gas nuclei can be observed in all samples. They are preferably located at the weakly X-ray attenuating AlMg50 particles. All four alloys exhibited a similar structure at this stage, although slight differences in the number of AlMg50 particles, corresponding to the different Mg contents, can be seen. In the second row, at 570 °C, the difference between the reference alloy AlSi8Mg4 and the others is obvious. While the reference alloy had already round and smoothly shaped bubbles, the other alloys developed irregularly shaped jagged bubbles. Nevertheless, at 587 °C, the bubble structures of the non-reference alloys seemed to have healed, so that the foam developed an acceptable structure, although with different pore sizes as can be seen later in Section 3.4.

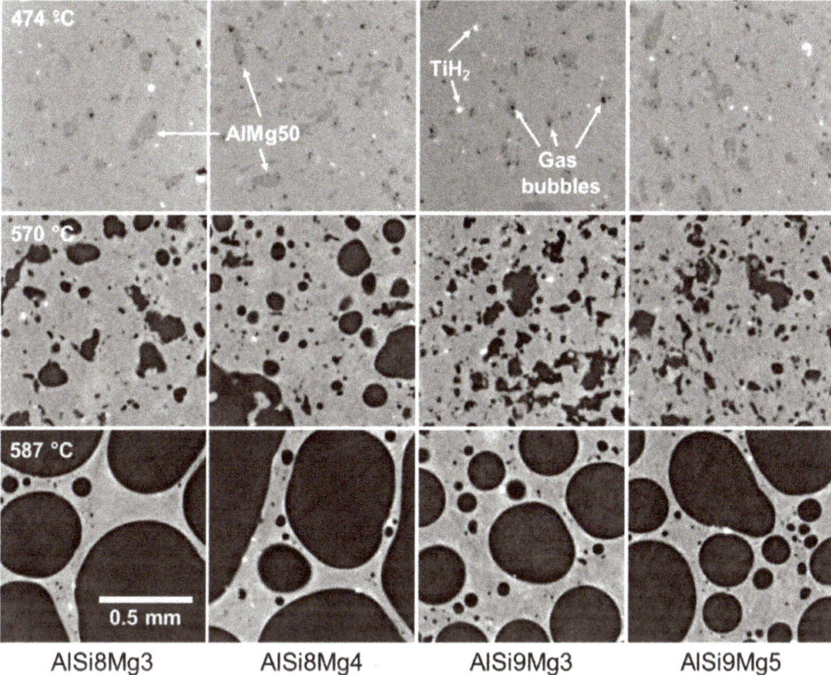

Figure 2. Representative tomographic slices extracted from tomograms recorded at three different temperatures for four different alloys. The first row shows the first nucleation stage, the second row the second nucleation stage, and the third row the early bubble growth stage. Gas bubbles, TiH_2, and AlMg50 particles are clearly visible in the first nucleation stage and marked with white arrows.

3.2. Expansion Kinetics

Figure 3 shows the projected area expansion curves for the selected representative samples of different alloy compositions as derived from laboratory radioscopy data and the corresponding temperature of the heating plate and the calibrated temperature of a dummy sample (a non-foamable piece of a similar aluminium alloy with the same weight and size of a real sample). We observed clear differences in the general course of the area expansion as well as in the maximum or end expansion values of the foams. While alloys AlSi9Mg3 and AlSi7Mg5 reached over 90% of their maximum expansion in ~25 s, AlSi8Mg4 and AlSi9Mg5 needed more than 200 s. Alloy AlSi7Mg5 reached its maximum expansion shortly after the expansion phase and shrunk slowly during the remaining holding time, while the other alloys continued growing slowly until the holding period was over. At the end of the holding period (t = 225 s), the heating power was switched off and after ~40 s (indicated by black arrows in Figure 3), the samples solidified. At this point, a small local minimum in the area expansion can be observed followed by a slight expansion stage. This is known as solidification expansion [34]. The final expansion of the solid foam is the so-called end expansion.

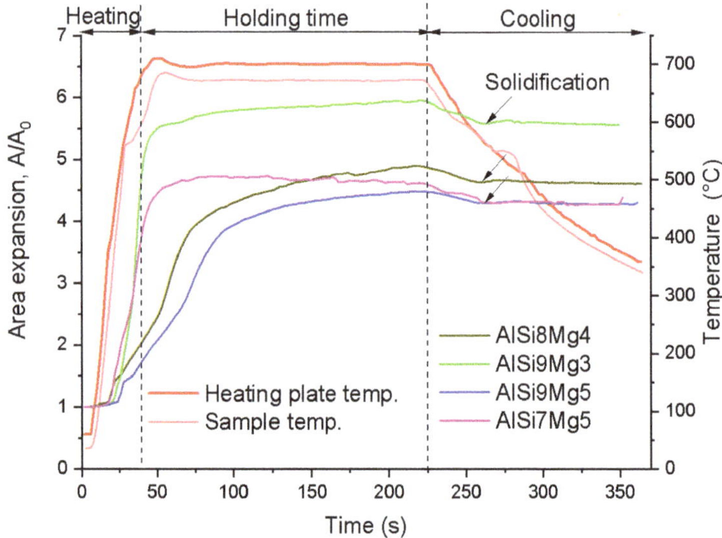

Figure 3. Area expansion curves for selected samples of different alloy composition and the common temperature profiles of the heating plate and the sample. Three experimental stages are indicated, namely, heating, holding for 180 s at 700 °C, heating plate temperature, and natural cooling after switching off the power. Black arrows indicate the solidification point in the expansion curve.

3.3. Maximum Area Expansion

Mg and Si variations of ±1 wt.% and ±2 wt.% around the reference alloy AlSi8Mg4 were evaluated. The impact of the variation of one element while keeping the other at a constant level on foam area expansion is given in Figure 4. In Figure 4a, we can observe that changes in the Mg content in the range 3–5 wt.% had little impact on expansion, but beyond this, range expansion is reduced. Figure 4b shows a flat maximum of expansion around 9 wt.% Si with a decrease in expansion for lower and higher values. The difference between the maximum and the end expansion expresses the tendency of shrinking (or of collapse in extreme cases) of the foams after solidification. Here, we can see for both variations that the difference was small in all cases.

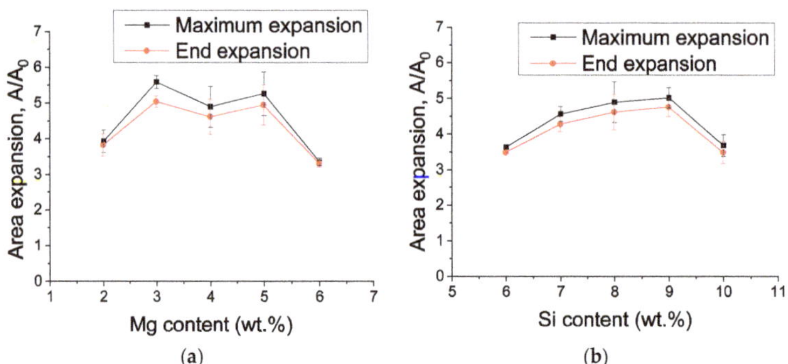

Figure 4. Maximum and end foam area expansion given as a function of (a) Mg content, with Si fixed at 8 wt.%, and (b) Si content, with Mg fixed at 4 wt.%.

In order to also describe simultaneous changes in Si and Mg composition, Figure 5 shows the results of a wider selection of compositions around the reference composition (in blue). Compositions in green show a higher maximum expansion and in red lower ones. The trend of higher expansions seems to be shifted to more Si and less Mg than for the reference alloy.

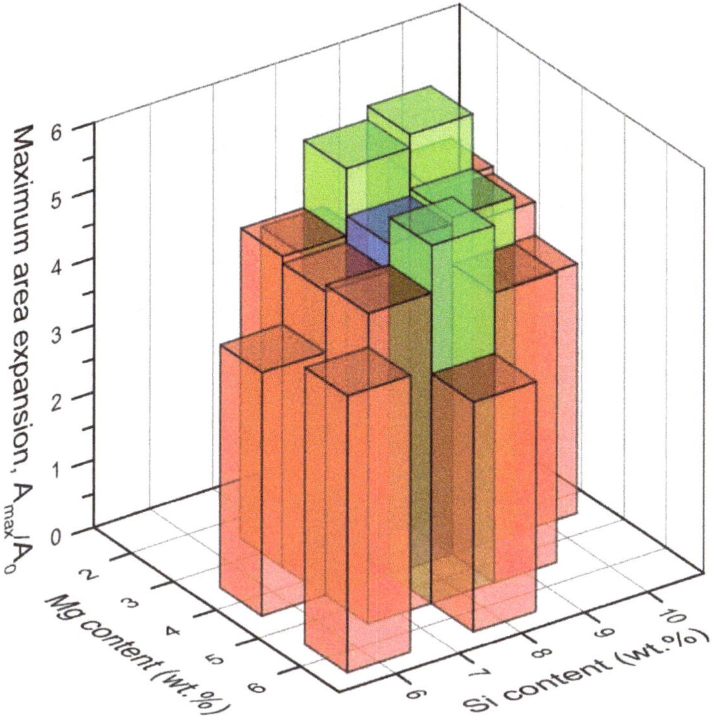

Figure 5. Maximum foam area expansion for varying Mg and Si content. Composition in blue is the reference alloy AlSi8Mg4. Compositions in green show a higher maximum expansion and in red lower ones. Only selected alloys around the reference alloy are plotted for a better overview.

3.4. Morphology

The evolution of the inner structure of liquid metal foams was elucidated by X-ray radioscopy. Examples of the nucleation stage, maximum expansion, and end expansion for alloys of composition AlSi8MgY (Y = 2–6 wt.%) and AlSiXMg4 (X = 6–10 wt.%) can be observed in Figures A3 and A4 of Appendix C. It is possible to observe how cracks induce later large bubbles as well as local coalescence and drainage are induced by gravity at the bottom part of the foams, especially for Si contents >8 wt.%.

The interior of solid foams can be assessed qualitatively from macrographs of sectioned foams (Figure 6). Obviously, minor changes in alloy composition lead to visible changes in pore size. Alloy AlSi7Mg5 (Figure 6a) shows the most irregular structure with a thick outer skin and large non-spherical pores. The reference alloy AlSi8Mg4 and alloy AlSi9Mg5 present the most homogeneous structure, combining a good expansion with spherical pores and a homogeneous pore size distribution. The remaining alloy AlSi9Mg3 still had a good expansion and an acceptable structure but with larger and more polyhedral pores. A ~1 mm thick dense layer can be observed at the bottom of alloys AlSi7Mg5 and AlSi9Mg5 caused by gravity-induced drainage.

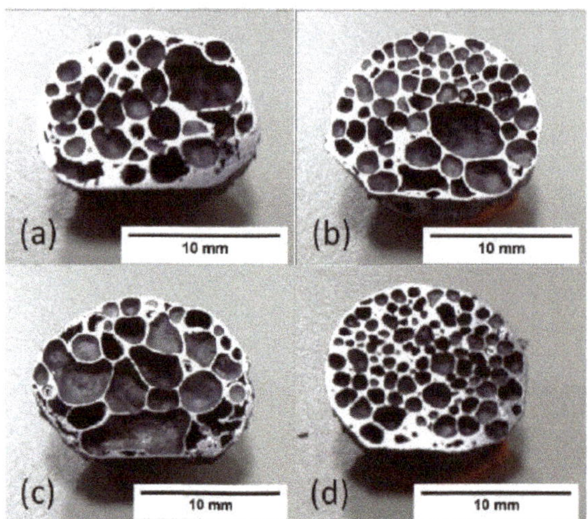

Figure 6. Cross-sections showing the pore structure of (**a**) AlSi7Mg5, (**b**) AlSi8Mg4, (**c**) AlSi9Mg3, and (**d**) AlSi9Mg5 after foaming and cutting into two pieces along a plane perpendicular to the sample substrate.

Drawing conclusions from individual foam samples, such as the ones shown in Figure 6, can be treacherous, as their volumes are small and statistical scatter among samples is large. Therefore, additional samples were quantitatively analysed by laboratory tomography, followed by pore segmentation as shown in Figure 7, where different colours indicate separated pores. The volume-weighted pore size distributions can be approximated by Gaussian functions. The mean pore size, sharpness of the distribution in terms of standard deviation, sample volume, and number of pores can be found in Table 2. The trends observed in Figure 6 are confirmed.

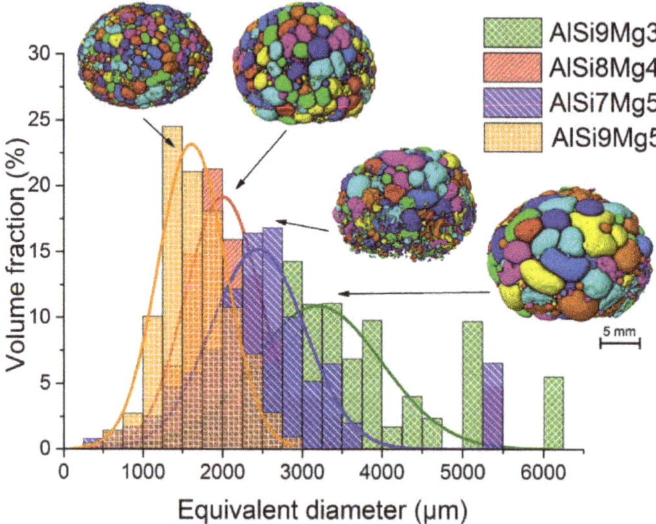

Figure 7. Volume-weighted pore size distribution histograms of four selected foams and corresponding Gaussian fits. Insets show rendered images of the segmented pore structures.

Table 2. Characteristic values of the foam structure obtained from tomographic analyses.

Alloy	Mean Pore Size (mm)	n	V (mm^3)	n/V (1/mm^3)
AlSi8Mg4	2.00 ± 0.47	1189	1704	0.69
AlSi7Mg5	2.44 ± 0.59	1086	1240	0.88
AlSi9Mg3	3.19 ± 0.81	448	2181	0.21
AlSi9Mg5	1.61 ± 0.42	1276	1355	0.94

3.5. Liquid Phase Evolution

To evaluate the evolution of the liquid phase during melting and solidification, the foaming process of the alloys was monitored by simultaneous in situ X-ray radioscopy and diffraction analysis. The results were compared with simulations of the solidification process carried out with the CALPHAD method.

Figure 8 depicts density maps of diffracted intensities (grey scale) dispersed in energy (in keV) of the reference alloy AlSi8Mg4 as a function of the time which is correlated with the temperature via the heating curve $T(t)$. Four selected radioscopic images of the evolving metal foam denote its structure and density at different positions. The phases are represented by the energy-dispersive peaks and labelled below these with the corresponding indices. The melting and solidification stadia showing peaks of solid phases mixed with a blurred area from the scattered intensities of the molten fraction are located between dashed blue and red lines (semi-solid phase). The fully molten stage is located between red dashed lines. It can be observed that the Al$_3$Mg$_2$ phase contained in the original AlMg50 particles disappears after melting and foaming, while the amount of the Mg$_2$Si phase increases after solidification (see also Figure A5).

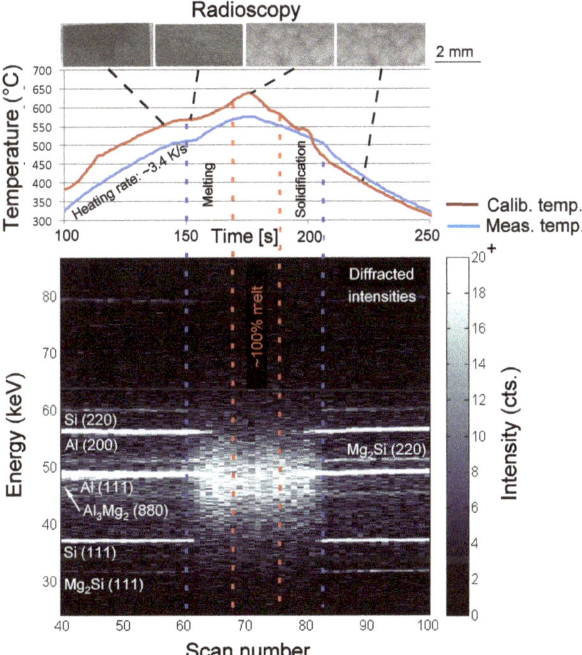

Figure 8. Density maps of diffracted intensities (grey scale) dispersed in energy (in keV) versus time correlated to the scan number and to the temperature profile and selected radioscopic images of an evolving AlSi8Mg4 foam.

The evolution of liquid fractions with varying temperature during foaming of four alloys was extracted from the density maps of diffracted intensities and plotted together with the calculated liquid fraction in Figure 9. The black curve represents the relative X-ray absorption of the samples extracted from the integrated intensities of the radiographies which correlates with their relative density via the Beer–Lambert equation [25]. The reference alloy AlSi8Mg4 shows a narrow melting interval starting from 580 °C. It is completely molten at 615 °C. We assumed that the absolute temperature in this case shifted to a higher temperature by ~20 K due to the insufficient sample contact to the heating plate. The AlSi7Mg5 presents the widest melt interval 555–625 °C and had a liquid fraction of 50% at 590 °C. The melting interval of alloy AlSi9Mg3 started at 560 °C and ended at relatively low 600 °C. The liquid fraction was 50% at 570 °C. The AlSi9Mg5 possessed the narrowest melting interval, and for temperatures >580 °C, only 5% solid phase remained.

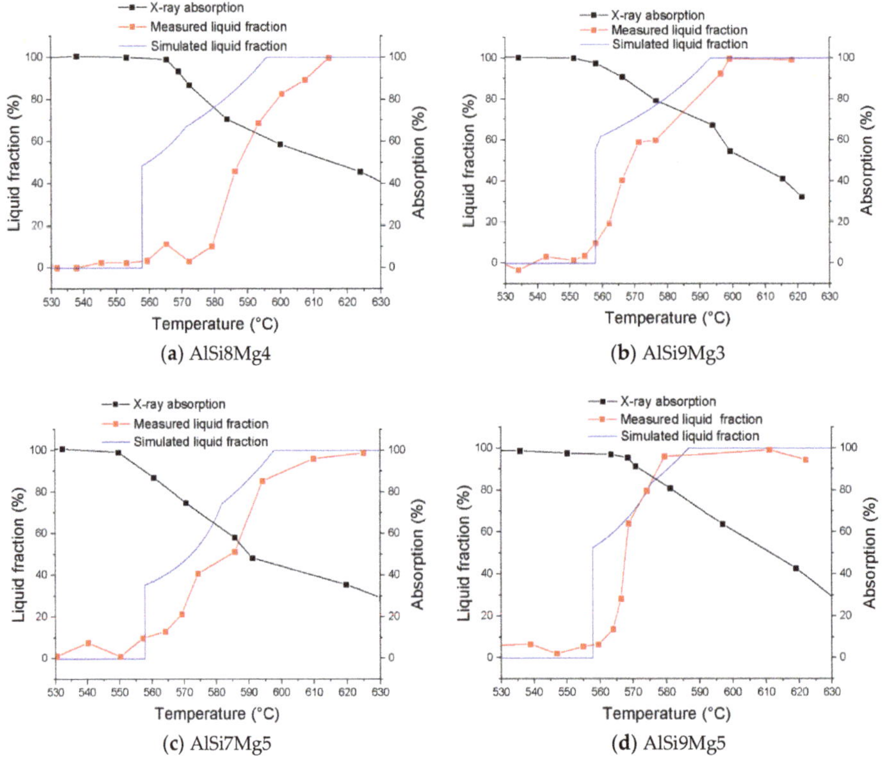

Figure 9. Development of the liquid phase over temperature extracted from the density maps of diffracted intensities (in red, see also Figure 8), calculated curves (in blue) and X-ray absorption of the samples related to the absorption of the precursors during foaming of (**a**) AlSi8Mg4, (**b**) AlSi9Mg3, (**c**) AlSi7Mg5, and (**d**) AlSi9Mg5 alloys.

4. Discussion

Melting of a foamable precursor is a mandatory step in the production chain of a metal foam following the PM route. It has been claimed that a narrow temperature range in which the liquid phase appears rapidly is beneficial [13]. To study the influence of the liquid phase on foam quality, we varied the alloy composition and found systematic changes. However, variations of composition may also affect other factors influencing foam quality including the quality of powder compaction. For satisfactory foaming, a high density and gas tightness have to be achieved. As shown in previous work on AlSiXMgY precursors, hot compaction at 400 °C and 300 MPa gives rise to good metallic bonding and relative densities above 96% which, in turn, suppresses crack formation in the early nucleation stages [20] as opposed to, for example, binary AlSiX alloys [4], where cracks occur that later lead to large pores and an inhomogeneous pore size distribution [35]. So far, no notable crack formation in the early stage was observed for all alloys studied here as shown in Figure 2, so that strong gas losses through open cracks all the way to the surface can be considered minimal. The increase or decrease in Si content does not have an influence on the density of the precursor, while an increase in Mg has a small positive effect on the compaction density, for example, from 96.1% relative density for AlSi7Mg3 to 97.5% for AlSi7Mg5 [36]. During compaction at 400 °C and 300 MPa, Mg atoms can diffuse into Al and react to form spinel $MgAl_2O_4$, break the Al_2O_3 oxide layer [37], and lead to improved metallic bonding and higher densities. On the other hand, this positive effect might be compensated by external oxidation of the entire foam thus hindering free foam expansion [10].

The rate of desorption from the blowing agent TiH_2 increases rapidly above 400 °C. An oxidising pre-treatment of the TiH_2 powder is known to be beneficial for foaming [19]. Gas nucleation starts already at ~450 °C close to or at AlMg50 particle surfaces, followed by a second nucleation step once the solidus temperature of 558 °C of the AlSiXMgY system has been reached [26]. The first nucleation step may be slightly influenced by differences in the amount of AlMg50 particles, although no significant influence is observed in Figure 2. After reaching the solidus temperature, the role of the amount of liquid fraction becomes visible, as it influences the second nucleation step and the following bubble growth as observed in Figure 2. In alloys like AlSi8Mg4 and AlSi9Mg5, which both have a short melting range, a large number of small pores appear in the initial state of the second nucleation step, leading later to foam structures with smaller pores (see Figure 6, Figure 7, and Figure A2).

The alloy AlSi7Mg5 (Figure 6a) exhibits an irregular pore structure and a thick outer skin caused by early shrinking and collapse as evident from Figure 3. The most homogeneous structures and a good expansion can be found for AlSi8Mg4. The AlSi9Mg5 exhibits a combination of good expansion with spherical pores and a sharper pore size distribution (see Table 2 and Figure 7). Alloy AlSi9Mg3 had the highest expansion of all tested alloys, as shown in Figure 3, but at the cost of an inferior pore structure, namely, larger polyhedral pores (see Figure 6c).

The liquid fractions in the melting and foaming alloys were measured by synchrotron diffraction experiments. An advantage of this method is that the real liquid fraction can be directly measured even if the system is not in thermodynamic equilibrium which is the case here. A disadvantage is that the quantification based on the peak area relies on several corrections as stated previously and is not exact but a reasonable qualitative approximation. Moreover, a direct temperature measurement is difficult because insertion of thermocouples into the foam would have an impact on foaming. Use of calibrated temperatures induces uncertainties.

The in situ diffraction method also allows for an identification of the phases involved. We could observe the disappearance of the Al_3Mg_2 phase present in the original precursors contained in the AlMg50 powders after melting and solidification as well as the increase of solid Mg_2Si phase after solidification which was already present in the precursors in lower amounts, indicating its formation already during hot compaction prior to foaming. Solid Mg_2Si exists in the melt as shown by Paes et al. [38]. Pronounced formation of solid Mg_2Si in the early stages of melting can reduce the amount of liquid fraction as was observed in alloy AlSi7Mg5, explaining the early foam stabilisation and the corresponding flat expansion profile in Figure 3, although the presence of Mg_2Si in the liquid stage could not be resolved by diffraction in Figure 8 due to the small quantities present and the strong scattering effect of the liquid. In all cases, the Si peaks disappeared abruptly between 560–580 °C, in the proximity to the eutectic temperature of AlSi, T_e = 577 °C. For AlSi8Mg4, the liquid fraction reached 100% at 615 °C instead of at the liquidus temperature, T_L = 595 °C, indicating that there was a shift of 20 K, probably caused by artefacts of the temperature measurement.

The highest expansions achieved corresponded to theoretical liquid fractions of 40–60% just after passing the solidus temperature. We assumed that the more liquid phase was formed in the early stages, the less gas losses due to the cracks take place, leading to better expansions due to the more effective use of the available gas and a more simultaneous nucleation. On the other hand, if too much liquid phase exists, the lack of solid particles and the corresponding low viscosity deteriorate foam stability, leading to coalescence and collapse and indirectly to a lower expansion. As a summary of the expansion performance of the alloys studied, Figure 5 clearly illustrates that some compositions (in green) showed higher maximum expansions, while others (in red) gave rise to lower expansions than the reference alloy (in blue). There was a tendency to higher expansions towards slightly higher Si and lower Mg contents than in the reference alloy. However, with 10 wt.% Si, only lower expansions could be achieved. It is known that Si lowers the viscosity of melts, whereas Mg increases it [39]. In Mg containing alloys, Mg_2Si particles are present in the melt [38], although they could not be detected in the in situ diffraction experiments, most likely due to the strong scattering by the molten phase. Low Mg contents lead to large pores due to the less stabilising oxides [3], fewer Mg_2Si particles, and lower viscosities as can be observed in Figures 6 and 7.

As a prediction and comparison, the amount of melt/liquid phase during melting and solidification was calculated using the CALPHAD method. The results obtained show the right tendency, as can be observed in Figure 9, but do not match the measurements exactly, because the method considers the amount of liquid phase during solidification of the already formed alloy. During melting of the foamable samples, the nominal composition and the corresponding phases in the thermodynamic equilibrium have still not formed, since the compacted samples mainly contain elemental powders (see Table 1), and diffusion during the short heating time does not allow for a full dissolution of the components. As shown in Figure 9, we can conclude that alloys with high Si contents show a narrower melting interval and lower temperatures for complete melting, while the influence of Mg is not clear, possibly due to the effect of a forming Mg_2Si phase.

A summary of the maximum area expansions achieved for foams with different alloy compositions as a function of the liquid fraction after passing the solidus temperature is presented in Figure 10. Most of the alloys studied show less expansion than the reference alloy, but some lead to a gain in maximum expansion. The latter determine the range of liquid fraction in which we can expect a good expansion. As it was shown, not only expansion is the parameter to take into account, but also foam structure, which is influenced by the amount of solid particles in the melt responsible for stabilisation. Therefore, a good compromise between high liquid fraction, facilitating expansion, and high solid fraction, enabling stabilisation, has to be found.

Finally, other process parameters, such as the foaming temperature and the temperature profile, could be further fine-tuned for each selected alloy. With all these improvements, foam quality can be positively influenced.

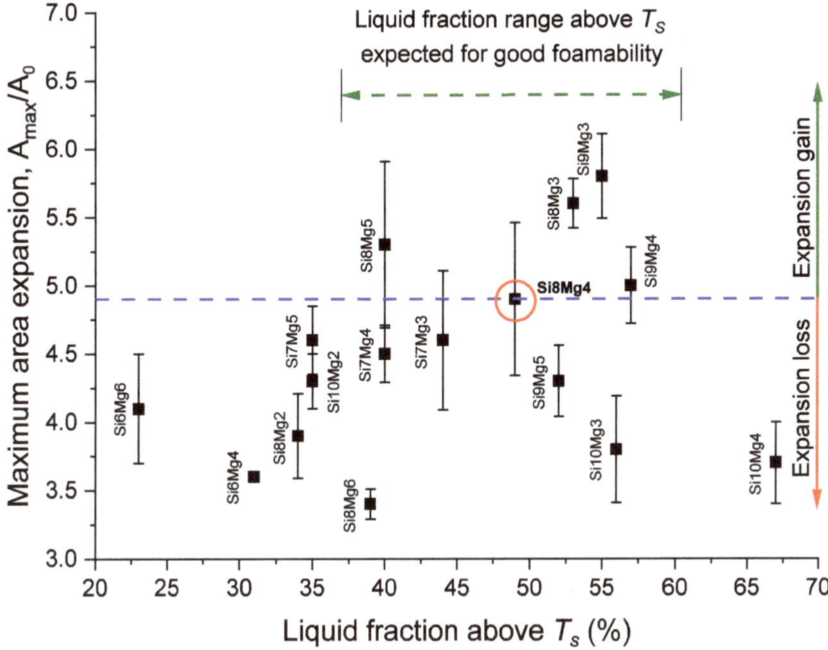

Figure 10. Maximum area expansion versus theoretical liquid fraction at the solidus temperature $T_S = 558\ °C$, displaying the expansion gain or loss found for different alloy compositions with respect to the reference system AlSi8Mg4. The range of expected good foamability is marked in green.

5. Conclusions

- An optimisation of alloy composition in the AlSiXMgY system for foaming was performed for compositions around the reference AlSi8Mg4.
- A tendency to higher expansions was observed for higher Si and lower Mg contents than in the reference system.
- A correlation between increased liquid fraction and high foam expansion was found up to a preferred range for the liquid fraction of 40–60% just above T_S. Even higher liquid fractions had an adverse effect on foamability.
- AlSi9Mg3 shows the highest expansion of all alloys with an acceptable foam structure, while AlSi9Mg5 has the best pore structure with an acceptable expansion. Therefore 4% Mg appears as a good compromise. Within the "island of good foaming" around AlSi8Mg4 cited in the introduction, the best point might lie slightly on the Si-rich side.

Author Contributions: Conceptualization, F.G.-M., J.B., L.A.R. and T.R.N.; methodology, F.G.-M., L.A.R., P.H.K., T.R.N., C.M.S. and M.K.; software, P.H.K. and C.M.S.; formal analysis, L.A.R. and M.K.; writing—original draft preparation, F.G.-M.; writing—review and editing, F.G.-M. and J.B. All authors have read and agreed to the published version of the manuscript.

Funding: This research was funded by the Deutsche Forschungsgemeinschaft through grants GA 1304/5-1 and BA 1170/40-1.

Acknowledgments: We acknowledge the Paul Scherrer Institut, Villigen, Switzerland for providing synchrotron radiation beam time at the TOMCAT beamline X02DA of the SLS.

Conflicts of Interest: The authors declare no conflict of interest.

Appendix A

The amount of liquid fraction above the solidus temperature of the AlSiXMgY system was calculated with Thermo-Calc. A variation of Si and Mg contents while keeping the other elements constant is shown in Figure A1. From these graphs, we can deduce that in this composition range, an increase of the Si content will increase the melt fraction above T_S, while an increase of Mg will reduce it.

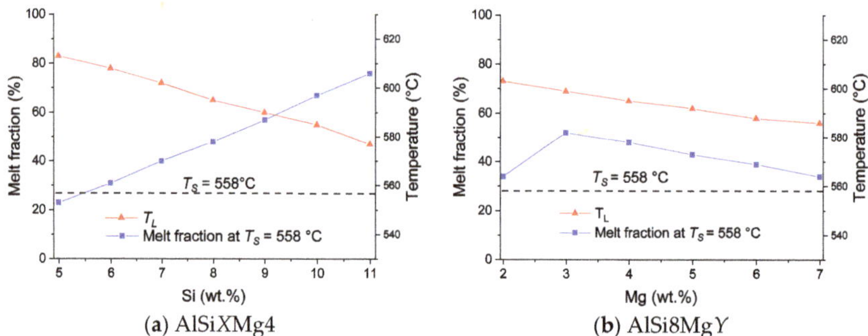

(a) AlSiXMg4   **(b)** AlSi8MgY

Figure A1. Amount of liquid fraction (in blue) above the solidus temperature T_S = 558 °C calculated with the software Thermo-Calc and the COST507B database for the alloys **(a)** AlSiXMg4 and **(b)** AlSi8MgY for different amounts of Mg and Si, respectively. The corresponding liquidus temperatures T_L are marked in red.

Appendix B

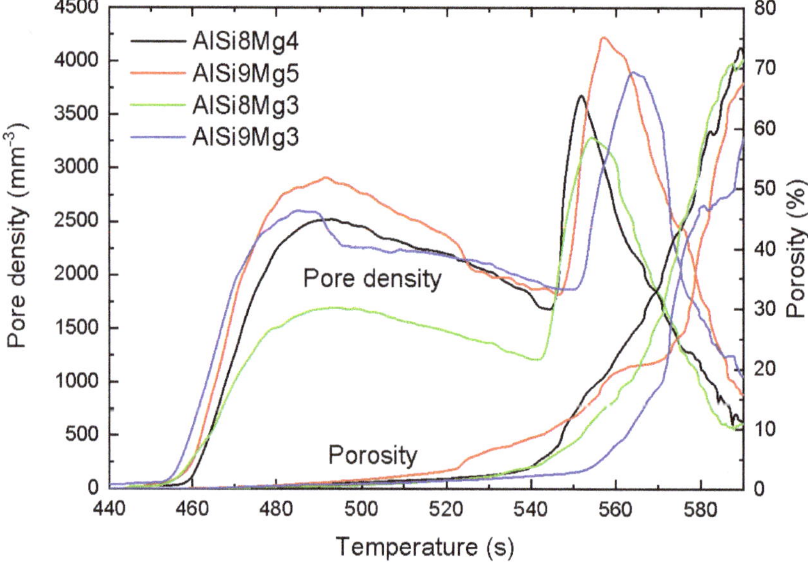

Figure A2. Pore density and porosity for selected samples of different alloy compositions over temperature corresponding to the samples shown in Figure 2.

Appendix C

X-ray radiographies of the evolving liquid metal foams allowed us to follow the structure development throughout foaming. Figures A3 and A4 show the nucleation, maximum expansion, and end expansion stages for a series of alloys of composition AlSi8MgY (Y = 2–6 wt.%) and AlSiXMg4 (X = 6–10 wt.%), respectively. Several features such as melt bubbles, cracks in early stages followed by big initial bubbles or drainage are marked in red.

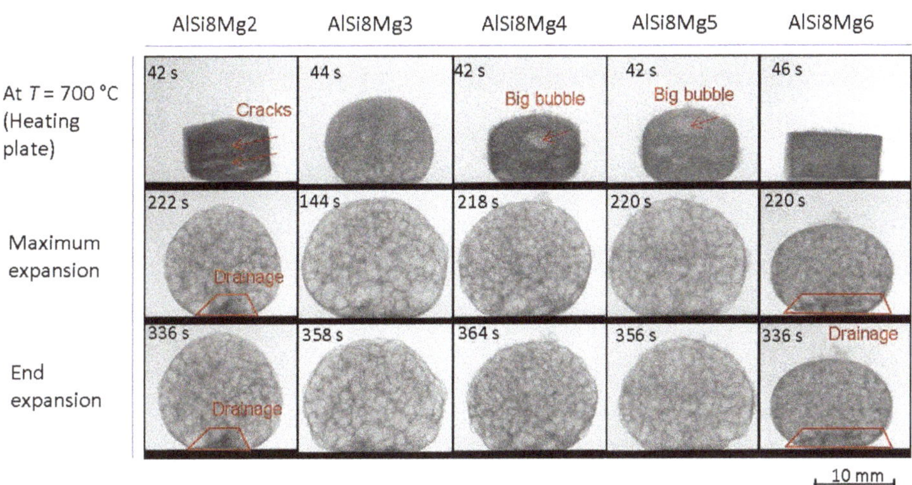

Figure A3. X-ray radiographies of evolving liquid metal foams of compositions AlSi8MgY (Y = 2–6 wt.%) extracted from an in situ radioscopic series.

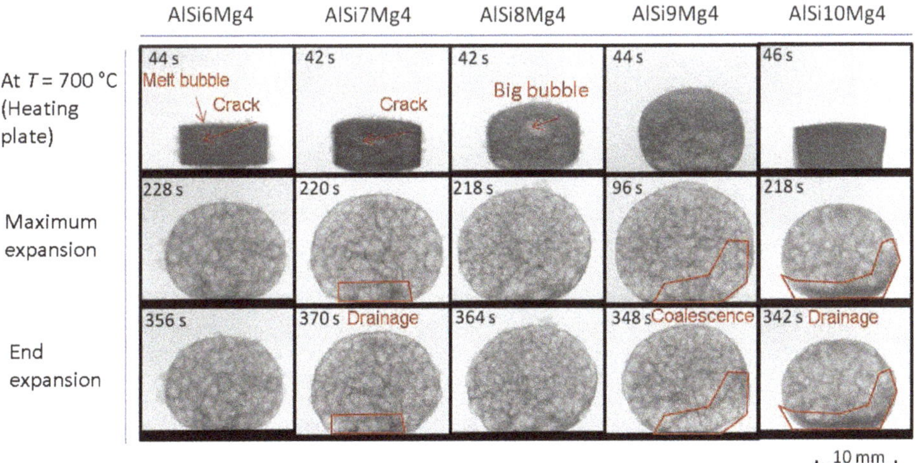

Figure A4. X-ray radiographies of evolving liquid metal foams of compositions AlSiXMg4 (X = 6–10 wt.%) extracted from an in situ radioscopic series.

Appendix D

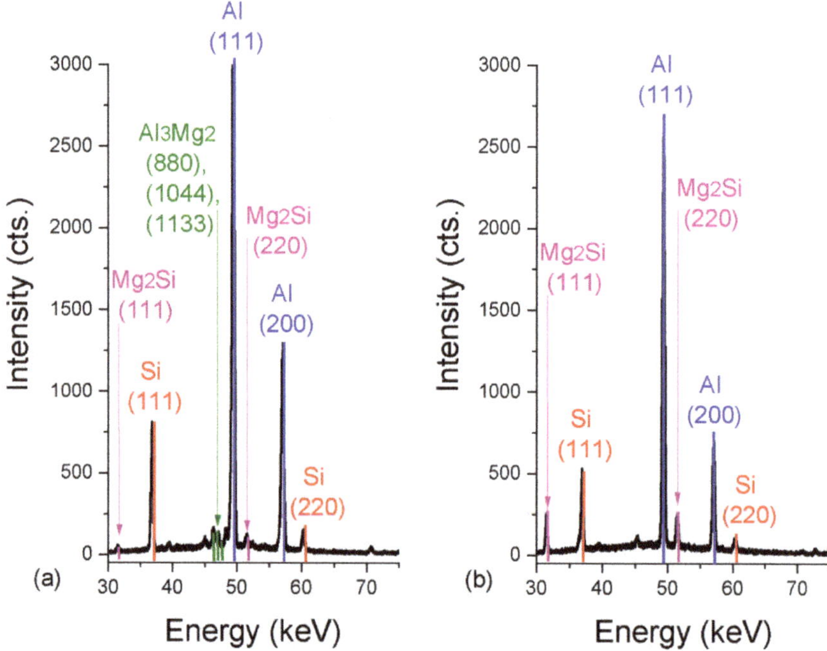

Figure A5. Diffraction pattern of an AlSi8Mg4 sample (**a**) before and (**b**) after foaming. The increase of the Mg_2Si phase after foaming is clearly observable.

References

1. García-Moreno, F. Commercial applications of metal foams: Their properties and production. *Materials* **2016**, *9*, 85. [CrossRef] [PubMed]
2. Banhart, J. Manufacturing routes for metallic foams. *JOM* **2000**, *52*, 22–27. [CrossRef]
3. Körner, C.; Arnold, M.; Singer, R.F. Metal foam stabilization by oxide network particles. *Mater. Sci. Eng. A* **2005**, *396*, 28–40. [CrossRef]
4. Rack, A.; Helwig, H.M.; Bütow, A.; Rueda, A.; Matijašević-Lux, B.; Helfen, L.; Goebbels, J.; Banhart, J. Early pore formation in aluminium foams studied by synchrotron-based microtomography and 3-D image analysis. *Acta Mater.* **2009**, *57*, 4809–4821. [CrossRef]
5. Kamm, P.H. Der Schäumprozess von Aluminiumlegierungen: Tomoskopische Untersuchung der Gasnukleation. Ph.D. Thesis, Technische University Berlin, Berlin, Germany, August 2017.
6. Neu, T.R. Einfluss der Kompaktierung auf das Schäumverhalten von pulvermetallurgisch hergestellten Metallhalbzeugen. Ph.D. Thesis, Technische Universität Berlin, Berlin, Germany, July 2018.
7. Duarte, I.; Banhart, J. A study of aluminium foam formation-kinetics and microstructure. *Acta Mater.* **2000**, *48*, 2349–2362. [CrossRef]
8. Weigand, P. Untersuchung der Einflussfaktoren auf die pulvermetallurgische Herstellung von Aluminiumschäumen. Ph.D. Thesis, RWTH Aachen University, Aachen, Germany, May 1999.
9. Lázaro, J.; Solórzano, E.; Rodríguez-Pérez, M.A.; Kennedy, A.R. Effect of solidification rate on pore connectivity of aluminium foams and its consequences on mechanical properties. *Mater. Sci. Eng. A* **2016**, *672*, 236–246. [CrossRef]
10. Simancik, F.; Behulova, K.; Bors, L. *Effect of Ambient Atmosphere on the Foam Expansion, Cellular Metals and Metal Foaming Technology: Metfoam, Bremen, 2001*; Banhart, J., Ashby, M.F., Eds.; MIT Publishing: Bremen, Germany, 2001; pp. 89–92.

11. Banhart, J.; Baumeister, J. Deformation characteristics of metal foams. *J. Mater. Sci.* **1998**, *33*, 1431–1440. [CrossRef]

12. Sahm, H. *Untersuchung des Einflusses von Magnesium auf die Schäumbarkeit von Aluminiumlegierungen*; Engineer Report; TU Bergakademie Freiberg: Freiberg, Germany, March 1994.

13. Helwig, H.M.; Garcia-Moreno, F.; Banhart, J. A study of Mg and Cu additions on the foaming behaviour of Al-Si alloys. *J. Mater. Sci.* **2011**, *46*, 5227–5236. [CrossRef]

14. Zhiqiang, G.; Donghui, M.; Xiaoguang, Y.; Xue, D. Effect of Mg addition on the foaming behaviour of AlSi7 based alloy prepared by powder metallurgy method. *Rare Metal Mater. Eng.* **2016**, *45*, 3068–3073. [CrossRef]

15. Lehmhus, D.; Busse, M. Potential new matrix alloys for production of PM aluminium foams. *Adv. Eng. Mater.* **2004**, *6*, 391–396. [CrossRef]

16. Helwig, H.M.; Banhart, J.; Seeliger, H.-W. Metal foams made of an aluminium alloy, use and manufacturing route. Patent EP 2 143 809, 14 September 2011.

17. Pohltec Metalfoam. Available online: http://metalfoam.de/ (accessed on 18 December 2019).

18. Nagelberg, A.S.; Antolin, S.; Urquhart, A.W. Formation of Al_2O_3/metal composites by the directed oxidation of molten aluminum-magnesium-silicon alloys: Part ii, growth kinetics. *J. Am. Ceram. Soc.* **1992**, *75*, 455–462. [CrossRef]

19. Jiménez, C.; Garcia-Moreno, F.; Pfretzschner, B.; Klaus, M.; Wollgarten, M.; Zizak, I.; Schumacher, G.; Tovar, M.; Banhart, J. Decomposition of TiH_2 studied in situ by synchrotron X-ray and neutron diffraction. *Acta Mater.* **2011**, *59*, 6318–6330. [CrossRef]

20. Helwig, H.M.; Hiller, S.; Garcia-Moreno, F.; Banhart, J. Influence of compaction conditions on the foamability of AlSi8Mg4 alloy. *Metall. Mater. Trans. B* **2009**, *40*, 755–767. [CrossRef]

21. Garcia-Moreno, F.; Fromme, M.; Banhart, J. Real-time x-ray radioscopy on metallic foams using a compact micro-focus source. *Adv. Eng. Mater.* **2004**, *6*, 416–420. [CrossRef]

22. Genzel, C.; Denks, I.A.; Gibmeier, J.; Klaus, M.; Wagener, G. The materials science synchrotron beamline EDDI for energy-dispersive diffraction analysis. *Nucl. Instrum. Methods Phys. Res. Sect. A* **2007**, *578*, 23–33. [CrossRef]

23. Jiménez, C.; Plaepow, M.; Kamm, P.H.; Neu, T.; Klaus, M.; Wagener, G.; Banhart, J.; Genzel, C.; García-Moreno, F. Simultaneous X-ray radioscopy/tomography and energy-dispersive diffraction applied to liquid aluminium alloy foams. *J. Synchrotron Rad.* **2018**, *25*, 1790–1796. [CrossRef] [PubMed]

24. Warren, B.E. *X-Ray Diffraction*; Dover: New York, NY, USA, 1990.

25. Garcia-Moreno, F.; Solorzano, E.; Banhart, J. Kinetics of coalescence in liquid aluminium foams. *Soft Matter* **2011**, *7*, 9216–9223. [CrossRef]

26. García-Moreno, F.; Kamm, P.H.; Neu, T.R.; Bülk, F.; Mokso, R.; Schlepütz, C.M.; Stampanoni, M.; Banhart, J. Using X-ray tomoscopy to explore the dynamics of foaming metal. *Nat. Commun.* **2019**, *10*, 3762. [CrossRef]

27. Fife, J.L.; Rappaz, M.; Pistone, M.; Celcer, T.; Mikuljan, G.; Stampanoni, M. Development of a laser-based heating system for in situ synchrotron-based X-ray tomographic microscopy. *J. Synchrotron Rad.* **2012**, *19*, 352–358. [CrossRef]

28. Mokso, R.; Schlepütz, C.M.; Theidel, G.; Billich, H.; Schmid, E.; Celcer, T.; Mikuljan, G.; Sala, L.; Marone, F.; Schlumpf, N.; et al. GigaFRoST: The gigabit fast readout system for tomography. *J. Synchrotron Rad.* **2017**, *24*, 1250–1259. [CrossRef]

29. Buhrer, M.; Stampanoni, M.; Rochet, X.; Buchi, F.; Eller, J.; Marone, F. High-numerical-aperture macroscope optics for time-resolved experiments. *J. Synchrotron Rad.* **2019**, *26*, 1161–1172. [CrossRef] [PubMed]

30. Paganin, D.; Mayo, S.C.; Gureyev, T.E.; Miller, P.R.; Wilkins, S.W. Simultaneous phase and amplitude extraction from a single defocused image of a homogeneous object. *J. Microsc.* **2002**, *206*, 33–40. [CrossRef] [PubMed]

31. Marone, F.; Stampanoni, M. Regridding reconstruction algorithm for real-time tomographic imaging. *J. Synchrotron Rad.* **2012**, *19*, 1029–1037. [CrossRef] [PubMed]

32. Andersson, J.O.; Helander, T.; Höglund, L.; Shi, P.; Sundman, B. Thermo-Calc & DICTRA, computational tools for materials science. *Calphad* **2002**, *26*, 273–312.

33. Schaffnit, P.; Stallybrass, C.; Konrad, J.; Stein, F.; Weinberg, M. A Scheil–Gulliver model dedicated to the solidification of steel. *Calphad* **2015**, *48*, 184–188. [CrossRef]

34. Mukherjee, M.; Garcia-Moreno, F.; Banhart, J. Solidification of metal foams. *Acta Mater.* **2010**, *58*, 6358–6370. [CrossRef]

35. Helfen, L.; Baumbach, T.; Stanzick, H.; Banhart, J.; Elmoutaouakkil, A.; Cloetens, P. Viewing the early stage of metal foam formation by computed tomography using synchrotron radiation. *Adv. Eng. Mater.* **2002**, *4*, 808–813. [CrossRef]
36. Radtke, L. Optimierung einer AlSiMg-Legierung für die Metallschaumherstellung. Master's Thesis, Technische Universität Berlin, Berlin, Germany, January 2017.
37. Lumley, R.; Sercombe, T.; Schaffer, G. Surface oxide and the role of magnesium during the sintering of aluminum. *Metall. Mater. Trans. A* **1999**, *30*, 457–463. [CrossRef]
38. Paes, M.; Zoqui, E.J. Semi-solid behavior of new Al–Si–Mg alloys for thixoforming. *Mater. Sci. Eng. A* **2005**, *406*, 63–73. [CrossRef]
39. Di Sabatino, M. Fluidity of aluminium foundry alloys. Ph.D. Thesis, Norwegian University of Science and Technology, Trondheim, Norway, September 2005.

Article

Effects of Eutectic Modification and Grain Refinement on Microstructure and Properties of PM AlSi7 Metallic Foams

Dirk Lehmhus [1,*], Daniela Hünert [2], Ulrike Mosler [3], Ulrich Martin [4] and Jörg Weise [1]

[1] Fraunhofer Institute for Manufacturing Technology and Advanced Materials IFAM, Wiener Straße 12, 28359 Bremen, Germany; joerg.weise@ifam.fraunhofer.de

[2] Rolls-Royce Deutschland Ltd. & Co KG, Eschenweg 11, 15827 Blankenfelde-Mahlow, Germany; daniela.huenert@rolls-royce.com

[3] Deutsche Bahn AG, DB Systemtechnik GmbH, Bahntechnikerring 74, 14774 Brandenburg-Kirchmöser, Germany; ulrike.mosler@deutschebahn.com

[4] TU Bergakademie Freiberg, Gustav-Zeuner-Straße 5, 09599 Freiberg, Germany; martin@ww.tu-freiberg.de

* Correspondence: dirk.lehmhus@ifam.fraunhofer.de; Tel.: +49-421-2246-7215

Received: 14 October 2019; Accepted: 13 November 2019; Published: 20 November 2019

Abstract: For AlSi7 foams, microstructure modification by variation of solidification rates and addition of Sr, B and $TiB_2/TiAl_3$ was investigated and its transfer to powder metallurgical metal foaming processes demonstrated. Microstructural characterization focused on grain size and morphology of the eutectic phase. Cooling rates during solidification were linked to secondary dendrite arm spacing, establishing a microstructure-based measure of solidification rates. Effects of refining and modification treatments were compared and their influence on foam expansion evaluated. Studies on foams focused on comparison of micro- and pore structure using metallographic techniques as well as computed tomography in combination with image analysis. Reference samples without additives and untreated as well as annealed TiH_2 as foaming agent allowed evaluation of pore and microstructure impact on mechanical performance. Evaluation of expansion and pore structure revealed detrimental effects of Sr and B additions, limiting the evaluation of mechanical performance to the TiB_2 samples. These, as well as the two reference series samples, were subjected to quasi-static compression testing. Stress-strain curves were gained and density-dependent expressions of ultimate compressive strength, plateau strength and tangent modulus derived. Weibull evaluation of density-normalized mechanical properties revealed a significant influence of grain size on the Weibull modulus at densities below 0.4 g/cm^3.

Keywords: aluminum foam; metal foam; aluminum alloys; grain refinement; modification; microstructure; mechanics of materials; metallurgy; melt treatment; powder metallurgy

---

1. Introduction

1.1. Foam Fundamentals

Metallic foams based on the powder compact melting or Fraunhofer process have by now reached sufficient levels of maturity to allow series production, e.g., for applications in the transport, machine tool and even the building industry [1–4]. The underlying process is based on hot compaction of a mixture of matrix metal, usually aluminum or an aluminum alloy, and foaming agent powders, usually TiH_2. The resulting precursor material is then heated above its melting point and expands to yield a liquid foam, which is stabilized by solidification. The method was patented in 1990 by Baumeister et al. and has since been described in several publications [5,6]. A recent overview contrasting this process in terms of foam structure, performance and cost with alternative approaches has recently been published by Lehmhus et al. [7].

As a natural consequence of the material's success, the call for statistically well founded design criteria as a prerequisite for making best use of the material's capabilities gains urgency. One major step to this end is an improved understanding of the role of different structural features in determining the mechanical properties, as well as their scatter. Both, however, are influenced by several characteristics, among which Mosler et al. and Martin et al. suggest the following hierarchy [8,9]:

(a)   global density
(b)   foam structure
(c)   matrix alloy
(d)   foam microstructure

Mu et al. introduce friction between cell walls as a further mechanism influencing behavior under compressive load, which, however, becomes effective only at high strain levels beyond the stress-strain curve's typical plateau [10]. Density as the dominating aspect may be subdivided into global density and systematic density variations such as density gradients as typically induced by foam drainage, or by solidification shrinkage. Mechanical testing parallel to such density gradients will cause the lowest density cross sections to fail first, leading to a lowered yield point and a steeper plateau region than observed in samples of matching average density either tested perpendicular to any density gradient, or showing a homogeneous overall structure [11].

To shed additional light on the relative importance of the above features in determining mechanical performance, the present study concentrates on microstructure variation achieved via modification and grain refinement of the foam matrix alloy, AlSi7. Part of the methodology is to eliminate as far as possible the effects of the global density (a) by normalizing the results of mechanical testing. This was done according to a Gibson-Ashby type formulation of strength and stiffness as a function of global density [12]. Foam structural characteristics (b) were documented for all samples subjected to mechanical testing based on computed tomography (CT) scans in order to have a further basis for explanation of potential outliers among compression test results. To independently study the significance of cell structure, a second reference sample series was introduced based on thermally treated TiH_2 as foaming agent. Treatments of this and similar kind have been demonstrated to allow tailoring of decomposition kinetics and can thus be employed to modify or improve pore structure and morphology [13–18]. Moreover, documentation of structural features is required for investigating any change induced by additives in this respect. The matrix alloy (c) itself is kept the same in all sample series, while the expression of the microstructure (d) is deliberately modified between series—either by means of appropriate additives (Sr, B, TiB_2) or by varying the cooling rate in solidification.

Background to this approach are earlier observations suggesting that in Al-Si foams, coarse forms of the eutectic (lamellar or needle-shaped Si phase) can lead to Al-Si interface planes similar in size to the typical cell wall thickness, the latter being approximately 80 μm for the alloy system in question [19]. Such interfaces have been suggested as preferred initial failure sites in aluminum alloy foams [20]. The notion that Si particles and their geometry influence failure is also supported by several fracture mechanical studies on bulk Al-Si and closely related alloys, though mostly focusing on failure under conditions of fatigue. Among these, Gall et al. suggest that fatigue cracks preferably grow along the Al-Si interface [21]. Su et al. also observed particle debonding under conditions of wear based on experimental and numerical studies, while alternative mechanisms include particle fracture and plastic deformation of the Al matrix. Spherical shapes of Si particles are shown to reduce susceptibility to fracture [22,23]. Lados et al. confirmed the role of both primary Al dendrite and Si phase shape and size in this respect [24]. Chan et al. suggested that crack growth in fatigue of B319 Al alloy is mostly via fractured and debonded Si particles, while interdendritic grain boundaries provide preferred fracture paths in later stages of rapid crack growth [25]. In contrast, Xia et al. derived quantitative values of considerable magnitude (namely an interface shear strength of 240 ± 6 MPa and a normal strength of

247 MPa) from nanoidentation-based experimental studies combined with finite element analysis and claimed a good match with certain atomistic simulations [26].

Grain refinement and modification are thus directed at eliminating potential weaknesses caused by the foam matrix alloy microstructure. The relevance of such investigations is stressed by the fact that due to their good processing characteristics, near-eutectic Al-Si alloys in a composition range from approximately 7 wt.% Si upwards and related systems, e.g., containing further additions of Mg or Cu, have retained their role as backbone both in classic metal foam and metal foam sandwich production [2,27] and in more recent developments such as Advanced Pore Morphology (APM) foams [28–31]. In its concentration on additive-based microstructure modification, the approach complements studies on heat treatment of foams [32–34] and variation of the matrix alloy [35–37]. Of these, when it comes to identifying the role of matrix alloy and specifically microstructure, the latter will suffer greatly from coincidental differences in expansion characteristics and thus pore structure, as has recently been shown in much detail by Helwig et al. [38]. The former, in contrast, allows control of matrix material properties at constant pore structure, but is limited to alloys that show a notable response to heat treatment, such as the 6000 and 7000 series Al alloys susceptible to precipitation hardening.

1.2. Grain Refinement and Modification of Al-Si Alloys

While grain refinement and modification of Al-Si and related alloys are established techniques in metal casting, neither has yet been evaluated in the context of powder metallurgically produced foams. Additions of TiB_2, a substance known for its refining capability, have as yet only been considered in the context of particle stabilization, e.g., by Kennedy et al. Grain refinement was also not investigated, nor was any such effect to be expected in these studies as a result of TiB_2 particle size, which greatly exceeded dimensions of nucleation sites effective in the microstructural refining effect [39].

Depending on the understanding specifically of grain refinement, transfer of this essentially liquid phase technique to powder metallurgy may seem contradictory, and, in fact, the relevant literature is mostly associated with casting processes during which the liquidus line is passed with a considerable margin. An exception to this rule is a study by Nafisi and Ghomashchi, who also considered the semi-solid region in their work on A356, i.e., AlSi7 Mg, alloys, though only with respect to casting billets for further, semi-solid processing. Thus, the effects of, e.g., inhomogeneous distributions of refiners in the semi-solid state, is not reflected in their investigation [40]. In any case, though metal foams of the type covered here do start as a powder metallurgical (PM) precursor, they cross solidus and usually also liquidus temperature for a limited time during foaming. It has repeatedly been shown that unless specific measures are taken to ensure that expansion occurs solely in the semi-solid region [38,41,42], the final microstructure of the foam is entirely formed during solidification [36]. A remaining concern is the question how effective treatments can be if modifiers are contained only in a limited Al powder fraction, which is diluted by addition of conventional Al and alloying element powders. For economic reasons (cost of specially prepared powders with refining/modifying agents), such a processing route seems mandatory if modification and refinement are to be established on a commercial basis.

Grain refinement generally relies on increasing the number of available nucleation sites in a melt. This can either be achieved by influencing the constitution of the melt in a way that the critical radius above which nuclei may grow is reduced, and thus using mechanisms of homogeneous nucleation, or via heterogeneous nucleation by offering additional nucleation sites. Current state of the discussion suggests that standard treatments with TiB_2/Al_3Ti and AlB_2 combine effects of both kinds, as is outlined in a dedicated review on the underlying principles of grain refinement mechanisms provided by Easton and St. John. According to them, the models proposed thus far fall into two main categories, termed the Nucleant and the Solute Paradigm. The Nucleant Paradigm itself encompasses nucleant particle theories, which stress the role of TiB_2 and isomorphous AlB_2 particles as primary nucleation sites, and phase diagram theories, which consider the properitectic Al_3Ti phase as the main nucleation site. The problem of the former theories is that based on crystallographic considerations and some experimental evidence, both TiB_2 and AlB_2 have to be considered poor nucleants and only really show their benefits in the presence of additional Ti. In comparison, Al_3Ti is a powerful nucleant due to a number of beneficial orientation relationships with α-Al and its peritectic reaction forming α-Al. However, pure phase diagram theories once again fail to deliver an explanation for the superior performance of combination of certain Ti levels with TiB_2, AlB_2 and $(Al,Ti)B_2$ as used in commercial grain refiners of the so-called 5-1 Ti-B-Al kind. This deficiency is reflected in theories combining aspects of both views, including a possible higher Al_3Ti stability in Al melts in the presence of borides or the peritectic hulk theory, which assumes a preferred nucleation of Al_3Ti on the melt-TiB_2 interface [43,44]. More recently, Schumacher et al. have demonstrated that the effectiveness of TiB_2 particles in fact relies on their being covered by Al_3Ti layers, which only form at Ti content levels exceeding 0.15 wt.% [45].

For modification of the eutectic Si phase, Sr- and Na-based treatments are common practice. Explanations of their effectiveness consider the influence of additives on Si nucleation and crystal growth. Among the latter, a distinction is made between kinetic effects caused by reduction of Si diffusion coefficients [46] and surface tension [47,48] in Sr-containing Al melts and blocking of Si lattice planes preferred in further crystal growth by local Na enrichment [49]. More recent studies by Srirangam et al. confirm significant influence of Sr additions on the liquid state structure of Al-Si alloys seen in simulations using synchrotron radiation, thus supporting a role of the additive in nucleation of the Si phase [50]. Further contributions to this topic by Timpel et al. based on atom-probe tomography and transmission electron microscopy suggest parallelism of two mechanisms, of which growth restriction of the eutectic Si phase is just one. Their measurements indicate the existence of two types of co-segregation of Sr with Al and Si, of which type I causes multiple twin formation in the Si crystal ("impurity-induced twinning"), which further facilitates multiple direction crystal growth, while larger size type II segregations restrict Si crystal growth [51].

Common to all theories, be they directed at grain refinement or eutectic modification, is that they require additional nucleation sites or solved additives in homogeneous distribution once solidification starts. Obviously, this is not necessarily a given thing in a PM material in which (a) refining/modifying agents may only be present in part of the metal powder components, and which (b) reaches the liquid state solely in the course of foam expansion and thus for seconds rather than minutes before solidification is initiated to stabilize the developing foam. Stability of the modified foam is an issue in itself, as earlier investigations indicate that B and Sr, but also TiB_2 additions alter characteristics such as surface tension and viscosity [50,52–54], parameters that are instrumental in determining a metal foam's susceptibility to detrimental phenomena such as cell wall rupture, coalescence or drainage [15,55,56]. For this reason, initial studies of refinement and modification have been performed on powder compacts without foaming agent and varied levels of additive-containing Al powders.

The present study will thus elucidate the possibility and the effects of transferring the aforementioned approaches for microstructural optimization to PM aluminum foams by evaluating microstructural features, expansion characteristics and mechanical behavior of such materials with and without additives.

2. Materials and Methods

As motivated above, the work described here is divided in three major sections, namely (a) the general evaluation of grain refinement and eutectic modification of powder compacts, (b) the evaluation of expansion characteristics of such powder compacts with addition of 0.5 wt.% TiH_2 as blowing agent and (c) the microstructural and mechanical evaluation of foams produced from these precursor materials.

In this, steps (a) and (b) turned out to effectively narrow down the set of samples suitable for detailed studies along the lines of step (c).

2.1. Materials and Sample Production

Production of precursor materials and foams followed the principles of the Fraunhofer or powder compact melting process. Alloying was based on mixing of elementary Al and Si powders. Table 1 lists the powders used and their composition, including, where applicable, levels of modifying agent content, as well as measures of particle size according to specification and measurement. Powders containing modifying or grain refining additions were specifically prepared for this purpose by Alpoco and match in constitution materials commonly employed in preparation of powder compacts used in melt treatment of aluminum casting alloys. The choice of Sr as modifying agent as well as B and TiB_2 for grain refinement was based on the dominant role of these additives in melt treatment in casting of aluminum in general, and specifically of hypoeutectic Al-Si alloys [57–60].

Particle sizes were measured using a Coulter LS 130 laser particle size analyzer (Beckman Coulter Inc., Brea, California, USA). Powders were dispersed in ethanol and subjected to ultrasonic agitation to break agglomerates before measurement.

Mixing of powders for hot pressing was done in a tumble mixing device. Compaction was realized as two-step process at a furnace temperature of 450 °C, using a Zwick 1474 universal testing machine with added furnace (ZwickRoell GmbH & Co. KG, Ulm, Germany), the initial load being 60 kN or 74.6 MPa during the first and a constant 90 kN (111.9 MPa) during the final stage. Holding time was 20 min for each step. The hardened steel dies with circular cross-section were filled at room temperature. During the first compaction stage, settling effects and plastic deformation in the powder bed led to a load decrease to below 20 kN. Hot extruded precursor material based on identical Al, Si and TiH_2 powders was acquired from Entwicklungsgemeinschaft Schunk-Honsel for use as reference. Powder mixtures were cold isostatically pressed to a relative density of approximately 0.75 to 0.85. The billets gained were heated to a temperature of 450 °C and extruded using a die with rectangular cross-section of 160 × 40 mm. The thickness of 40 mm allowed producing cylindrical samples for expansion measurements and foaming of specimens for compression tests from the extruded precursor material with the central axis oriented perpendicular to the extrusion direction and the larger transversal extension of the extruded geometry.

Axially compacted samples were turned to 31 mm diameter and 28.5 mm height, corresponding to a maximum global sample of 0.66 g/cm^3 considering the foaming molds used. Higher levels of foam density were achieved by interrupting foam expansion prior to complete filling of the mold. Both extruded and hot pressed materials show a preferred direction of expansion. In both types of samples, this direction coincides with the cylinder axis.

Foaming of all samples took place in a specially designed furnace configuration with an attached cooling device allowing both water and forced convection cooling. Furnace and attached cooling rig were designed to be raised and lowered automatically to avoid moving the then unstable sample prior to solidification. Foaming molds were machined from high temperature oxidation resistant steel grades (1.4713, 1.4828) with internal dimensions of D45 × 55 mm. The setup, including a foaming mold, is depicted in Figure 1.

Table 1. Composition, production method and supplier of powder variants.

Powder	Supplier Information						Measurement	
	Name	Production Method	Purity [wt.%]	Particle Size [μm]	Modifying Agent (MA)	MA Content [wt.%]	Particle Size, D_{mean} [μm]	Particle Size, D_{50} [μm]
Al-1	Ecka [1] grade AS71	air atomized	99.7	<160	-	-	77.38	70.08
Al-2	Alpoco	inert gas atomized	99.7	<150	-	-	39.40	25.41
Al/Sr	Alpoco	inert gas atomized	n. a. [2]	<150	Sr	0.05	39.61	28.00
Al/B	Alpoco	inert gas atomized	n. a. [2]	<150	B	0.007	57.30	41.63
Al/TiB$_2$	Alpoco	inert gas atomized	n. a. [2]	<150	Ti/TiB$_2$	0.05 Ti, as 5/1 TiBAl	42.11	29.71
Si	Ölschläger	milled	98.5	60–100	n.a.	n.a.	25.71	20.25
TiH$_2$	Chemetall grade N	milled	99	< 63	n.a.	n.a.	18.14	16.04

[1] Ecka Granules. [2] Not applicable.

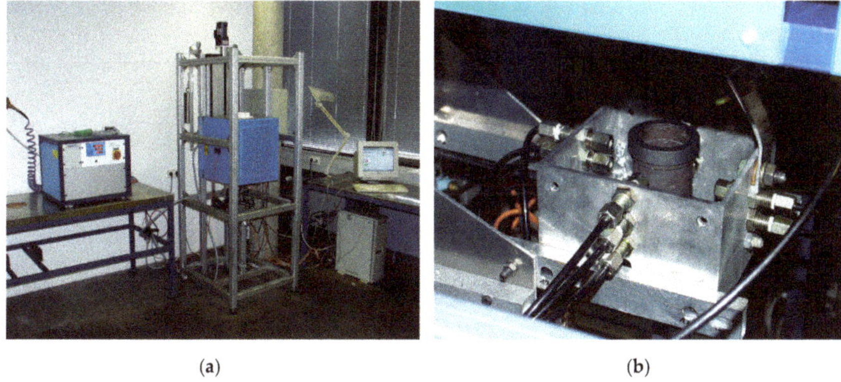

Figure 1. Foaming furnace used for sample production—(**a**) general layout including control units for heating (left) and movement (right), (**b**) nozzle arrangement for forced convection (FC) cooling (below furnace, with cylindrical foaming mold).

For initial evaluation of refining and modification effects achievable in powder compacts, altogether nine different types of samples without blowing agent were produced containing three different levels of B, TiB_2 and Sr, respectively. These specimens, as well as reference samples containing neither grain refining nor modifying additives, were subjected to thermal treatments at 660 and 680 °C, meant to simulate the foaming process. For those powder compacts representing the medium level of additives, a further variation of the quenching step following thermal treatment was foreseen. Full details of all sample series are reported in Table 2.

Table 2. Overview of powder compact sample series for evaluation of grain refinement and modification of the Al-Si eutectic.

Sample Series		1			2		
Holding temperature	[°C]	660			680		
Holding time	[min.]	5			10		
Cooling method		water quenching (WQ)	forced convection (FC)	natural convection (NC)	water quenching (WQ)	forced convection (FC)	natural convection (NC)
Reference AlSi7					n.a.		
	[wt%]	0.007	-	-	-	-	-
B grain refined		0.014	0.014	0.014	0.014	0.014	0.014
		0.028	-	-	-	-	-
	[wt%]	0.026	-	-	-	-	-
TiB_2 grain refined [1]		0.035	0.035	0.035	0.035	0.035	0.035
		0.0465	-	-	-	-	-
	[wt%]	0.013	-	-	-	-	-
Sr modified		0.025	0.025	0.025	0.025	0.025	0.025
		0.0465					

[1] Ti as Al-5Ti-1B.

Evaluation of the expansion characteristics of a subset of the above sample types, including all additives, led to three different compositions of foamed samples being produced for structural, microstructural and mechanical evaluation. These are listed in Table 3. Further foam samples were produced to substantiate the observations on microstructural features made on powder compact samples.

Table 3. Overview of foam sample series for mechanical testing. As also implied by the series' designations, forced convection cooling was employed in all cases.

		Sample Series		
		Al-FC	Al/TiB$_2$-FC	Al-TiH$_2$-ht-FC
Number of compression test samples	[-]	60	38	20
Compaction method	hot …	extrusion	axial pressing	axial pressing
Al powder		Al-1	Al-2; Al/TiB$_2$	Al-1
Foaming agent		TiH$_2$ as received	TiH$_2$ as received	TiH$_2$, treated 4h@500 °C, air
Ti level (as 5/1 TiBAl)	[ppm]	-	465	-
min./max. density of compression test samples	[g/cm^3]	0.18/0.78	0.17/0.61	0.19/0.92
cooling method		forced conv. [1]	forced conv.	forced conv.

[1] Forced convection using compressed air.

2.2. Microstructural Characterization

Foam sample preparation followed general guidelines laid down by Müller et al. and Mosler et al. [61,62]. To clearly distinguish microstructural features such as the individual primary Al grains, Barker etching was employed. The method provides grain-level contrast under polarized light based on local crystallographic orientation via this feature's influence on the growth of thin oxide layers during anodic etching [63]. For quantitative image analysis, the software package A4i-Analysis developed by Acquinto was used.

Grain size was established via two alternative approaches. For Barker etched samples, the largest extension of several grains' was measured on approximately 100 grains per sample. Since larger grains showed better contrast, grain size tends to be slightly overestimated in this case. As an alternative, scanning electron microscopy and electron backscatter diffraction (EBSD) were employed for image acquisition. The method proved critical in eutectic regions, where presence of several Al-Si transitions in close vicinity did not allow for unambiguous detection of orientations. To compensate for this effect, a dilation of clearly identified Al grains was performed, leading to a single phase microstructure with polyhedral grains. General consideration of the alloy's characteristics suggests that these approximate to the circumference of the complex shaped dendritic grains which make up the true microstructure. As a consequence, it is possible to derive the grain sizes of the true microstructure from its processed counterpart. In the present study, this has been done by measuring linear intercepts of grain boundaries for approximately 250 grains per sample. The second advantage, besides the increased database, is the fact that the method is capable of distinguishing and thus evaluating even very small grains, while the light microscopy approach effectively stressed larger grains showing better contrast due to the subjective selection step involved. In the following text, whenever grain sizes are given, the fundamental principle used in determining these is given.

Analysis of the degree of modification was based on metallographic sections images of which were acquired using light microscopy. In following Ohser and Lorz's recommendation, the specific interface area S_V between Al and Si phase in the eutectic as well as the integral of the average curvature of the Al-Si boundary M_V in the two-dimensional (2D) section were determined [64].

As additional parameter, values of secondary dendrite arm spacing were established relying on light microscopy images of metallographic sections. Two methods were employed, namely selection of dendrites followed by interactive measurement of arm spacing and an automated approach based on the assumption that the thickness of dendrite arms matches the arm spacing. The latter method was used on anodically etched bright field light microscopy images providing high levels of contrast between the primary aluminum grains and the eutectic phase, thus allowing straightforward binarization. Evaluation first singled out isolated dendrite arms not connected (in the respective 2D section) to a dendrite stem, then determined and averaged the minimal Feret diameter of these. Generally, good agreement between both methods was observed; however, less effort and greater numbers of dendrite arms included in the analysis make the latter the principle of choice.

2.3. Pore Structure and Morphology

For evaluation of the pore structure, both cutting of foam samples to produce photographic images and computer tomography (CT) were employed. In the former case, preparation relied on wire EDM cutting as a means to avoid exerting mechanical loads on foam structural members. Part of the CT measurements was executed by Dr. Illerhaus at the Bundesanstalt für Materialforschung und -prüfung (BAM). Further CT investigations were performed on reference compression test samples using a Procon CT-MINI device, and on foamed samples using modified Yxlon equipment. Resolution was 60 μm, 65.2 μm and approximately 200 μm, respectively. The latter exceeds minimum cell wall thickness by an approximate factor of two, but still allows reconstruction of cellular structure based on averaging effects.

2.4. Foam Expansion Measurements

Foam expansion characteristics were determined using the so-called mechanical expandometer to simultaneously measure furnace and sample temperature as well as volume expansion of the foam by means of Ni-CrNi thermocouples and position encoders based on inductive sensors [65]. Samples tested were of 29 mm diameter and 9 mm height. Per sample type, at least three measurements were performed. Furnace temperature control was set to 750 °C, leading to an approximate 770 ± 5 °C within the furnace just beside the steel tube containing the sample. To eliminate side effects attributable, e.g., to thermal expansion of the device itself, a correction curve was calculated based on three measurements under the same conditions but without sample. This curve was subtracted from each individual expansion measurement. A further, temperature based correction relied on matching of solidus temperatures determined by differential scanning calorimetry (DSC, Netzsch STA 409 C) with temperature profiles measured during foam expansion.

2.5. Mechanical Testing and Evaluation of Test Results

General concerns about viability of mechanical test results in cellular materials were put to the test by Andrews et al. and more recently Yu et al. Andrews et al., as a consequence, proposed the principal dimensions of samples to be at least seven times the average pore size [66] and Yu et al. suggested specimen heights to exceed this parameter at a factor of six [67]. Alkheder and Vural basically confirmed these observations in a recent numerical study and add considerations of boundary layer influence, which they found to be insignificant once sample edge length equaled 10 times the average pore size [68]. With representative pore size values for the type of foam studied here between 2 and 3 mm, a diameter of 30 mm at a height of 25 mm was considered acceptable. Compliance with the criterion was checked, with positive results, based on CT measurements.

To eliminate effects of the foams' typical solid skin, samples for mechanical testing were machined to final dimensions from foam cylinders originally measuring 45 mm in diameter and 55 mm in height by means of turning or wire EDM cutting. The main axis of these specimens coincides with that of the original foam cylinder, the main direction of expansion and gravity during foaming and the direction of the applied compressive load. Samples tested parallel to the direction of expansion have previously been shown to exhibit lower strength than those tested perpendicular to it, most likely due to a density gradient of identical direction [11]. In the present case, limitation of sample height to 25 mm allowed selecting a region of constant density in sampling.

Compression tests were performed in the quasi-static regime at a constant strain rate of 0.1 s^{-1} and stopped once densification had clearly been reached for all density levels, i.e., at a total strain of at least 80%, or at a load exceeding 100 kN (corresponding to 203.7 MPa). Modified foams were tested at TU Bergakademie Freiberg using a MTS 810 universal testing system. Reference samples were tested at Fraunhofer IFAM using a Zwick 1476 universal testing device and a 100 kN load cell.

As principal characteristics of the stress-strain response on which further evaluation was based, tangent modulus (defined as the maximum slope in the elasto-plastic region), ultimate compressive strength and plateau strength were selected. The latter was determined as the stress value associated with the intersection of a linear fit to the plateau region with a tangent to the elasto-plastic region of the stress-strain curve at the point of maximum inclination.

The strain interval for linear plateau region fits has an arbitrarily chosen lower boundary at 0.1 engineering or technical strain, while for the upper limit, a density dependent formulation was chosen in accordance to the initiation of densification as proposed by Gibson and Ashby [12]:

$$\varepsilon < (1 - 1.4 \cdot \rho_{rel}) \cdot (1 - D^{-1}) \tag{1}$$

The value of the parameter D depends on, e.g., the characteristics of the matrix material. For metallic foams, Gibson and Ashby suggested a value of 2.3 [12], which has been taken over for the present study. Both definitions ascertain that for the whole density range studied neither initial stress peaks nor densification affect plateau stress values.

Ultimate compressive strength (UCS) was defined as the peak stress reached immediately after the elasto-plastic region. Identification of such a peak, which does not occur in very ductile foams [32], proved possible for all samples.

Results of this kind were evaluated using Weibull statistics [69]. For this purpose, all mechanical characteristics derived were normalized based on Gibson's and Ashby's fundamental models of strength and elastic modulus as function of density, which reads as follows, in its simplified form [12]:

$$\sigma_{UCS/pl} = C_{UCS/pl} \cdot \rho_{rel}^{3/2} \tag{2}$$

$$E = C_E \cdot \rho_{rel}^{2} \tag{3}$$

Normalization can be realized by dividing the respective property value by the relative density to the applicable power, i.e., 1.5 for plateau and ultimate strength and 2 for the tangent modulus, seen here as elastic materials characteristic. In the present study, a twofold deviation from this straightforward approach was followed: first, the dependence of density on strength was formulated based on absolute rather than relative density. Both descriptions are equivalent when assuming that in the former case, the respective power of the matrix density's reciprocal forms part of a redefined constant C_E. Second, instead of using the standard values introduced in Equations (2) and (3), normalization relied on the exponent of foam density found when fitting a power law function to the experimental density-dependent compression test data.

3. Results

3.1. Microstructural Modification of Powder Compacts

Figure 2 summarizes data on the influence of cooling rate and levels of different additives on grain size. Results from sample series 1 (holding temperature/time 660 °C/5 min.) are included. The diagram confirms that refining and modifying additions can reduce grain size significantly even at the highest cooling rate that could be realized by water quenching. Refining effects are visible even at lower additive levels, i.e., for an initial constitution which contains additives only in part of the aluminum powder fraction. Nevertheless, highest additive levels lead to smallest grains, but in size these still retain an order of magnitude which slightly exceeds the minimum cell wall thickness.

Figure 3 shows selected microstructures complementing the above diagram. In these, Barker etching is employed to achieve distinction between grains. Lines hold reference, Sr-, TiB$_2$- and B-containing samples from top to bottom, while from column 1 to 3 the cooling rate increases from natural via forced convection to water quenching. Additive content levels are 250, 350 and 140 ppm for the Sr, B and TiB$_2$ treated samples, respectively. Specifically the water-cooled Sr-based samples appear to show an exceedingly fine microstructure, additional quantitative data on which is related in Figure 4. A similar observation is made in Figure 5, which contrasts non-etched metallographic sections corresponding to the same sample types. The finding is, however, not directly reflected in the quantitative data gathered in Figure 2; grain sizes are almost identical for water-quenched samples containing additives, with Sr assuming an intermediate position between B and TiB$_2$. A possible explanation for smaller grain sizes to be found in Sr-containing samples is the fact that Sr boosts the level of undercooling in solidifying Al alloys, and thus, promotes formation of additional crystallization nuclei [70].

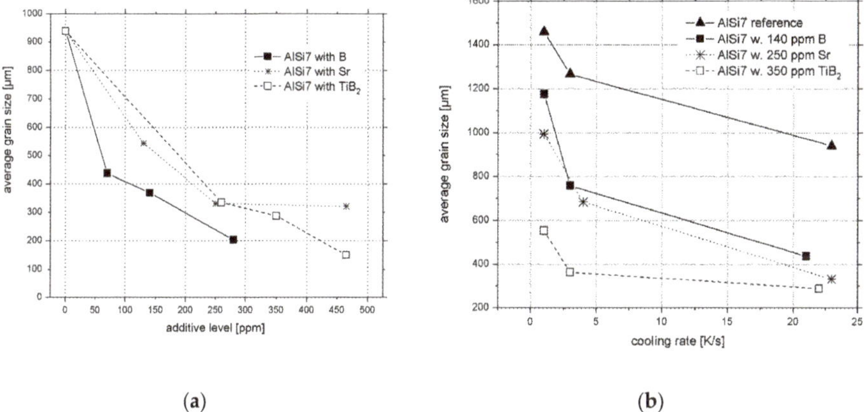

(**a**) (**b**)

Figure 2. Average grain size of powder compacts, (**a**) as a function of Sr, B and TiB$_2$ additive content when quenched in water, (**b**) depending on cooling rate for reference material without additives and samples containing medium levels of Sr, B and TiB$_2$ addition as given in the diagram.

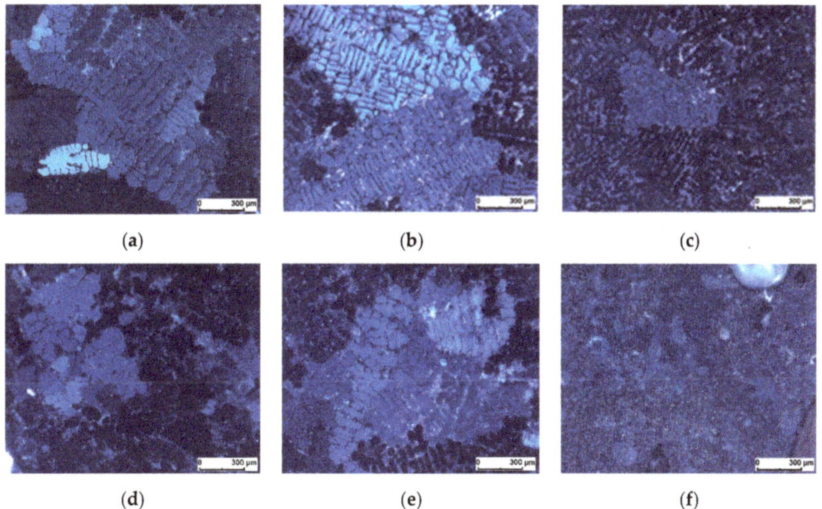

(**d**) (**e**) (**f**)

Figure 3. *Cont.*

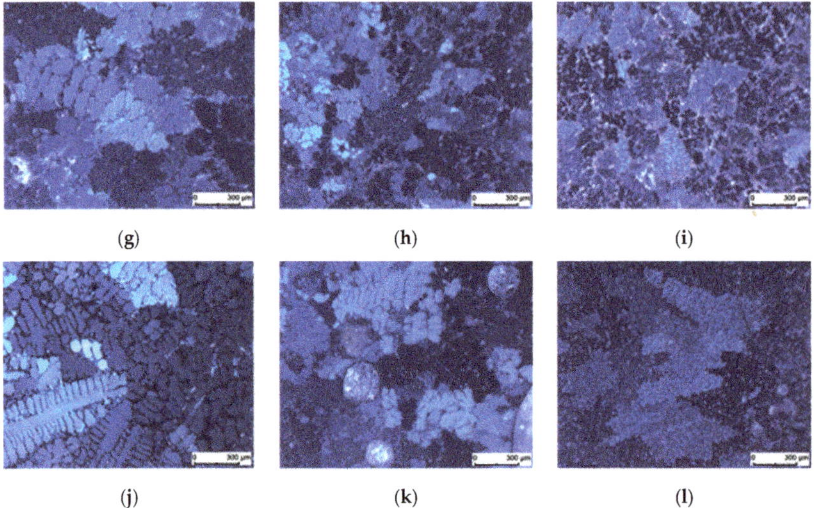

Figure 3. Metallographic sections of powder compacts containing (**a–c**) no additives (AlSi7 reference material), (**d–f**) Sr, (**g–i**) TiB$_2$ and (**j–l**) B additions after thermal treatment (melting) for 5 min at 660 °C and cooling via (**a,d,g,j**) natural convection (NC), (**b,e,h,k**) forced convection (FC) and (**c,f,I,l**) water quenching (WQ), revealing differences in grain size and morphology (Barker etching). Content levels of Sr, TiB$_2$ and B are 250, 350 and 140 ppm, respectively.

Figure 4 displays graphically the influence of Sr, TiB$_2$ and B additions on the morphology of the eutectic phase for the two holding times and temperatures compared. For B-refined samples, only data for sample series 1 is available. The diagrams show that both TiB$_2$ and B additions have a coarsening effect on the eutectic phase. With rising additive content, both the integral of average curvature and the specific boundary layer increasingly fall below the reference data set by the additive-free samples. For Sr, the situation is different, as a coarsening is observed only for the lowest content level, whereas specifically the data from sample series 2 treated for 10 min at 680 °C testifies to the expected effect, i.e., a significant rise of both parameters scrutinized. Nevertheless, the focus of the investigation remains on sample series 1, as the analysis of temperature and expansion versus time curves for the various materials suggests that the respective conditions are a better approximation of the thermal history of foam than the increased time and temperature values adopted for sample series 2.

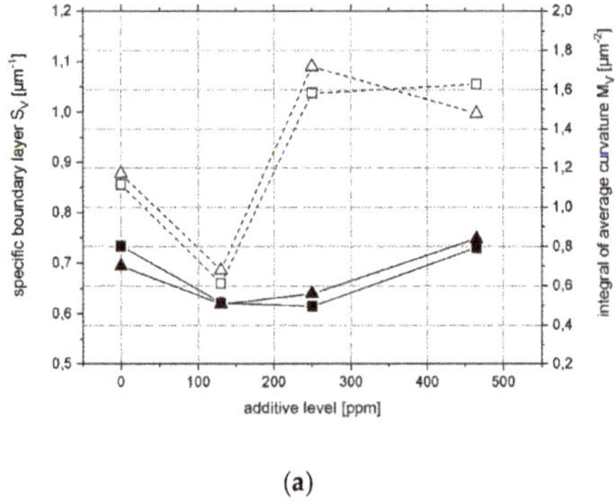

(**a**)

Figure 4. *Cont.*

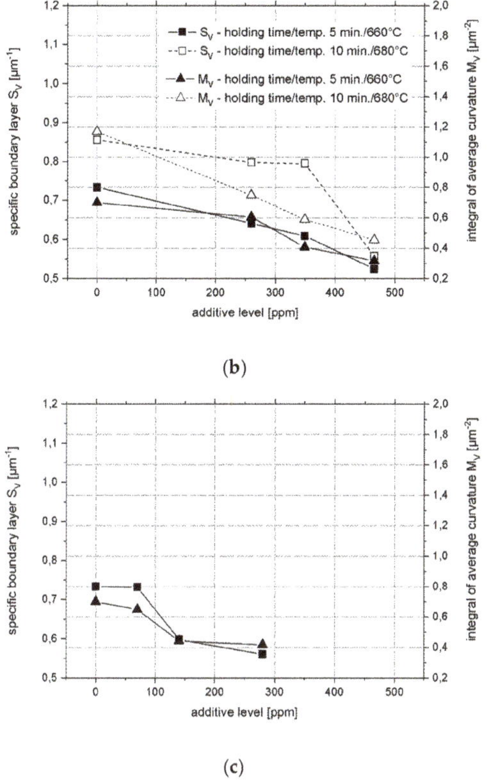

(**b**)

(**c**)

Figure 4. Influence of holding time and temperature on expression of geometrical characteristics of the eutectic phase in powder compacts after melting and resolidification, (**a**) with Sr additions, (**b**) with TiB$_2$ and (**c**) with B additions. Cooling is done by water quenching in all cases. For samples containing B additions, experimental data is available only for materials held for 5 min at 660 °C (series 1). The legend provided in Figure 4b refers to all diagrams.

Micrographs corresponding to the diagrams in Figure 4 are presented in Figures 5 and 6. While the former reflects the influence of cooling rate without as well as at the medium additive levels considered, the latter relates the impact of the amount of additives under conditions of water quenching, and thus, directly corresponds to the quantitative data presented above.

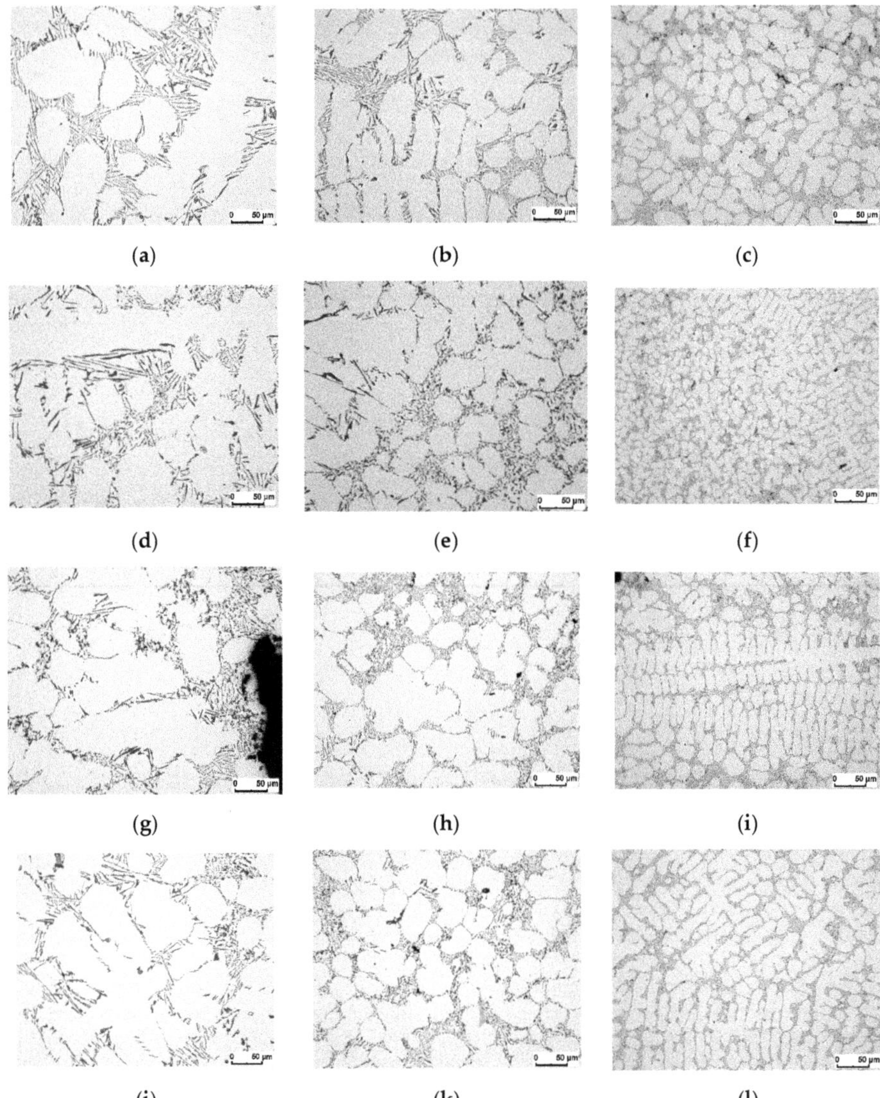

Figure 5. Powder compacts, morphology of the eutectic phase, influence of cooling rate and additives. Columns from left to right correspond to (**a,d,g,j**) natural convection, (**b,e,h,k**) forced convection and (**c,f,I,l**) water quenching, lines from top to bottom showing (**a–c**) reference samples without additives, followed by (**d–f**) 250 ppm Sr, (**g–i**) 350 ppm TiB$_2$ and (**j–l**) 140 ppm B. All images taken at identical magnification.

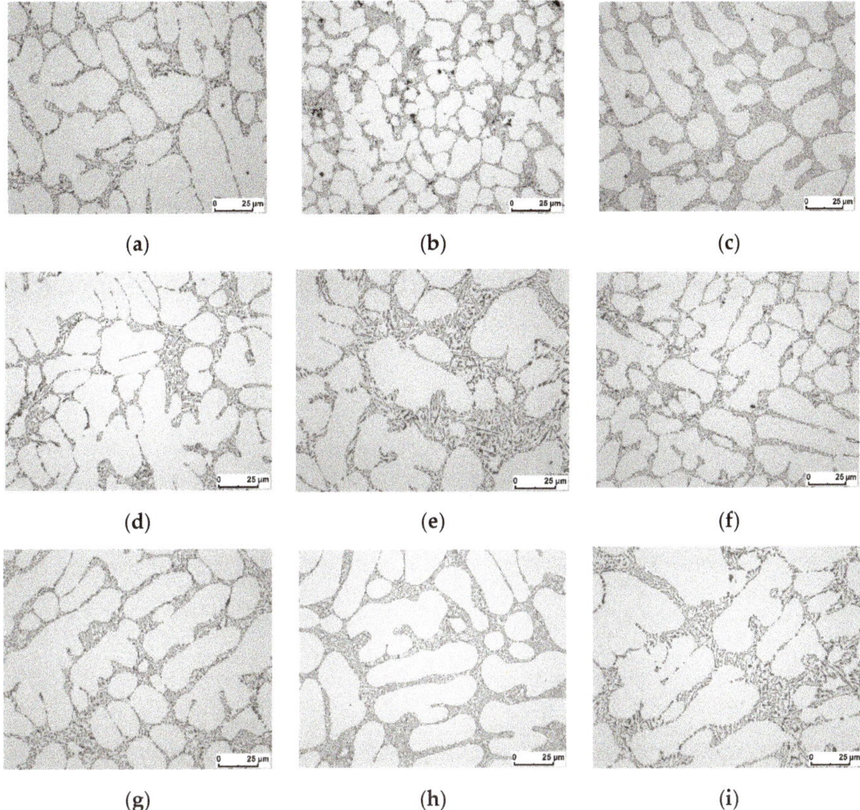

Figure 6. Powder compacts, morphology of the eutectic phase, influence of additive content at high cooling rates. In columns from left to right, additive level is stepped up from 130 to 250 to 465 ppm (Sr, top row, (**a–c**)), 260 to 350 to 465 ppm (TiB$_2$, center row, (**d–f**)) and 70 to 140 to 280 ppm (B, bottom row, (**g–h**)). Note the twofold increase in magnification compared to Figure 5.

Figure 5 underlines the dominant effect of cooling rate on grain refinement as opposed to that of moderate additive levels. Contrast between the third column and the two others exceeds the difference between the reference sample in the top row and the additive-containing ones at matching cooling conditions. Still, both a refining and a modification effect are discernable: Specifically at forced convection, i.e., at slightly elevated cooling rates, the eutectic phase seems finest in B and TiB$_2$-containing samples. Figure 6 optically confirms the data represented in Figure 4, as specifically the coarsening of the eutectic structures is clearly visible in lines 2 (d–f) and 3 (g–i), which correspond to TiB$_2$ and B additions, while the finest structure is revealed in Figure 6c for highest Sr levels.

Since direct measurement of cooling rates in a solidifying foam is difficult, as a result, it may differ significantly depending on whether or not a thermocouple was in actual contact with the matrix material during data acquisition, and literally impossible if the macroscopic structure development must not be influenced (which is the case when using thermocouples within the sample); thus, alternative ways of establishing correct values have to be developed. Correlating microstructural features with solidification rates thus becomes attractive. Secondary dendrite arm spacing (SDAS) can be used for this purpose, with an expression of the type:

$$\lambda_a = A \, v^{-1/3} \tag{4}$$

Describing the relationship between both parameters according to Sahm et al., where λ_a describes the actual SDAS value in µm and v the cooling rate in Ks^{-1} [71]. While the overall size of grains, be they dendritic in nature or not, is clearly influenced by refining additions, literature shows that secondary dendrite arm spacing (SDAS) is not. This allows SDAS measurements to be performed on reference samples containing neither additives nor foaming agents. For the current study, calibration experiments have been performed based on the melting of powder compacts without foaming agent and cooling via natural convection, forced convection using a pressurized air supply and a water jet. During all experiments, cooling rates were measured using thermocouples, while associated SDAS values were determined metallographically. Following suggestions from Schumann, the measured "solidification rate" was determined as the average cooling rate between 550 °C and 400 °C [63]. As a result, the value A = 45.75 has been gained by fitting an equation of the aforementioned type to the measured points, see Figure 7 as well as Equation (6). In doing so, the original equation has been modified to the two parameter form given in Equation (5):

$$\Lambda_a \ [\mu m] = A \cdot v^n \tag{5}$$

$$\Lambda_a \ [\mu m] = 45.75 \cdot v^{-0,3295}, R^2 = 0.94211 \tag{6}$$

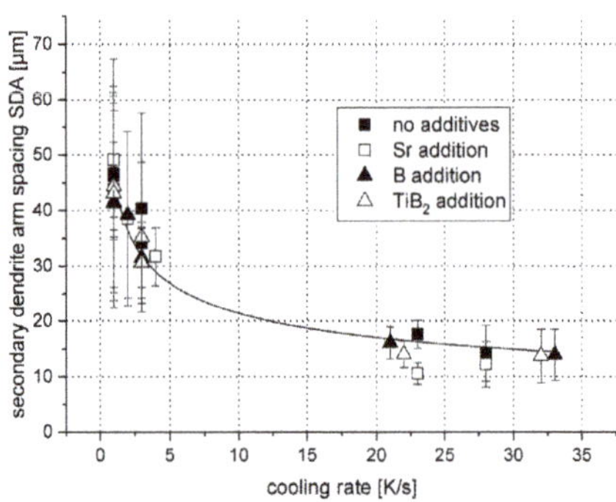

Figure 7. Measured values for secondary dendrite arm spacing (SDA) and cooling rate and fit curve based on Sahm's equation (Equation (4), [71]).

For the second parameter, the exponent n, regression results in a value of −0.3295, which almost matches the suggested theoretical value of −1/3, and thus, supports the viability of the underlying measurements. Based on this, it is possible to ascribe actual cooling rates to the foamed compression test samples via a microstructural feature.

Figure 7 underlines the fact that SDAS is independent of Sr, B or TiB_2 additions, as no systematic deviation of the various sample series from the fit curve can be asserted. Study of the powder compacts has provided insights into refining and modification effects of the various additives considered. Furthermore, the investigations allowed to compare influence of additives with the effect of cooling rate specifically on grain size. The findings provide a microstructure-based measure of cooling rate which can be transferred from fully dense materials to foams, as well as a quantification thereof (see Equation (6)).

3.2. Expansion Characteristics and Pore Structure of Precursor Material Variants

Figure 8 sums up the expansion measurements performed on different material variants. Measurements on B-containing samples have not been included based on the results of the previous evaluation of grain refinement performance, which favors further study of TiB_2 in the composition ranges accessible. The error bars denote the standard deviation observed within the sample series tested (three to six samples evaluated in each case). Comparison of the maximum porosities achievable shows that both Sr and B additions adversely affect foamability. The effect is most evident for B additions, where levels as low as 70 ppm suffice to reduce maximum expansion from an average of 438.8% to 342.2%, and thus, by 22.0%, whereas Sr addition levels of 250 ppm and 465 ppm lead to a decrease of 8.5% and 21.8%, respectively.

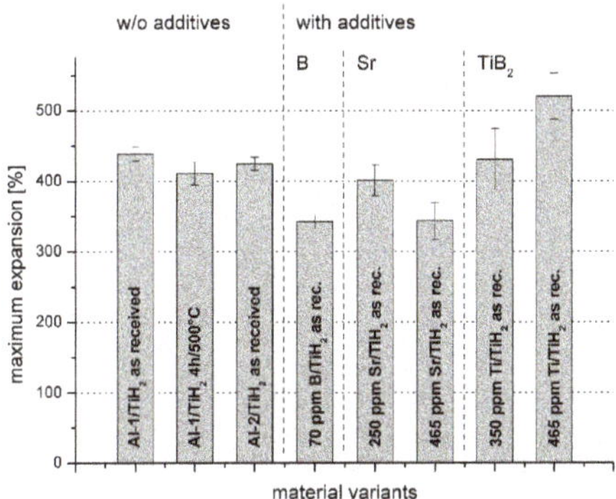

Figure 8. Maximum expansion of various precursor material variants, highlighting the influence of additive type and content.

This finding is contrasted by improvements in maximum achievable porosity achieved by addition of TiB_2. Ti contents of 350 ppm result in a maximum expansion which equals both the standard material (sample series Al-1/TiH_2 as received) at an average expansion of 430.9% and an otherwise identical material based on pure aluminum powders with reduced oxygen content (sample series Al-2/TiH_2 as received, maximum expansion 425.1%), the latter matching the respective values of the additive-containing powders. This sample series has explicitly been included to rule out a major influence of the oxygen content of Al powders on deviations in expansion characteristics. The parameter as such is known to significantly influence this property, as has, e.g., been shown by Weigand, who determined the effective oxygen content range for pure Al alloys to be between 0.2 and 0.72 wt.% [72], while Asavavisichai reported high expansion for the same material, though in a cold rather than hot compacted state, at oxygen contents 0.24, 0.3 and 0.33 wt.%, with a decrease observed for 0.73 wt.% [73]. The fact that the oxygen content of all powders used falls into this range supports the observation that no significant differences in foamability were found between the two standard sample series without additives and different oxygen content levels. Thus, the results of this comparison underline that for the present study, any influence of oxygen content on expansion is secondary to phenomena associated with the various additives. This point is stressed by the fact that a further rise in Ti content to 465 ppm brings about a major increase in expansion by 18.5% to 520.1%. This superior performance of the TiB_2 sample series matches similar observations by Kennedy et al. on pure Al foam, though TiB_2 addition was studied in terms of stabilization and strengthening instead of grain refinement, and thus, used a much higher content level of 10 wt.% TiB_2 [39].

Sr additions to Al-Si alloys differ in their effects on characteristics such as melt viscosity and surface tension based on the Si content. Limited reduction in viscosity is reported for eutectic compositions, while an increase has been observed for hypoeutectic alloys according to Song et al. [54].

Lower melt viscosity is generally considered to adversely affect foam stability. Static metal foam stabilization concepts suggest increasing melt viscosity, e.g., by means of adding non-surface-active ceramic particles, to suppress cell wall drainage, thinning and rupture of cell walls [55,56,74]. In contrast, concepts of foam stability based on dynamic influences indicate that low viscosity might favor healing of local defects which could otherwise develop into rupture sites during stretching of these membranes in foam expansion [15]. The latter point is of interest when considering the superior performance exhibited by TiB_2 based samples. For hypoeutectic compositions and specifically AlSi7, as investigated here, Yan et al. suggest a reduction in viscosity at least for the semi-solid state when $TiB_2/TiAl_3$ is employed for grain refinement [53].

On the other hand, no effect of TiB_2 additions on surface tension is known, whereas Sr is reported by [50] to reduce this properties' value in Al-Si alloys. Considering once again static foam stability criteria, a decrease of surface tension should influence stability to some advantage, since it would mean a reduction of free surface energy. However, the dynamic concept suggested by Lehmhus assumes that high surface tension increases the driving force for compensation of neck formation during stretching of cell walls in the course of foam expansion, and may thus increase foam stability in conditions where dynamic effects represent the major threat to foam stability [15]. Similarly, Nadella et al. correlate reduced stability in Al-Si-Mg foams with the lowering effect of Mg additions on the surface tension [75].

The overall result of expansion measurements is fundamental for the decision to concentrate on Sr and TiB_2 as grain refining/modifying agent in the course of the present study, and solely on TiB_2 in terms of sample production for foam microstructure evaluation, and specifically for mechanical testing.

Figure 9 contrasts the cell morphology of samples containing no additives and as received and thermally treated titanium hydride as foaming agent with that of a sample containing untreated TiH_2 in combination with TiB_2 additions for grain refinement. The images reveal no noticeable influence of this additive on the cellular structure, whereas it is clearly visible that thermal treatment of the foaming agent leads to a more regular structure with increased average pore size.

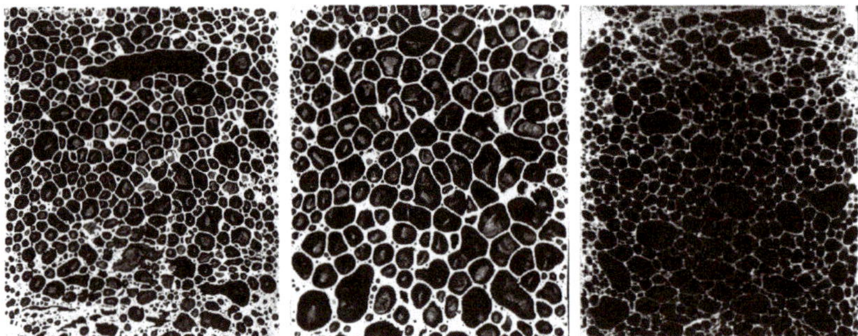

Figure 9. Foam pore structure, alloy variants AlSi7 with "as received" TiH_2 as foaming agent (left), thermally treated TiH_2 as foaming agent (center), "as received" TiH_2 as foaming agent and Ti/TiB_2 modification (right). Left and center image represent scans of cut samples, while the right image is derived from computer tomography (CT) data.

3.3. Microstructural Modification of Foams and Influence on Mechanical Properties

Microstructural evaluations of foams have been performed for materials containing no additives, 465 ppm of Sr and 465 ppm of Ti as TiB_2. The parameters evaluated include the SDAS value, from which actual cooling rates could be derived. Data was gathered in top, central and bottom locations within the foam. The findings suggest that top and center locations experience similar thermal conditions, whereas the bottom of the foam samples is subjected to slower cooling. This observation may be linked to effects of drainage, the influence of the thermal mass associated with the mold support, and the shielding of this part of the mold from forced convection cooling and water quenching. Table 4 lists the cooling rates determined from SDAS in conjunction with Equation (6) for the different positions and sample compositions.

Table 4. Local cooling rates determined for foam samples containing no as well as Sr and Ti as TiB_2 additives (465 ppm each) determined via evaluation of SDAS.

Sample Series	Cooling	Cooling Rate Ks^{-1} at Position Within Foam Sample		
		Top	Center	Bottom
Reference AlSi7	forced convection	17.79	11.83	7.10
	water quenching	47.51	82.94	69.13
AlSi7 with 465 ppm Sr	forced convection	11.21	14.40	25.05
	water quenching	28.17	67.85	55.36
AlSi7 with 465 ppm Ti as TiB_2	forced convection	19.86	15.79	9.77
	water quenching	63.03	92.04	79.50

The listing in Table 4 shows that cooling rates reached within the foam may even exceed the values actually measured in solid powder compacts. This may be explained by reductions in heat capacity per unit volume which coincide with the increased porosity. However, the finding must be seen critical as it implies that the derived cooling rates are in many cases based on an extrapolation of Equation (6) to values not covered by the original data set, which only extends to about 33 K/s (see Figure 7).

Figure 10 summarizes the studies on grain refinement in foams and should be matched with Figures 2 and 4 for comparison with results obtained on powder compacts. Figure 11 adds the corresponding metallographic sections. The top line contains images of Barker etched samples giving an indication of grain size, while the bottom line images allow evaluation of the eutectic structure.

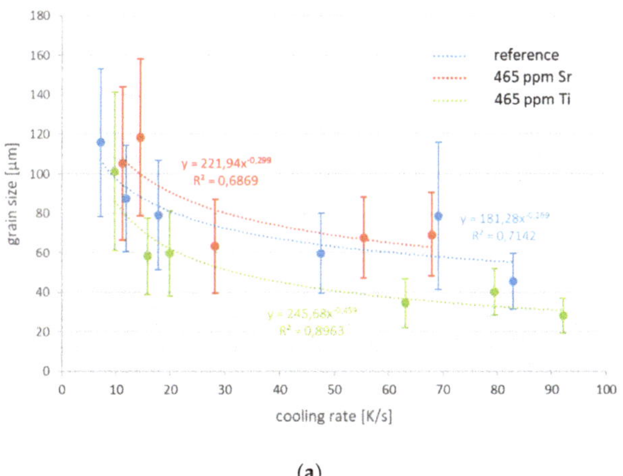

(**a**)

Figure 10. *Cont.*

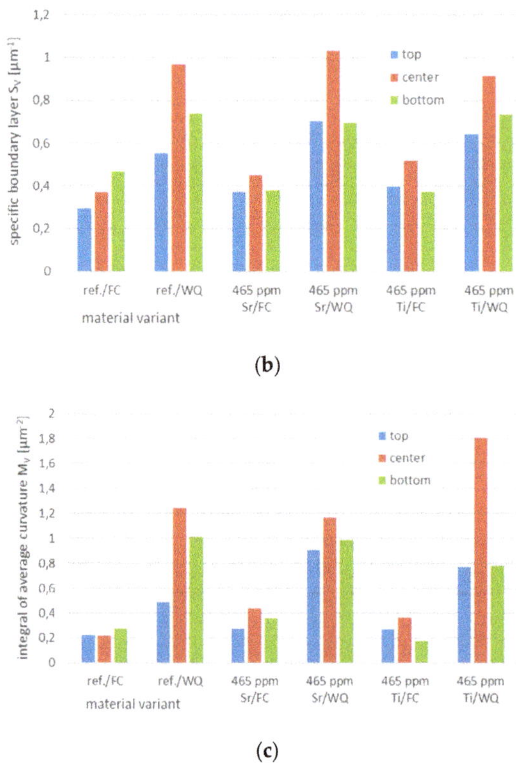

(b)

(c)

Figure 10. Foam microstructure, diagrams: (**a**) grain size as function of additive level and (**b**, **c**) geometrical characteristics of the eutectic phase as a function of cooling rate.

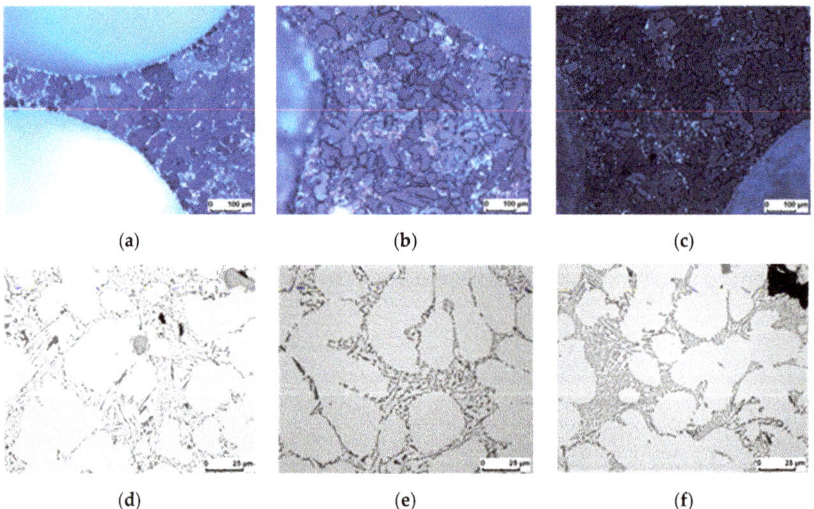

(a)   **(b)**   **(c)**

(d)   **(e)**   **(f)**

Figure 11. Foam microstructure, (**a–c**) Barker etching emphasizing grain sizes, (**d–f**) expression of the eutectic phase. (**a,d**) Reference material AlSi7 with conventional, as received foaming agent, (**b,e**) with 465 ppm Sr and (**c,f**) with 465 ppm Ti as TiB$_2$ added.

Data in Figure 10a confirms that grain refining additions are effective and further add to the effect of cooling rate. TiB_2 containing samples undercut the reference material by a margin of approximately 20% at low and in excess of 30% at high cooling rates, at which grain sizes well below 40 μm are observed. For Sr additions, given the prevailing level of scatter, no effect on grain refinement can be substantiated. It is noteworthy that these results fall below those of measurements on powder compacts as summarized in Figure 2 by approximately one order of magnitude.

In contrast, Figure 10b,c reveal that under conditions of foaming, in stark divergence from comparable findings based on non-foamed powder compacts and irrespective of the cooling rate, addition of Sr has very limited influence on the expression of the eutectic. Values of both the specific boundary layer S_V and the integral of average curvature M_V roughly match corresponding measurements on powder compacts, but show a more significant dependency on cooling rate than on additive content.

Figure 11 illustrates the findings expressed in the diagrams: Sr-, but even more so TiB_2-containing samples show regions with fine and coarse eutectic morphologies that do not differ greatly from those observed in the reference samples.

For all density levels compared, foams produced with the thermally treated foaming agent show highest levels of strength. Moreover, they are characterized by a clearly defined stress peak immediately following the elasto-plastic region (encircled in Figure 12). As Alkheder and Vural pointed out, based on general considerations and 2D simulation results, such peaks can be interpreted as expression of stored elastic energy, which is released once collapse of the structure starts. Naturally, occurrence of this effect is favored in regular structures, in which the capacity of resistance to an external load is distributed more homogeneously, and thus localization of failure only arises at higher levels of global load [68,76].

When comparing grain refined and non-refined material variants based on untreated foaming agents, ultimate compressive strength levels nearly match for densities of approximately 0.3 and 0.4 g/cm^3. At higher densities, the non-refined foams outperform the refined ones. Initial peaks can be identified, though they are less clearly distinguished in these sample series, and specifically so in the grain refined one.

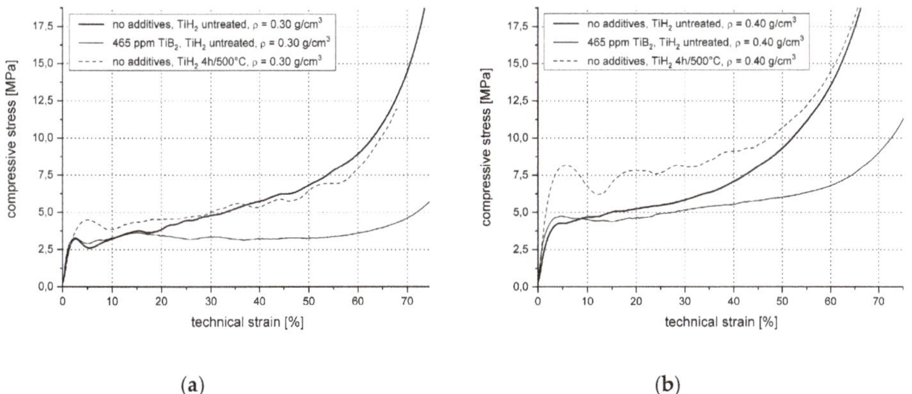

(a)　　　　　　　　　　　　　　　(b)

Figure 12. *Cont.*

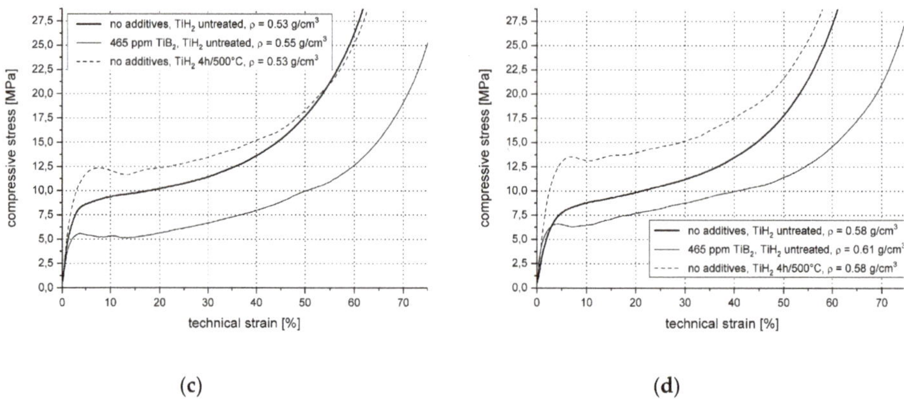

(c)

(d)

Figure 12. Compression test results—comparison between stress-strain curves associated with different density levels: 0.3 g/cm³ (**a**), 0.4 g/cm³ (**b**), 0.53-55 g/cm³ (**c**), 0.58-0.61 g/cm³ (**d**). Note the difference in the scaling of the x- (i.e., stress-) axis in Figure 12 (**a,b**) compared to (**c,d**).

A notable distinction between refined and non-refined samples is a postponed onset of densification observable in the former.

Figure 13 shows exemplarily the density dependence of plateau strength, ultimate compression strength and tangent modulus for the two unmodified reference series samples and the grain refined sample series. The diagram includes the fit curves derived for the various properties and specimen series using a variant of the standard Gibson/Ashby approach for describing the dependence between density and mechanical characteristics in foams already introduced as Equation (2) (plateau and ultimate compressive strength) and 3 (tangent modulus) above. Fitting is done according to a least squares method with variation of the factor C and the exponent n. Table 5 sums up all values derived for the parameters C and n and the three sample series. As all evaluations are based on the absolute density of the foam, the parameter C as given in the table differs in definition from the form introduced in Equations (2) and (3).

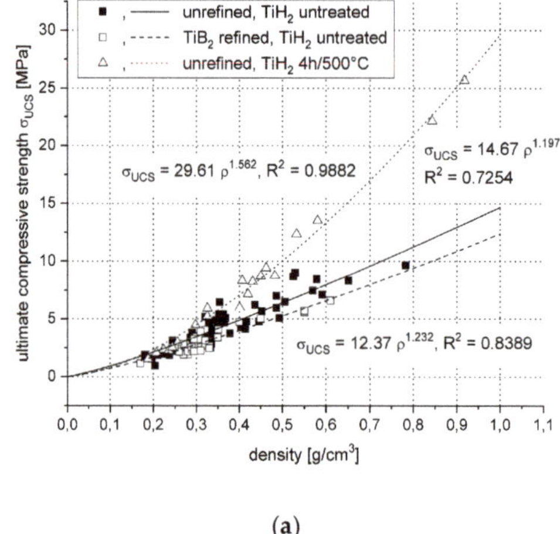

(a)

Figure 13. *Cont.*

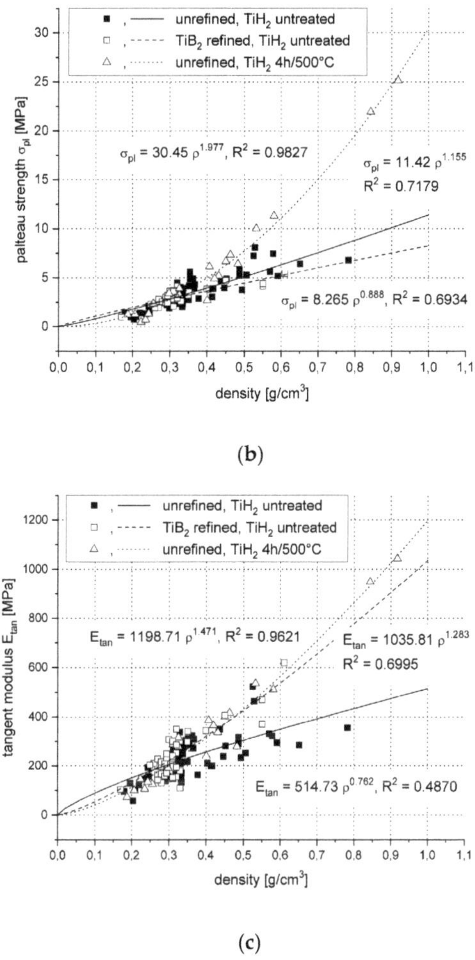

(b)

(c)

Figure 13. Compression test results: ultimate compressive strength (**a**), plateau strength (**b**) and tangent modulus (**c**) versus density.

Table 5. Compression test results, comparison of fit curve parameters for density dependence of plateau and ultimate strength as well as tangent modulus for different sample series (standard series with TiH_2 as received and thermally treated, TiB_2 grain refined with TiH_2 as received—all series produced using forced convection cooling with air).

Property	Sample Series	$\rho_{Matrix}^{-n} \cdot C_{pl/UCS/Et}$	n	R^2
Plateau Strength (C_{pl}) [1]	AlSi7/Ref-FC	11.415	1.155	0.7179
	AlSi7-TiB$_2$-FC	8.2648	0.8882	0.6934
	AlSi7-TiH$_2$/ht-FC	30.45	1.977	0.9827
Ultimate Compressive Strength (C_{UCS}) [1]	AlSi7/Ref-FC	14.672	1.197	0.7254
	AlSi7-TiB$_2$-FC	12.373	1.232	0.8389
	AlSi7-TiH$_2$/ht-FC	29.609	1.562	0.9882
Tangent Modulus (C_{Et}) [1]	AlSi7/Ref-FC	514.73	0.7620	0.4870
	AlSi7-TiB$_2$-FC	1035.8	1.283	0.6995
	AlSi7-TiH$_2$/ht-FC	1198.7	1.471	0.9621

[1] Indicates the associated constant quantified in column 3.

These data clearly show that among all sample series, those based on the thermally treated foaming agent show by far the best adherence to the general dependency of strength and stiffness on density postulated by Gibson and Ashby, even though the values of the fitting curve's exponent do neither meet expectations for stiffness (with an expected value of 2 compared to 1.47 for the respective sample series) nor for strength (1.5 versus 1.98 and 1.56, respectively).

Thus, since this is the major difference specifically between both sample series without modification, but different foaming agents, it must be assumed that this phenomenon is based on the respective expression of the pore structure, which has already been shown to be favorable in the case of thermally treated TiH_2 in Figure 9 as well as various earlier publications [13,14,16,17,77].

Closer observation, however, reveals that the expression of this structural deviation is alleviated at higher levels of porosity. In the present case, a boundary of this kind can be identified at a density of about 0.4 g/cm^3, which corresponds roughly to the earlier observations related to the stress strain curves in Figure 12, which also tend to deviate more significantly at lower porosity and level of expansion. The observation as such is in line with earlier studies suggesting a healing of initial deficiencies in pore structure following an extended period of expansion, i.e., an expansion to a higher final porosity [14,77]. This finding is of considerable importance, as it may serve to define a limit in density above which pore structure can be effectively controlled to adapt global mechanical performance, which in turn implies that microstructural effects to this end should primarily be sought for—and aimed at—below this density level.

A fundamental assumption of the present study was that failure of foam is considerably affected by local failure initiation sides either in the form of Al-Si interfaces or grain boundaries approaching in their dimensions those of cell walls and struts. Such a dependence of failure initiation on local, randomly distributed microstructural weaknesses suggest a description of strength based on a Weibull distribution function. General recommendations for this type of analysis suggest a number of at least 30 samples/data points. At 38 and 60 for grain refined and untreated reference samples, this condition is fulfilled in both cases. Moreover, approximation via a Weibull distribution function is considered justified if an asymmetric shape of the probability density function is observed. This has qualitatively been verified for the three parameters plateau strength, ultimate compressive strength and tangent modulus via the observed histograms of these characteristics in their normalized form. The respective diagrams are contrasted in Figure 14, which contains data for non-refined, reference and TiB_2-refined foams. Essentially, the non-refined foams do show a slight asymmetry, while the grain refined samples tend to exhibit a more symmetric distribution. Note that the parameters compared here are normalized using the observed density dependence as described in Section 2 of this manuscript.

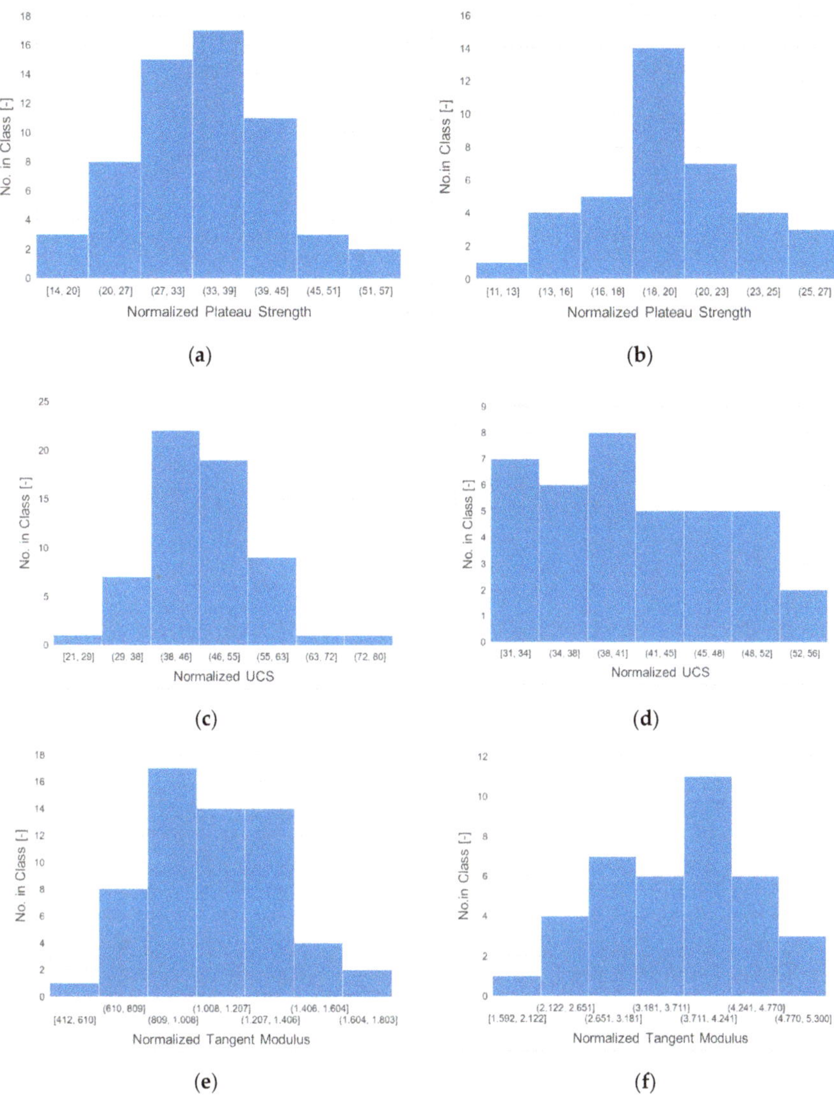

Figure 14. Distributions of normalized properties for the reference material AlS7 expanded using "as received" TiH$_2$ and the same with added TiB$_2$ grain refiner—(**a**,**b**), plateau strength, (**c**,**d**) ultimate compressive strength and (**e**,**f**) tangent modulus.

The following Figures 15–17 depict Weibull evaluation plots of the same compressive properties—namely plateau strength, ultimate compressive strength and tangent modulus. The Weibull distribution parameters for derived from this evaluation for both sample series are given in Table 4. The underlying cumulative distribution function is given by:

$$F(\rho) = 1 - \exp(-(\sigma_0 \cdot \rho^{-1})^m) \tag{7}$$

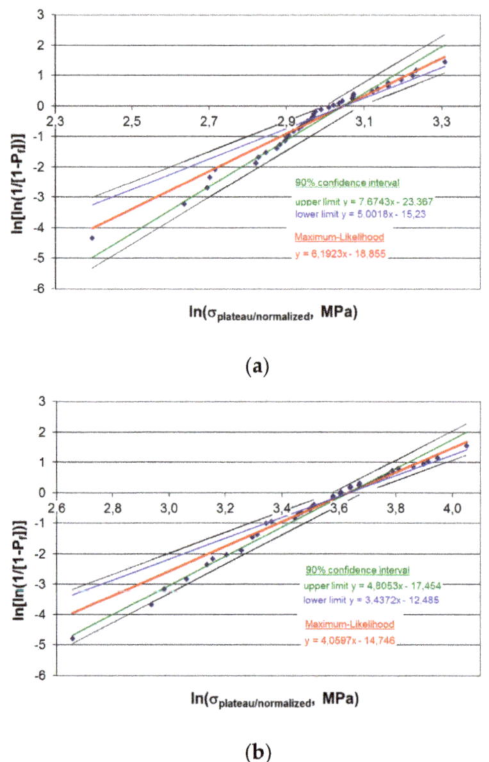

(**a**)

(**b**)

Figure 15. Compression test results—Weibull evaluation of plateau strength for AlSi7 with (**a**) and without TiB2 grain refiner (**b**), as-received TiH$_2$ as foaming agent.

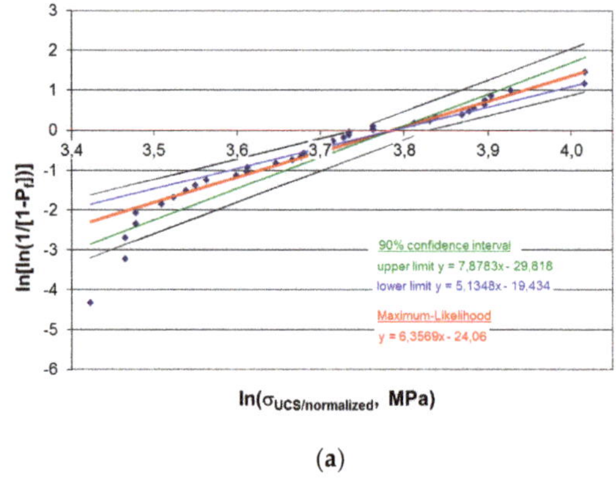

(**a**)

Figure 16. *Cont.*

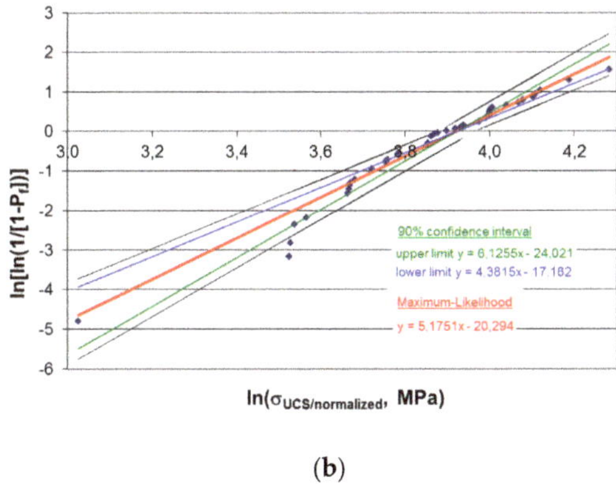

(**b**)

Figure 16. Compression test results—Weibull evaluation of ultimate compressive strength for AlSi7 foam with (**a**) and without TiB2 grain refiner (**b**), as-received TiH_2 as foaming agent.

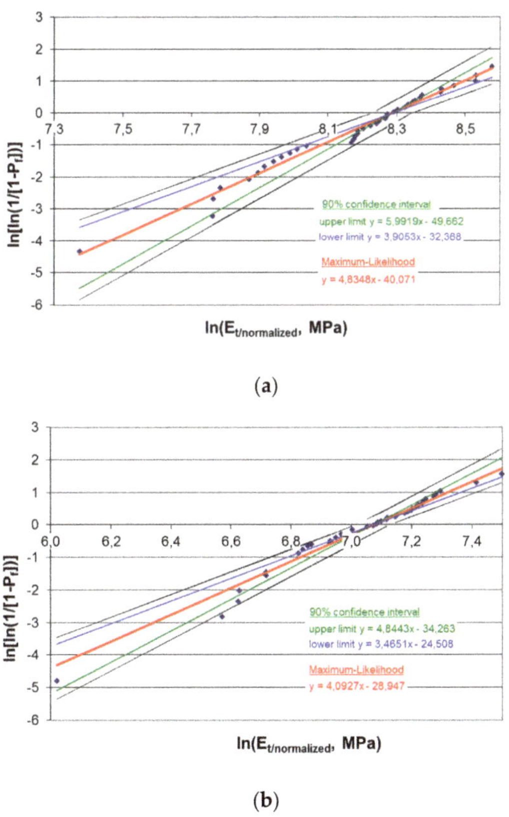

(**a**)

(**b**)

Figure 17. Compression test results—Weibull evaluation of tangent modulus for AlSi7 foam with (**a**) and without TiB_2 grain refiner (**b**), as-received TiH_2 as foaming agent.

In general, the plots in Figures 15–17 show a good match of the experimental data with the expected linear relationship in the double-logarithmic plot and thus support the possibility of expressing material performance via a Weibull distribution. For both plateau and ultimate compressive strength, the respective scale parameter reaches higher values in the case of the non-refined reference material. The tangent modulus, however, clearly deviates from this behavior (Table 6). Common to all material properties, however, is the fact that the shape parameter, i.e., the modulus m, is increased by the refinement. As higher values of this parameter indicate a narrower distribution, the conclusion is that grain refinement can indeed reduce the level of scatter observed.

Table 6. Compression test results, comparison of Weibull distribution parameters for AlSi7 foam with and without TiB_2 grain refiner with as received TiH_2 as foaming agent.

Property	Weibull Parameter	Reference AlSi7	Refined AlSi7-TiB_2
Plateau Strength	scale parameter $\sigma_{0,\ plateau/normalized}$	37.8	21.0
	shape parameter/ modulus $m_{plateau}$	4.06	6.19
Ult. Comp. Strength	$\sigma_{0,\ UCS/normalized}$	50.5	44.0
	m_{UCS}	5.18	6.36
Tangent Modulus	$E_{0,\ t/normalized}$	1179.4	3976.2
	m_{Et}	4.09	4.83

4. Discussion

The present study shows that grain refinement of PM aluminum foams made from hypoeutectic Al-Si matrix alloys is a viable approach. Both the microstructural evaluation of powder compact samples without foaming agent as well as the corresponding studies on actual foam samples show a significant influence of the refining agent chosen, TiB_2. However, contrasting data gathered from powder compacts and foam samples shows that though the relative influence of the refining treatment on grain size is similar in both cases, the absolute grain sizes differ significantly. For powder compacts, values roughly between 150 and 900 μm were found, while foam samples show grains of approximately 40 to 150 μm. Differences in cooling rate may be excluded as underlying cause of this phenomenon, as the evaluation of secondary dendrite arm spacing (SDAS) and the establishment of a correlation between this microstructural feature and the cooling rate on powder compacts has facilitated the association of cooling rates with the grain sizes measured on foams, too. As expected, the respective data do not support any large deviation of the cooling rates prevalent in foams from the range covered in the experiments performed on powder compacts. However, there is still room for several possible explanations: first, Ti content in foam samples is naturally increased, as the blowing agent, TiH_2, will add to it. The result may be a secondary grain refining effect superimposed to that of the deliberately added TiB_2. Second, the topology of the foam itself, and specifically the fact that the dimensions of its main structural elements, cell walls and struts fall short of the grain sizes measured on powder compacts by up to one order of magnitude, will prohibit the formation of large grains in a foam. This sterical hindrance is amplified further by the fact that the smooth surface of the solidified foam's cell walls suggest that solidification and thus nucleation of grains occurs not only within these membranes, but may even start at the interface between liquid and gas, potentially helped by the fact that, e.g., oxide particles present in the matrix alloy [78] can accumulate here and act as heterogeneous nucleation sites, leaving once again less room for grains to grow within the cell walls and struts.

In contrast to grain refinement, modification of the eutectic structures has not been achieved in foams. The reason for this maybe deduced from the comparison of the two powder compact sample test series, which differ in temperature and in the time of exposure to the respective temperature. As mentioned earlier, comparison of initial powder compact sample series clearly shows that successful modification of the eutectic structure in a material produced by compaction of a powder mixture in which only one component brings in the modifying agent apparently requires an extended process window for distribution of the dissolved modifier. 10 min at 680 °C serve this purpose, while 5 min at 660 °C do not entirely; however, the latter is a better representation of the actual conditions during foaming as performed in the present study, which uses unmodified foaming agents in additive-containing samples. Since these conditions cannot be altered without putting the stability of the foam at risk—which, as this study has shown, is compromised by Sr addition anyway—the only way to achieve Sr-based eutectic modification in a metal foam of the type studied here would seem to be choosing an Al powder that contains the optimum level of modifying agent from the start. This would effectively eliminate the need for distributing Sr by the time- and temperature-dependent process of diffusion within the matrix material. In the present study, this approach was not adopted, since working with a master alloy-like powder containing increased levels of Sr allowed facile production of samples covering a range of Sr contents rather than a single value. In any case, the adverse effect of Sr on foam stability practically ruled out the production of compression test samples from this material variant and thus limited the studies on modification to microstructural investigations.

Comparison of compression test data between samples with and without grain refiner but identical, non-optimized foaming agents suggests that macroscopic deficiencies dominate mechanical response. The effect of grain refinement is thus somewhat obscured in stress-strain curves as well as strength-versus-density plots. This finding is stressed further by the obvious increase in strength observed for the third sample series based on thermally treated foaming agents. As a consequence, further studies on grain refinement should be based on samples with comparable regularity in macroscopic structure to facilitate a clearer discrimination of the influence of refinement and modification. Meanwhile, the fact that the optimization of the macroscopic foam structure achieved by means of using a thermally treated, oxidized and depleted blowing agent [14,77] not only results in the highest strength of all sample types compared, but also shows least scatter (see fit quality as expressed by R^2 in table) and best adherence to the values of the exponent of the strength-density relation theoretically predicted by Gibson and Ashby [12] underlines that it is in fact this macroscopic structure which primarily controls a metal foam's mechanical performance. In this respect, the results of the present study confirm the hierarchy of influencing factors put forward in the introduction based on authors like Mosler et al., Martin et al. or Mu et al. [8–10].

Furthermore, it is apparent that additional improvements in grain refining efficiency are required to bring grain sizes down to levels which clearly fall below the cross-sectional dimensions of the main structural elements of an aluminum foam of the type considered here. This relates to cell walls, but predominantly to the struts as the primary structural members; their size at least should ideally be undercut by one order of magnitude. At present, with typical membrane thicknesses of approximately 80 μm and struts approaching two–three times this value [19], this has not yet been achieved. The expected benefit of this would be a shift of cell wall failure mechanisms to regions in which the quasi-isotropic behavior of polycrystal dominates. As of now, the ratio between cell wall thickness and typical grain size implies that single crystal plasticity effects, and namely anisotropy of grains with unfavorable orientation relative to the loading direction, may act as additional weak spots within the foam's structure. Moreover, as the studies of microstructural features of the powder compacts have shown, it must be assumed that any beneficial effect of grain size reduction is partly obscured by the observed coarsening effect of TiB_2 and B on the expression of the eutectic phase. In this context, it must be considered unfortunate that the insufficient expansion characteristics of the Sr-containing materials did not allow stable production of sufficient numbers of samples for mechanical testing and specifically Weibull evaluation. Despite these facts, the Weibull evaluation does show a beneficial effect of grain refinement in the increased values of the Weibull modulus m in samples treated, a characteristic that indicates a reduction in scatter of the respective properties.

5. Conclusions

Though in principle successfully applicable, the results obtained show that grain refinement of the type employed here cannot lead to grain sizes that match or undercut the dimensions of typical cell walls. Thus, it remains questionable whether this approach alone can have significant influence on the mechanical performance of the foams. A different picture is observed for modification treatments affecting the eutectic phase. Here, both the influence of the modifying agents as well as secondary effects of the refining treatment can shift the size of the respective microstructural features to levels below typical cell wall thickness levels. This is of major relevance in view of the initial argument that specifically the interface area between Si and Al phases within the eutectic can act as weak spot at which cell wall failure is initiated. However, transferring this effect to foams requires a base material, i.e., matrix material powder, which already contains the modifying agent in homogeneous distribution and thus eliminates the need to achieve just this by diffusion during the foaming process; the present study has shown that the combination of time and temperature typical for foam expansion cannot guarantee this. Moreover, Sr additions introduced for eutectic modification turned out to adversely affect expansion characteristics—hence, Sr-modified samples could not be included in the evaluation of the foams' mechanical performance.

Comparison of density-dependent strength and stiffness shows modified foams as weaker compared to both their untreated counterparts. A possible explanation may be that on the whole, more ductile matrix combined with an irregular structure, possibility of ductile deformation promotes premature macroscopic yield when compared to the fracture-based failure associated with the initial assumption that Si phases might be preferred fracture planes in cell walls.

Thus, the investigations clearly show that the influence of microstructural features is limited when it comes to average strength levels determined for samples identical in terms of composition—except for the respective modifying agents—and processing conditions. The scatter of these average strength values, however, is noticeably reduced by grain refinement. For designing with a material, the consequences of this achievement are similar to an increase in strength, since reduced scatter means that higher stress levels can be accepted at a given level of safety.

Furthermore, it is of great importance for metal foam production that the equivalence of two different concepts to achieve the intended microstructural changes could in principle be demonstrated; both an increase in cooling rate and the use of grain refining agents yield similar results. The fact that the effectivity of the former approach seems more pronounced does not devaluate the latter, since in foam production, local cooling rates are limited specifically where large parts are concerned due to effects such as the reduced thermal conductivity within the foam. Thus, grain refinement based on $TiB_2/TiAl_3$ additions may prove a valuable alternative.

Comparisons based on non-foamed powder compacts have shown that a limited increase in foaming time and final temperature will positively affect the efficacy of treatments; specifically for treatments based on solution of active substances, this finding is less than surprising. This is an additional argument for combining modification and grain refinement with the use of thermally treated foaming agents, which naturally induce a shift of processing conditions in the required direction. Moreover, such a combination, though not tested in the present study, must seem attractive based on the chance to combine improvements in micro- and pore structure.

Future investigations should have a closer look at conditions of adding grain refining substances, especially in terms of increasing levels of concentration of $TiB_2/TiAl_3$-containing Al powders, since using large proportions of such specialty powders for precursor material production would have unwanted economic implications. However, basing refinement on such an emphasized master alloy approach may cause local variations in the availability of nucleation sites, especially when assuming that the nucleant paradigm correctly describes the process, as in this case, non-soluble constituents, which cannot widely be redistributed during foam expansion, take the stress in initiating nucleation. The question of the ideal level of addition also needs to be answered, since in the current study maximum values were set by the powders available. Moreover, a further study of size effects should be envisaged to clarify to what degree microstructural modification will affect results of the kind seen by Blazy et al. [79].

Finally, the ongoing search for alternative grain refiners and modifying agents should be taken into account; recent developments in this area have been summarized by Easton et al. as well as Liu [58,60]. These, and other studies such as that of Xu et al., have highlighted potential new approaches in grain refinement, e.g., scandium addition [80]. Beyond a switch to new refining and modifying agents, combinations of both processes may also be considered [81], the more so since the present study has shown that Sr- and B-based modification on the one hand and TiB_2-based grain refinement have opposed effects on foam expansion, Thus, combinations might partially cancel out negative side effects of Sr- and B-based treatments. In general, eutectic modification should in itself not be neglected—specifically if solutions can be identified with less detrimental effect on foam expansion. In the end, it is not only grain size that may reach the dimensions of cell walls, but also the size of eutectic silicon structures, which may just as well act as preferred fracture planes.

In conclusion, we believe the present study has shown that grain refinement of aluminum foam has both promises and challenges. Of the latter, not all have been successfully addressed yet. There is room for improvement, and we believe that our results provide a sound basis for such future research, for which we have suggested several promising paths.

Author Contributions: Conceptualization, U.M. (Ulrike Mosler), D.L., D.H., J.W. and U.M. (Ulrich Martin); methodology, U.M. (Ulrike Mosler), D.H. and D.L.; validation, D.H., U.M. (Ulrike Mosler) and D.L.; formal analysis, D.H., U.M. (Ulrike Mosler) and D.L.; investigation, D.H., U.M. (Ulrike Mosler) and D.L.; data curation, D.H., U.M. (Ulrike Mosler) and D.L.; writing—original draft preparation, D.L.; writing—review and editing, D.H., U.M. (Ulrike Mosler), U.M. (Ulrich Martin), J.W. and D.L.; visualization, D.H. and D.L.; supervision, U.M. (Ulrike Mosler), D.L. and U.M. (Ulrich Martin).

Funding: Part of this research was funded by the German Research Foundation (Deutsche Forschungsgemeinschaft, DFG) within SPP 1075, "Cellular Metallic Materials".

Acknowledgments: CT measurement facilities were made accessible by B. Illerhaus, Bundesanstalt für Materialprüfung (BAM), Berlin, and J. Gudat, ProCon GmbH, Hannover, whose help is greatly appreciated. Illerhaus also gave support in the evaluation of CT data. Furthermore, Ray Cook, Alpoco, provided Al powders and contributed to the discussions on grain refinement and modification in general.

Conflicts of Interest: The authors declare no conflict of interest.

References

1. Baumeister, J.; Rausch, G.; Stöbener, K.; Lehmhus, D.; Busse, M. Aluminium foam composites–applications in railroad manufacturing. *Mater. Werkst.* **2007**, *38*, 939–942. [CrossRef]

2. Banhart, J.; Seeliger, H.-W. Aluminium Foam Sandwich Panels: Manufacture, Metallurgy and Applications. *Adv. Eng. Mater.* **2008**, *10*, 793–802. [CrossRef]

3. Neugebauer, R.; Hipke, H. Machine tools with metal foams. *Adv. Eng. Mater.* **2006**, *8*, 858–863. [CrossRef]

4. Garcia-Moreno, F. Commercial Applications of Metal Foams: Their Properties and Production. *Materials* **2016**, *9*, 85. [CrossRef]

5. Baumeister, J. Verfahren Zur Herstellung Poröser Metallkörper. German Patent No. DE 40 18 360 C1, 8 June 1991. (date of filing, patent granted 29 May 1991).

6. Banhart, J. Manufacture, characterization and application of cellular metals and metal foams. *Prog. Mater. Sci.* **2001**, *46*, 559–632. [CrossRef]

7. Lehmhus, D.; Vesenjak, M.; de Schampheleire, S.; Fiedler, T. From Stochastic Foam to Designed Structure: Balancing Cost and Performance of Cellular Metals. *Materials* **2017**, *10*, 922. [CrossRef]

8. Martin, U.; Mosler, U.; Lehmhus, D.; Müller, A.; Heinzel, G. Hierarchical structure of aluminium alloys and relation to compression behavior. *High Temp. Mater. Process.* **2007**, *26*, 291–296. [CrossRef]

9. Mosler, U.; Martin, U.; Losseva, N.; Oettel, H. A statistic approach to estimate the compression strength of aluminium foams. In *Cellular Metal: Manufacture, Properties, Applications*; Banhart, J., Fleck, N.A., Mortensen, A., Eds.; MIT Publishing: Bremen, Germany, 2003; pp. 387–392.

10. Mu, Y.; Yao, C.; Liang, L.; Luo, H.; Zu, G. Deformation mechanisms in closed-cell aluminum foam in compression. *Scr. Mater.* **2010**, *63*, 629–632. [CrossRef]

11. Avalle, M.; Lehmhus, D.; Peroni, L.; Pleteit, H.; Schmiechen, P.; Belingardi, G.; Busse, M. AlSi7 metallic foams-aspects of material modelling for crash analysis. *Int. J. Crashworth.* **2009**, *14*, 269–285. [CrossRef]

12. Gibson, L.J.; Ashby, M.F. *Cellular Solids*; Cambridge University Press: Cambridge, UK, 1997.

13. Kennedy, A.R.; Lopez, V.H. The decomposition behaviour of as-received and oxidized TiH_2 foaming-agent powder. *Mater. Sci. Eng. A* **2003**, *357*, 258–263. [CrossRef]

14. Lehmhus, D.; Rausch, G. Tailoring titanium hydride decomposition kinetics by annealing in various atmospheres. *Adv. Eng. Mater.* **2004**, *6*, 313–330. [CrossRef]

15. Lehmhus, D. Dynamic Collapse Mechanisms in Metal Foam Formation. *Adv. Eng. Mater.* **2010**, *12*, 465–471. [CrossRef]

16. Matijasevic, B.; Banhart, J. Improvement of aluminium foam technology by tailoring of blowing agent. *Scr. Mater.* **2006**, *54*, 503–508. [CrossRef]

17. Matijasevic-Lux, B.; Banhart, J.; Fiechter, S.; Görke, O.; Wanderka, N. Modification of titanium hydride for improved aluminium foam manufacture. *Acta Mater.* **2006**, *54*, 1887–1900. [CrossRef]

18. Proa-Flores, P.M.; Drew, R.A.L. Production of Aluminum Foams with Ni-coated TiH_2 Powder. *Adv. Eng. Mater.* **2008**, *10*, 830–835. [CrossRef]

19. Stanzick, H.; Wichmann, M.; Weise, J.; Helfen, L.; Baumbach, T.; Banhart, J. Process control in aluminium foam production using real-time X-ray radioscopy. *Adv. Eng. Mater.* **2002**, *4*, 814–823. [CrossRef]

20. Mosler, U.; Heinzel, G.; Martin, U.; Oettel, H. Microstructure and deformation behaviour of aluminium foams. *Mater. Werkst.* **2000**, *31*, 519–522. [CrossRef]

21. Gall, K.; Yang, N.; Horstemeyer, M.; McDowell, D.L.; Fan, J. The debonding and fracture of Si particles during the fatigue of a cast Al-Si alloy. *Metall. Mater. Trans. A* **1999**, *30*, 3079–3088. [CrossRef]

22. Su, J.F.; Nie, X.; Stoilov, V. Characterization of fracture and debonding of Si particles in AlSi alloys. *Mater. Sci. Eng. A* **2010**, *527*, 7168–7175. [CrossRef]

23. Su, J.F. Characterization of Fracture and Debonding of Silicon Particles in Aluminium Silicon Alloys. Master Thesis, University of Windsor, Windsor, ON, Canada, 2009.

24. Lados, D.A.; Apelian, D. Relationships between microstructure and fatigue crack propagation paths in Al–Si–Mg cast alloys. *Eng. Fract. Mech.* **2008**, *75*, 821–832. [CrossRef]

25. Chan, K.S.; Jones, P.; Wang, Q. Fatigue crack growth and fracture paths in sand cast B319 and A356 aluminum alloys. *Mater. Sci. Eng. A* **2003**, *341*, 18–34. [CrossRef]

26. Xia, S.; Qi, Y.; Perry, T.; Kim, K.-S. Strength Characterization of Al/Si interfaces: A hybrid method of nanoindentation and finite element analysis. *Acta Mater.* **2009**, *57*, 695–707. [CrossRef]

27. Banhart, J.; Seeliger, H.-W. Recent Trends in Aluminum Foam Sandwich Technology. *Adv. Eng. Mater.* **2012**, *14*, 1082–1087. [CrossRef]

28. Stöbener, K.; Lehmhus, D.; Avalle, M.; Peroni, L.; Busse, M. Aluminum foam-polymer hybrid structures (APM aluminum foam) in compression testing. *Int. J. Solids Struct.* **2008**, *45*, 5627–5641. [CrossRef]

29. Lehmhus, D.; Baumeister, J.; Stutz, L.; Schneider, E.; Stöbener, K.; Avalle, M.; Peroni, L.; Peroni, M. Mechanical characterisation of particulate aluminium foams-strain-rate, density and matrix alloy vs. adhesive effects. *Adv. Eng. Mater.* **2010**, *12*, 596–603. [CrossRef]

30. Vesenjak, M.; Gacnik, F.; Krstulovic-Opara, L.; Ren, Z. Behavior of composite advanced pore morphology foam. *J. Compos. Mater.* **2011**, *45*, 2823–2831. [CrossRef]

31. Duarte, I.; Vesenjak, M.; Krstulović-Opara, L.; Ren, Z. Compressive performance evaluation of APM (Advanced Pore Morphology) foam filled tubes. *Compos. Struct.* **2015**, *134*, 409–420. [CrossRef]

32. Lehmhus, D.; Banhart, J.; Rodriguez-Perez, M.A. Adaptation of aluminium foam properties by means of precipitation hardening. *Mater. Sci. Technol.* **2002**, *18*, 474–479. [CrossRef]

33. Lázaro, J.; Solórzano, E.; Escudero, J.; de Saja, J.A.; Rodríguez-Pérez, M.A. Applicability of Solid Solution Heat Treatments to Aluminum Foams. *Metals* **2012**, *2*, 508–528. [CrossRef]

34. Khan, K.L.A.; Kumar, G.; Prasad, R. Effect of Age Hardening and Quenching Media on Aluminium Foams. *Int. J. Appl. Eng. Res.* **2018**, *13*, 245–248.

35. Duarte, I.; Banhart, J. A study of aluminium foam formation–kinetics and microstructure. *Acta Mater.* **2000**, *48*, 2349–2362. [CrossRef]

36. Lehmhus, D.; Busse, M. Potential new matrix alloys for production of PM aluminium foams. *Adv. Eng. Mater.* **2004**, *6*, 391–396. [CrossRef]

37. Campana, F.; Pilone, D. Effect of wall microstructure and morphometric parameters on the crush behaviour of Al alloy foams. *Mater. Sci. Eng. A* **2008**, *479*, 58–64. [CrossRef]

38. Helwig, H.-M.; Garcia-Moreno, F.; Banhart, J. A study of Mg and Cu additions on the foaming behaviour of Al–Si alloys. *J. Mater. Sci.* **2011**, *46*, 5227–5236. [CrossRef]

39. Kennedy, A.R.; Asavavisichai, S. Effect of TiB_2 particle addition on the expansion, structure and mechanical properties of PM Al foams. *Scr. Mater.* **2004**, *50*, 115–119. [CrossRef]

40. Nafisi, S.; Ghomashchi, R. Grain refining of conventional and semi-solid A356 Al–Si alloy. *J. Mater. Process. Technol.* **2006**, *174*, 371–383. [CrossRef]

41. Helwig, H.-M.; Hiller, S.; Garcia-Moreno, F.; Banhart, J. Influence of Compaction Conditions on the Foamability of AlSi8Mg4 Alloy. *Metall. Mater. Trans. B* **2009**, *40*, 755–767. [CrossRef]

42. Banhart, J.; Garcia-Moreno, F.; Heim, K.; Seeliger, H.-W. Light-weighting in transportation and defence using aluminium foam sandwich structures. In Proceedings of the International Symposium on Light Weighting for Defence, Aerospace and Transportation, Indian Institute of Metals, Goa, India, 11 November 2017.

43. Easton, M.; St. John, D. Grain Refinement of Aluminium Alloys: Part, I. The Nucleant and Solute Paradigms–A Review of the Literature. *Metall. Mater. Trans. A* **1999**, *30*, 1613–1623. [CrossRef]

44. Easton, M.; St. John, D. Grain Refinement of Aluminium Alloys: Part II. Confirmation of, and a Mechanism for, the Solute Paradigm. *Metall. Mater. Trans. A* **1999**, *30*, 1625–1633. [CrossRef]

45. Schumacher, P. Keimbildungsmechanismen während der Kornfeinung von Al-Si-Legierungen. *Giess. Rundsch.* **2003**, *50*, 228–230.

46. Müller, K. *Möglichkeiten der Gefügebeeinflussung Eutektischer und Naheutektischer Aluminium-Silizium-Gusslegierungen unter Berücksichtigung der Mechanischen Eigenschaften*; Fortschrittsberichte VDI, Series 5; VDI Verlag GmbH: Düsseldorf, Germany, 1996.

47. Chen, X.-G. Kristallisation des Aluminium-Silizium-Eutektikums und Anwendung der Thermischen Analyse zur Kontrolle der Veredelung. Ph.D. Thesis, RWTH Aachen, Aachen, Germany, 1990.

48. Chen, X.-G.; Ellerbrok, R.; Engler, S. Über die eutektischen Körner von Aluminium-Silizium-Legierungen. Teil 1. Veredeltes Eutektikum. *Giessereiforschung* **1990**, *42*, 1–10.

49. Qiyang, L.; Qingchun, L.; Qifu, L. Modification of Al-Si alloys with Sodium. *Acta Metall. Mater.* **1991**, *39*, 2497–2502. [CrossRef]

50. Srirangam, P.; Kramer, M.J.; Shankar, S. Effect of strontium on liquid structure of AlSi hypoeutectic alloys using high-energy X-ray diffraction. *Acta Mater.* **2011**, *60*, 503–513. [CrossRef]

51. Timpel, M.; Wanderka, N.; Schlesiger, R.; Yamamoto, T.; Lazarev, N.; Isheim, D.; Schmitz, G.; Matsumura, S.; Banhart, J. The role of Strontium in modifying lauminium-silicon alloys. *Acta Mater.* **2012**, *60*, 3920–3928. [CrossRef]

52. Dahle, A.K.; Tondel, P.A.; Paradies, C.J.; Arnberg, L. Effect of Grain Refinement on the Fluidity of Two Commercial AI-Si Foundry Alloys. *Metall. Mater. Trans. A* **1996**, *27*, 2305–2313. [CrossRef]

53. Yan, M.; Luo, W. Effects of grain refinement on the rheological behaviors of semisolid hypoeutectic Al–Si alloys. *Mater. Chem. Phys.* **2007**, *104*, 267–270. [CrossRef]

54. Song, X.; Bian, X.; Zhang, J.; Zhang, J. Temperature-dependent viscosities of eutectic Al–Si alloys modified with Sr and P. *J. Alloys Compd.* **2009**, *479*, 670–673. [CrossRef]

55. Banhart, J. Metal Foams: Production and Stability. *Adv. Eng. Mater.* **2006**, *8*, 781–794. [CrossRef]

56. Haibel, A.; Rack, A.; Banhart, J. Why are metal foams stable? *Appl. Phys. Lett.* **2001**, *78*, 154102. [CrossRef]

57. Chen, Z.; Kang, H.; Fan, G.; Li, J.; Lu, Y.; Jie, J.; Zhang, Y.; Li, T.; Jian, X.; Wang, T. Grain refinement of hypoeutectic Al-Si alloys with B. *Acta Mater.* **2016**, *120*, 168–178. [CrossRef]

58. Easton, M.; Qian, M.; Prasad, A.; St. John, D. Recent advances in grain refinement of light metals and alloys. *Curr. Opin. Solid State Mater. Sci.* **2016**, *20*, 13–24. [CrossRef]

59. Lee, C. Effect of Ti-B addition on the variation of microporosity and tensile properties of A356 aluminium alloys. *Mater. Sci. Eng. A* **2016**, *668*, 152–159. [CrossRef]

60. Liu, Z. Review of Grain Refinement of Cast Metals Through Inoculation: Theories and Developments. *Metall. Mater. Trans. A* **2017**, *48*, 4755–4776. [CrossRef]

61. Müller, A.; Mosler, U. Gefügeinterpretation von Aluminiumschäumen. In *Sonderbände der Praktischen Metallographie*; Petzow, G., Ed.; Werkstoff-Informationsgesellschaft mbH: Frankfurt, Germany, 2001; Volume 32, p. 279.

62. Mosler, U.; Heinzel, G.; Martin, U.; Oettel, H. Quantitative Charakterisierung der zellularen Struktur von Aluminium-Schäumen. In *Sonderbände der Praktischen Metallographie*; Petzow, G., Ed.; Werkstoff-Informationsgesellschaft mbH: Frankfurt, Germany, 2001; Volume 32, p. 121.

63. Schumann, H. *Metallographie*; Wiley-VCH Verlag: Weinheim, Germany, 2004.

64. Ohser, J.; Lorz, U. Quantitative Gefügeanalyse: Theoretische Grundlagen und Anwendung. In *Freiberger Forschungshefte, B 276: Metallurgie und Werkstofftechnik, Werkstoffeinsatz*; Deutscher Verlag für Grundstoffindustrie: Leipzig, Germany, 1994.

65. Weber, M. Herstellung von Metallschäumen und Beschreibung der Werkstoffeigenschaften. Ph.D. Thesis, Technical University of Clausthal, Clausthal, Germany, 1995.

66. Andrews, E.W.; Gioux, G.; Onck, P.; Gibson, L.J. Size effects in ductile cellular solids. Part II: Experimental results. *Int. J. Mech. Sci.* **2001**, *43*, 701–713. [CrossRef]
67. Yu, H.; Guo, Z.; Li, B.; Yao, G.; Luo, H.; Liu, Y. Research into the effect of cell diameter of aluminium foam on its compressive properties. *Mater. Sci. Eng. A* **2007**, *454*, 542–546. [CrossRef]
68. Alkheder, M.; Vural, M. Mechanical response of cellular solids: Role of cellular topology and microstructural irregularity. *Int. J. Eng. Sci.* **2008**, *46*, 1035–1051. [CrossRef]
69. Weibull, W. A statistical distribution function of wide applicability. *J. Appl. Mech.* **1951**, *18*, 293–297.
70. Chen, X.; Geng, H.; Li, Y. Study on the Eutectic Modification Level of Al-7Si Alloy by Computer Aided Recognition of Thermal Analysis Cooling Curves. *Mater. Sci. Eng. A* **2006**, *419*, 283–289. [CrossRef]
71. Sahm, P.R.; Egry, I.; Volkmann, T. *Schmelze, Erstarrung, Grenzflächen–Eine Einführung in die Physik und Technologie Flüssiger und Fester Metalle*; Viehweg-Verlag: Wiesbaden, Germany, 2003.
72. Weigand, P. Untersuchung der Einflussfaktoren auf die Pulvermetallurgische Herstellung von Aluminiumschaum. Ph.D. Thesis, RWTH Aachen, Aachen, Germany, 1999.
73. Asavavisichai, S.; Kennedy, A.R. The role of oxidation during compaction on the expansion and stability of Al foams made via a PM Route. *Adv. Eng. Mater.* **2006**, *8*, 568–572. [CrossRef]
74. Körner, C.; Arnold, M.; Singer, R.F. Metal foam stabilization by oxide network particles. *Mater. Sci. Eng. A* **2005**, *396*, 28–40. [CrossRef]
75. Nadella, R.; Sahu, S.N.; Gokhale, A.A. Foaming characteristics of Al-Si-Mg (LM25) alloy prepared by liquid metal processing. *Mater. Sci. Technol.* **2010**, *26*, 908–913. [CrossRef]
76. Tekoglu, C.; Gibson, L.J.; Pardoen, T.; Onck, P.R. Size effects in foams: Experiments and modelling. *Prog. Mater. Sci.* **2011**, *56*, 109–138. [CrossRef]
77. Lehmhus, D.; Busse, M. Mechanical performance of structurally optimized AlSi7 aluminum foams—An experimental study. *Mater. Werkst.* **2014**, *45*, 1061–1071. [CrossRef]
78. Asavavisichai, S.; Kennedy, A.R. The effect of oxides in various aluminium powders on foamability. *Procedia Eng.* **2012**, *32*, 714–721. [CrossRef]
79. Blazy, J.-S.; Marie-Louise, A.; Forest, S.; Chastel, Y.; Pineau, A.; Awade, A.; Grolleron, C.; Moussy, F. Deformation and fracture of aluminium foams under proportional and non-proportional multi-axial loading: Statistical analysis and size effect. *Int. J. Mech. Sci.* **2004**, *46*, 217–244. [CrossRef]
80. Xu, C.; Ma, C.; Sun, Y.; Hanada, S.; Lu, G.; Guan, S. Optimizing strength and ductility of Al-7Si-0.4 Mg foundry alloy: Role of Cu and Sc addition. *J. Alloys Compd.* **2019**, *810*, 151944.
81. Riestra, M.; Ghassemali, E.; Bogdanoff, T.; Seifeddine, S. Interactive effects of grain refinement, eutectic modification and solidification rate on tensile properties of Al-10Si alloy. *Mater. Sci. Eng. A* **2017**, *703*, 270–279. [CrossRef]

 *metals*

Article

Automated Continuous Production Line of Parts Made of Metallic Foams

Isabel Duarte [1,*], Matej Vesenjak [2] and Manuel J. Vide [3]

[1] Department of Mechanical Engineering, Centre for Mechanical Technology and Automation (TEMA), University of Aveiro, Campus Universitário de Santiago, 3810-193 Aveiro, Portugal

[2] Faculty of Mechanical Engineering, University of Maribor, Smetanova ul. 17, SI-2000 Maribor, Slovenia; matej.vesenjak@um.si

[3] M.J.Amaral-Equipamentos Industriais Lda, Apartado 14, Bouça da Aguincheira-São Pedro de Castelões, 3730-038 Vale de Cambra, Portugal; geral@mjamaral.pt

* Correspondence: isabel.duarte@ua.pt; Tel.: +350-234-370-830

Received: 20 April 2019; Accepted: 6 May 2019; Published: 8 May 2019

Abstract: The paper presents an automated continuous production line (7 m × 1.5 m × 1 m) of high-quality metallic foams using a powder metallurgical method. This continuous production line was used to obtain metal foam parts and/or components by heating the foamable precursor material at melting temperatures close to the temperature of the metallic matrix and cooling the formed liquid metallic foam (in liquid state), which then results in a solid closed-cell metallic foam. This automated continuous production line is composed of a continuous foaming furnace, a cooling sector and a robotic system. This installation has enabled a technological breakthrough with many improvements solving some technical problems and eliminating the risks and dangers related to the safety of workers due to the high temperatures involved in this process. The whole process becomes automatic without any need for human intervention.

Keywords: aluminum alloy foams; powder metallurgy; continuous production; mechanical properties

---

1. Introduction

Metallic foams are one of the most interesting lightweight materials that have a high potential to be used in large-scale in various areas of engineering. They are multifunctional, lightweight, recyclable and non-flammable with reduced environmental impact. The fact that they may be formed by open-cells and/or closed-cells, makes it possible for them to be widely used in the near future [1]. The open-cell metallic foams are mainly used as functional materials for chemical and biomedical applications (e.g., treatment of water, heat exchangers) [2]. The closed-cell foams are used as structural materials for commercial, industrial and military applications (e.g., energy absorbing structures, ballistic protection) [1,3–5]. The latter exhibit an excellent structural efficiency and excellent capacity to absorb (impact and sound) energy, together with noise and vibration damping in structures such as houses, cars, trains and other devices and equipment [6]. These special advantages are a combination of properties derived from their porous cellular structures along with the properties of the metal they are made of. These foams are considered as one of the future materials that must be lightweight, durable, ecofriendly and economical. When compared to the other foams made of polymer and ceramic, these foams have more potential. Polymer foams [7] are flexible, but cannot support high temperatures, while metallic ones are extremely tough and can plastically deform and possess a high capacity of energy absorption. Also, metallic foams are easily recyclable unlike some polymeric foams, which cannot be recycled or, if they are, it is quite expensive. Moreover, polymeric foams increase fire hazards, releasing smoke and toxic gases, which does not happen with the non-flammable metallic foams. Ceramic foams [8] are relatively stiff, endure high thermal shocks and are more stable in harsh environments compared with metallic and polymer foams. However, ceramic foams have a brittle behavior that limits their uses, while metallic foams have a ductile behavior that is more appropriate for engineering purposes. Despite the advantages of metallic foams, they have not been used in a large scale yet. The main reason is the fact that the current manufacturing processes to prepare the closed-cell foams do not allow a meticulous control of the cellular structures in terms of their pores' shape and size. The control of cellular structures of these foams will make it possible to predict the mechanical, thermal and electrical behavior.

Two main strategies have been adopted to achieve cellular metals with regular structures. The first strategy consists of improving the most viable existing manufacturing processes, as the case of the powder metallurgy that allows the preparation of the ductile metallic foams to be used as structural materials. Powder metallurgy method consists of heating a foamable precursor material to a temperature close to the melting temperature of the metallic matrix in which this foamable precursor material is prepared by the hot compaction of a mixture of the metal and blowing agent powders [9]. The improvement was achieved by using pre-treated blowing agent powder (e.g., titanium hydride) instead of untreated powder, ensuring that the initial decomposition temperature is close to the melting temperature of the metallic matrix [10]. The second strategy is to develop new classes of the cellular metals with closed pores using methods that allow the control of the size and shape of their pores. Advanced pore morphology (APM) metallic foam elements [11], syntactic foams [12] and hybrid foams [13] are some of these examples. The APM metallic foam elements are simple spheres of metal closed-cell foams, which are obtained by the powder metallurgy method using the smallest pieces of the precursor material (e.g., 1–2 mm in size) [11]. The small sample sizes ensure that surface tension forces during the melting process are relatively large compared to the hydrostatic pressure, forming near spherical shapes with an easily reproducible unit cell [14]. Syntactic foams are prepared by infiltrating a metal melt into the interstice's spaces of a packed hollow spheres made of metal or ceramic or porous particles [15,16]. Hybrid foams are prepared by impregnating an open-cell metal foam with a polymer [17]. Despite these developments, the closed-cell metallic foams prepared by the traditional powder metallurgy method have several advantages in comparison with other new cellular metals. For example, the syntactic foams have high density (>2 g/cm^3) when compared to the typical closed-cell metal foams (<1 g/cm^3) that limits their application, especially for lightweight constructions [6]. Also, in the case of hybrid foams the use of polymers increases the risk of fire hazards. Furthermore, the powder metallurgy method allows to fabricate parts with complex geometries, to fill the hollow structures with a formed liquid metal foam during its formation [18] and to easily incorporate metallic inserts (e.g., screws) [19] into the metal foam during its formation, promoting a metallic bonding. For example, aluminum alloy closed-cell foams have been extensively tested as core [20] and filler materials [21,22] of sandwich structures and thin-walled tubes, mainly to be used as energy absorbing structures for vehicles in which the joining between the solid thin-walled structure and the closed-cell aluminum alloy foam is made during the foam formation. However, since the metallic foam component is prepared using a batch furnace, the quality of the components depends on several factors including manufacturing parameters (e.g., temperature and time) and the experience of the workers. In order to have a strict control over the process, without the need of human intervention in dangerous tasks, an automated continuous production line was developed to obtain parts and/or components of metallic foams. This article presents this production line, illustrating several case-studies as well as the potentiality of this newly developed installation.

2. Materials and Methods

Two foamable precursor materials with 160×20 mm (Figure 1a) and 20 mm $\times 5$ mm (Figure 1b) in cross-section, made of AlSi7 alloy with 0.5 wt. % of titanium hydride were prepared by a combination of cold isostatic pressing and hot extrusion as described in [23]. Table 1 presents characteristics of the used powders. The powder mixture of aluminum alloy and titanium hydride were first compacted to cylindrical slugs with 70–80% of theoretical density by cold isostatic pressing (CIP). After this, the cylindrical slug was pre-heated to 360 °C and extruded to rectangular bars of 160 × 20 mm and 20 mm × 5 mm in cross-section through a horizontal 25 MN direct extrusion machine. Closed molds made of S235JR carbon steel (0.17% C, 1.40% Mn, 0.045% P, 0.045% S, 0.009% N) were made using steel plates and thin-walled tubes. Square (outer width: 25 mm; wall thickness: 2 mm) and cylindrical (outer diameter: 30 mm; wall thickness: 1.5 mm) thin-walled tubes made of AA 6060 T66 having an outer and inner diameter of 30 mm and 26 mm, respectively, were cut to the required length (e.g., 150 mm). Carbon steel bars (diameter: 4 mm) made of S235JRG2C (0.20% C, 1.40% Mn, 0.045% P, 0.045% S) were used as reinforcements to prepare the in-situ reinforced foams.

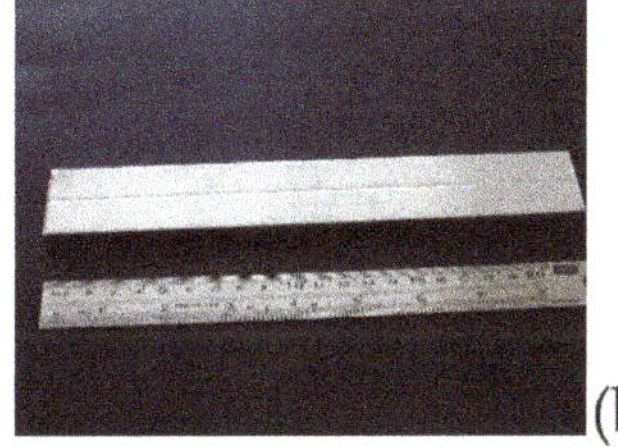

Figure 1. Foamable precursor materials made of aluminum alloys and titanium hydride with 160 × 20 mm (**a**) and 20 mm × 5 mm (**b**) in cross-section.

Table 1. Properties of the powders.

Powders	Purity (%)	D90 (μm)	D50 (μm)	D10 (μm)	Medium Diameter (μm)	Oxygen (%)
Aluminum	99.7	128	57	17	67	0.7
Silicon	99.5	91	30	4.4	48	0.5
6061-alloy	–	250	116	59	140	1.1

The compressive and three-point bending properties presented in this work were measured using a 50 kN servo-hydraulic dynamic INSTRON 8801 testing machine (Instron, Norwood, MA, USA) under quasi-static loading conditions (crosshead loading rate: 0.1 mm/s).

3. Results

The development of this automated continuous production line was based on the thorough knowledge acquired in the fabrication of metallic foams using laboratory batch furnaces [9,23]. In the traditional process, a foamable precursor material is placed into the cavity of a closed mold. The worker places the mold containing the precursor into a pre-heated furnace at high temperatures. This is a dangerous task for the worker due to exposure to extremely high temperatures (700–800 °C for the case of aluminum alloys). The total weight of the mold with the foamable precursor material should also be appropriate for the easy handling by the worker.

After that, the heating of the foamable precursor material leads to a formation of the liquid metallic foam due to the thermal decomposition of the blowing agent and the melting of the metal. The precursor expands, filling the cavity of the mold. During this step, the worker should supervise the exterior of the mold to follow the filling process of the mold. When the first melt leaves the mold, the worker must quickly yet carefully remove the mold containing the formed liquid metallic foam. In addition to this, the formed liquid foam must be solidified immediately to prevent the collapse mechanisms (drainage and coalescence) and must be done in a very steady manner to avoid any turbulent movements that could damage the foam [24]. This is another dangerous task in which the worker is, once again, exposed to extremely high temperatures. To perform the dangerous tasks, the workers should use personnel protective equipment, such as heat protection clothing, gloves and safety glasses. In the traditional procedure, the cooling is not controlled since it is the worker who cools the surrounding of the mold with a hose of compressed air until room temperature is obtained. Results show that formation of defects and structural imperfections within solid foams are created due to the fact, that the solidification is not uniform [25]. All of these aspects could increase the number of rejected metallic foam pieces. Automated continuous production line was developed to eliminate the dangerous tasks performed by the workers, to ensure that the quality of the foam is not compromised by the experience of the worker and to guarantee a controlled solidification. The main components of the continues production line are the continuous furnace, cooling sector and robotic system that will be described below.

3.1. Continuous Foaming Furnace

Figure 2 shows a scheme (Figure 2a) and photos (Figure 2b–d) of the continuous foaming furnace developed to fabricate 3D components of metallic foams. This continuous furnace has five zones, which are the feed zone of the foamable precursor material into the molds (open or closed) or into the thin-walled structures, followed by the three heating zones and the exit zone. The loading and movement of the foamable precursor material within the furnace is achieved by means of a conveyor belt, as shown in Figure 2d. The length of the feed zone is approximately 0.7 m (Figure 2b), while the length of the exit zone is approximately 0.3 m (Figure 2c).

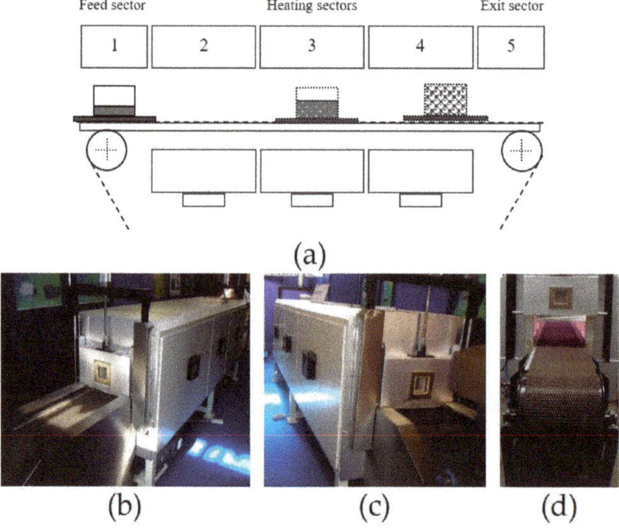

Figure 2. Scheme (**a**) and photos of the continuous foaming furnace belt, showing the feed (**b**) and exit (**c**) sectors and the conveyor belt (**d**).

The furnace has three heating zones respective to the three foaming stages until the maximum expansion value is reached [26]. The first foaming stage accounts for temperatures below the solidus temperature of the metallic matrix, where the expansion is very small. In the second foaming stage, a rapid increase of expansion is observed due to the metal being in liquid state and due to the releasing of the gas into the metal from by the thermal decomposition of the blowing agent, simultaneously. The third foaming stage also represent a rapid increase in expansion as the temperature increases until the maximum expansion is reached. After that, gas is no longer released, and the formed liquid foam starts to collapse due to the coalescence and drainage mechanisms. Thus, the formed liquid metallic foam should be removed from the furnace once it is close to the maximum expansion to avoid its collapse. The three heating zones are of the same dimensions (size: 0.3 m in height and 0.4 m in width) and could be controlled individually according to the thermal foaming cycle.

All of the heating zones could be operated from room temperature up to 1273 K. This furnace is heated by means of electrical resistances, which were installed at the top of the furnace and under the conveyor belt. The conveyor belt is made of stainless-steel chains to support the weight of the foamable precursor material and molds or hollow structures, as well as the thermal fatigue during the thermal foaming cycle.

Two panel doors were installed in the furnace to decrease the temperature variation throughout the heating zones, as well as to reduce the turbulence airflows effect. One of them is located at the entrance of the first heating zone (Figure 2b) and the other is located at the end of the third heating zone of the furnace (Figure 2c). Both panel doors have windows for observation of the furnace's interior. Each heating zone (Figure 2b,c) has also got a window located on the sidewall of the front furnace for visualizing and controlling the foaming process. The movement of the foamable precursor material within the furnace is associated with the movement of the conveyor belt created by a set of engines and transmission systems. The maximum linear speed of the conveyor belt is approx. 0.5 m/s. The temperature of the three heating zones and speed rate of the conveyor belt should be adjusted according to the characteristics of the foamable precursor material and the geometry and dimensions of the component to be fabricated. Figure 3 shows two examples of rejected foam parts (Figure 3a,b) and the block foam prepared with adjusted parameters (Figure 3c). The use of a high speed of the conveyor belt and/or low temperatures of the heating zones will not allow enough time for the precursor material to expand and filling the cavity of the mold, as shown in Figure 3a. On the other hand, a very low speed rate of the conveyor belt promotes the collapse of the foam since the formed liquid metallic foam remains at high temperatures for a long time, leading to a high loss of the liquid metal, as can be seen in Figure 3b. For each foam part, the parameters of the furnace should be adjusted to obtain a high-quality foam as shown in Figure 3c.

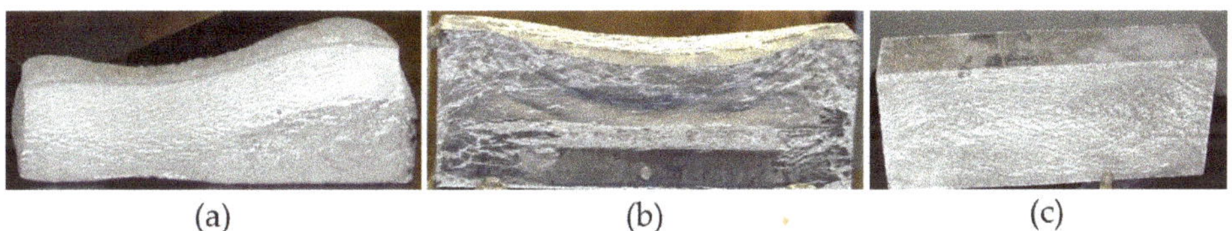

(a)   (b)   (c)

Figure 3. Rejected foam parts using a steel mold ($200 \times 80 \times 50$ mm^3) due to no filling the cavity of the mold (**a**) and collapsed foam (**b**) and foam block with properly adjusted parameters (**c**).

3.2. Cooling Sector and Robotic System

A cooling sector was designed and built to control the solidification of the liquid metallic foam, decreasing the number of rejected foam parts. In fact, uncontrolled solidification of the liquid metallic foam causes defects and structural imperfections, decreasing the mechanical properties of the resulting solid metal foams. Figure 4 shows rejected parts and nearly perfect foam parts, which were prepared with uncontrolled (Figure 4a) and controlled (Figure 4b) cooling, respectively.

This cooling sector was installed immediately after the continuous foaming furnace. The robotic system was also built and installed in the cooling sector to extract the mold or the hollow structures filled with the liquid metallic foam from the furnace at high temperatures to the cooling sector, making it an automatic operation without any human intervention. Figure 5 shows a 3D model of the cooling sector (Figure 5a) composed by the robotic system (Figure 5b), the cooling systems and the conveyor belt (Figure 5c) designed by SolidWorks software. Figure 6 shows real photos of the cooling sector and an overview of the final automated continuous production line.

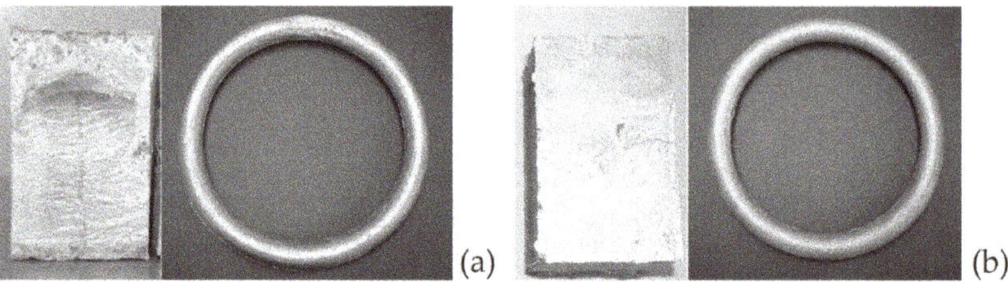

Figure 4. Foam block ($80 \times 50 \times 50$ mm³) and a complex foam part (outer diameter: 250 mm) prepared by the uncontrolled cooling (**a**) and controlled cooling (**b**).

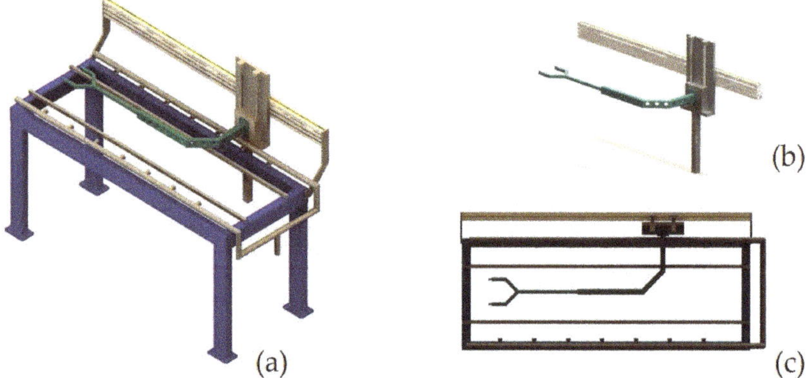

Figure 5. Scheme (**a**) and photos of the continuous foaming furnace belt, showing the feed (**b**) and exit (**c**) sectors and the conveyor belt.

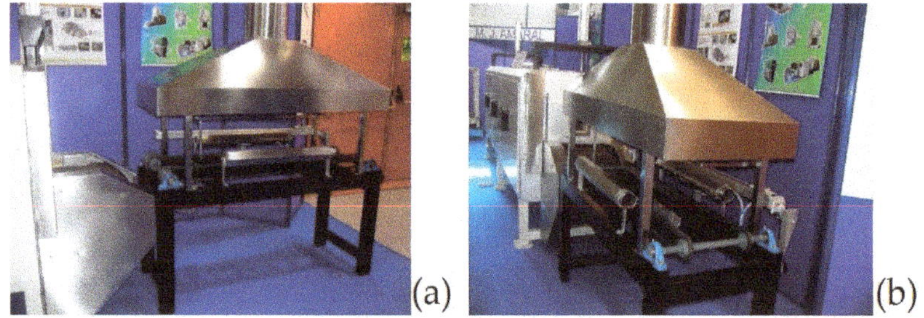

Figure 6. Cooling sector (**a**) and the automated continuous production line developed (**b**) for fabricating parts made of metallic foams.

The cooling sector is a metallic structure equipped with a high-pressure turbine connected to two manifolds drilled and mounted on both sides of a conveyor belt. The angle of the manifolds drilled varies according to the dimensions of the required component. A robotic system was developed to extract the mold containing the formed liquid metallic foam from the furnace at a high temperature to the cooling sector, which makes it possible to operate it without any human intervention required. This robotic system is installed in the cooling sector. The mechanical arm is built from stainless steel AISI 304 and is mounted at the vertical rail made of aluminum and a horizontal rail made of the same material, as shown in Figure 6. The vertical drive is controlled by a pneumatic cylinder, while the horizontal one is controlled by a motor reducer with the possibility of adjusting the speed rates and positions. The movement of this robotic system is associated with the command of the opening and closing of the furnace door at the end of the third heating zone. The mechanical arm system receives a trigger signal, starts to move towards the furnace door, while the furnace door also opens. This signal is triggered when the mold touches the suspended metal chain in the third heating zone. Briefly, the sequence steps during this operation are:

- The signal is emitted, triggering the start of two immediate mechanisms: the opening of the furnace door and the course of the mechanical arm into the furnace;
- The mechanical robotic system arm dovetails in the tray containing the mold with the formed liquid foam;
- The mechanical robotic system arm extracts the tray containing the mold with the formed liquid foam (third heating zone) from the furnace, and puts it into the cooling sector;
- The mechanical system arm puts the tray on the second conveyor belt and the furnace door is closed, simultaneously.

This last operation is associated to the velocity of the conveyor belt of the cooling sector, which is different from the first conveyor belt of the continuous furnace. Two conveyor belts were installed to avoid any interference between the foaming process and the solidification of the formed liquid foam to solid foam. The first conveyor belt moves the precursor material from the loading sector through the continuous furnace to the exit sector, where the expansion of the precursor occurs. The second conveyor belt is placed in the cooling sector, where the foam is solidified. Both carrier conveyor belts are made of stainless-steel chains. The steel chains were designed to support temperatures up to 1000 °C in continuous operation, to assure a good thermal fatigue resistance, a good corrosion resistance, good dimensional stability and a good resistance to weld to the workpiece surface. The movement of the precursor material (with or without the mold) is associated to the movement of the conveyor belt created by the engine/motors and the drums. This is equipped by drive motors that ensure a continuous motion through the furnace and the rate variations according to the materials to be expanded. The control of the velocity of the conveyor belt and temperatures of each heating zones are programmed according to the required thermal cycle of foaming.

3.3. Operation Mode

The operation mode of the automated continuous foaming production line is as follows:

- The loading of the mold (open or closed) or the hollow structures containing the foamable precursor material is placed into a stainless-steel tray in the feed sector of the continuous foaming furnace on the conveyor belt at the room temperature.
- The first furnace door is opened and the tray with the mold and foamable precursor material enters the furnace.
- The first furnace door is closed immediately after the tray enters the furnace.
- The tray containing the mold and the precursor material moves through the conveyor belt by a controlled motion through the three heating zones. The foamable precursor material expands, filling the mold completely.
- The mold reaches the three-heating zones and touches the suspended metallic chains. This physical contact activates a signal to start the operation of extracting the mold with the formed liquid metallic foam by the robotic system, which is associated with the command of opening/closing the second furnace door. The mechanical arm system activates the movement towards the furnace and the furnace door at the end of the continuous furnace starts to open.
- The mechanical system arm enters in the third heating sector of the furnace and dovetails in the tray, which was developed for this purpose.
- The mechanical arm system begins the transportation of the mold with the liquid metallic foam and places it on the conveyor belt of the continuous foaming furnace that is programmed to ensure the uniform solidification of the formed foam up to the room temperature.
- The worker can open the mold and extract the resulting component made of solid metallic foam.

The command and control parameters of the entire installation are performed through one electric console. For example, it controls the thermal foaming cycle associated to each heating zone, the velocities of the conveyor belts, as well as the movement of the doors.

3.4. Case-Studies

Simple and complex parts made of aluminum alloy foams were fabricated using this automated continuous production line. The closed mold is built according to the dimensions and geometry of the foam part to be fabricated. Figure 7 illustrates an example to fabricate a complex foam part, showing the 3D model designed in SolidWorks software (Figure 7a), the closed mold made of stainless steel constructed for this purpose (Figure 7b) and finally, the components made of aluminum alloy (Figure 7c).

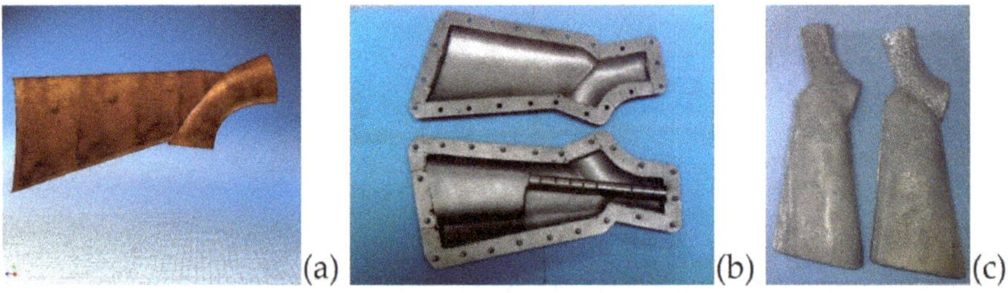

Figure 7. 3D model of the complex foam prototype designed in SolidWorks software (**a**). Stainless steel closed-mold (**b**). Complex foam parts (approx. 290 mm in height) (**c**).

The molds could be fabricated using screw system or clamping system, as illustrated in Figure 8a,b respectively. The use of molds with clamping system makes it much easier to extract the resulting solid foam components compared to those with screw system, which require a special tool, reducing the global production time. Figure 9 shows examples of real molds with screw system (Figure 9a) and with clamping system (Figure 9b), which were constructed to prepare aluminum alloy foam parts.

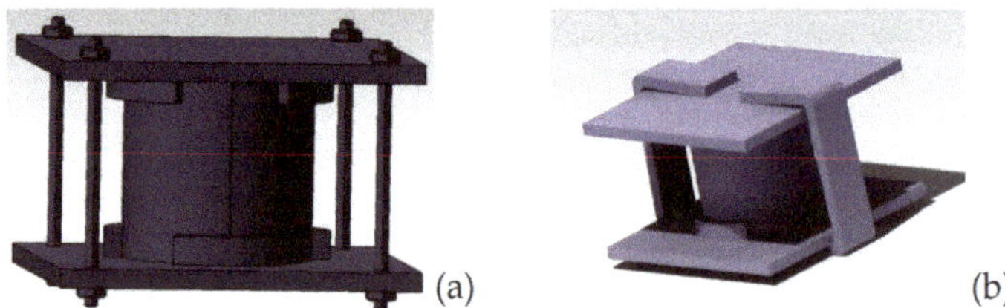

Figure 8. Molds with screw system (**a**) and with clamping system (**b**).

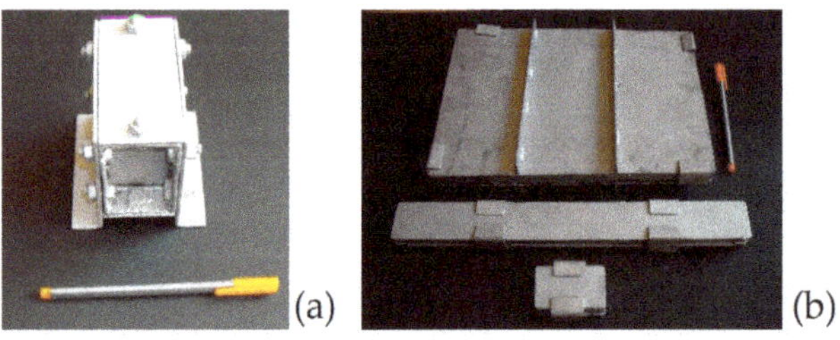

Figure 9. Real molds with screw system (**a**) and with clamping system (**b**).

Figure 10 shows simple and complex foam parts made of aluminum alloys fabricated using this automated continuous production line with stainless steel molds constructed for each component. The foamable precursor material is selected according to the dimensions and geometry of the respective component. For example, the foam blocks and the complex foam parts were prepared using the foamable precursor material with 160 mm × 20 mm in cross-section (Figure 1a). Simple thin foam panels and sandwich foam panels were prepared using the foamable precursor material with 20 mm × 5 mm in cross-section (Figure 1b). The rectangular block of the foamable precursor material (160 mm × 20 mm in cross-section) is usually cut according to the characteristics of the foam component to be fabricated (volume, geometry and dimensions). However, foam parts could be fabricated using various small precursor pieces (Figure 11a) instead of a single precursor [27], as illustrated in Figure 11.

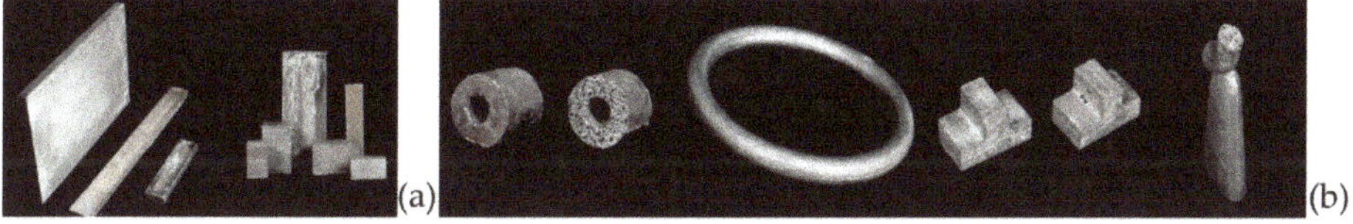

Figure 10. Simple thin foam panels (e.g., $210 \times 250 \times 15$ mm^3 and $300 \times 35 \times 15$ mm^3) and foam blocks (e.g., $200 \times 50 \times 50$ mm^3 and $80 \times 50 \times 50$ mm^3) (**a**) and complex foam parts (e.g., steering wheel 250 mm in outer diameter) (**b**).

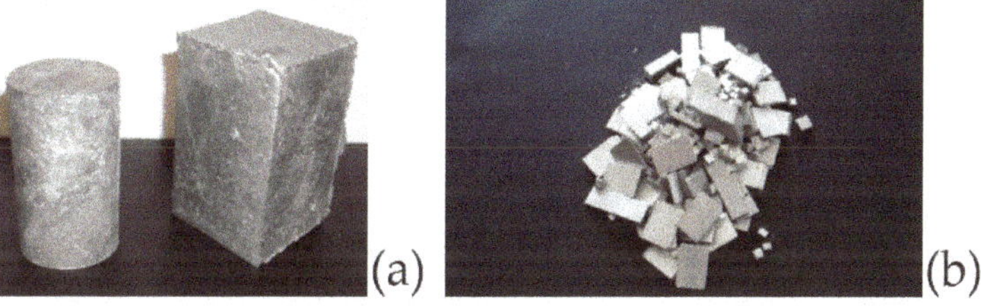

Figure 11. Foam parts (rectangular block: $80 \times 50 \times 50$ mm^3; cylindrical foam: 30 mm in diameter and 60 mm in height) (**a**) fabricated using rejected precursor pieces (**b**).

This manufacturing process has several advantages when compared to other existing processes to produce metallic foams. It enables the production of parts and components of complex geometry of good quality in a simple, effective and practical way, as illustrated in Figures 7 and 10b. Another advantage is that it joins the metal foams with other materials during its foam formation, promoting a metallic bonding. This facilitates and simplifies its application, since it eliminates the expensive joining step, leading to highly competitive products. Moreover, the application of the common joining techniques [28–30], such as welding and brazing, damages the cellular structures of the foams. The aluminium alloy foams are usually used as a core or a filler of sandwich panels and thin-walled structures, respectively. Results have shown that it is possible to fabricate in-situ foam filled tubes [18,20] (Figure 12a) and in-situ sandwich foam panels [21] (Figure 12b) using aluminium alloy tubes or aluminium alloy sheets by applying this manufacturing process. The manufacturing conditions could be adjusted to ensure the structural integrity of these sheets or tubes when exposed to high foaming temperatures that are close to the melting temperature of the alloy. Results have also demonstrated that the thermal treatment that is submitted to these tubes or sheets at high temperatures during the foam formation is beneficial to obtain a predictable and stable mechanical behavior of the resulting in-situ foam structures. The ductility of these structures increases, leading to an efficient crashworthiness without formation of cracks and abrupt failure when subjected to compressive and bending loads. Results also show that the global weight of the resulting in-situ foam structures can be controlled by reducing the wall thickness of the tubes or sheets [18]. Moreover, the metallic inserts could be incorporated into the foams [31], resulting in in-situ reinforced foams, as shown in Figure 12c. This could be an advantage for the industry. For example, fasteners or other standard parts used in vehicles, like screws, bolts and pin rivets could be incorporated, minimizing the discontinuity, slip and fracture under loading utilized in the processes such as screwing.

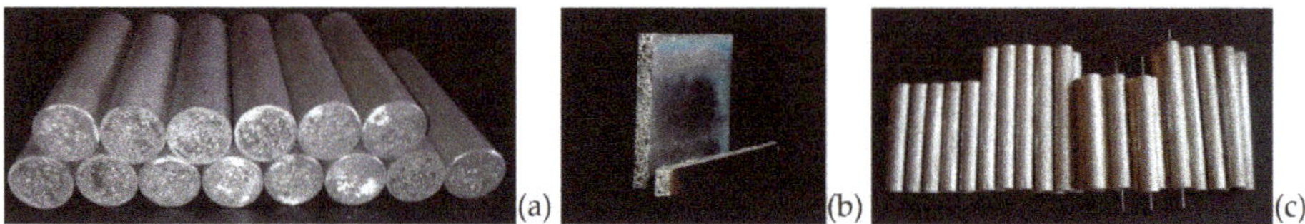

Figure 12. In-situ foam filled tubes (150 mm in length, 30 mm in outer diameter) (**a**), in-situ sandwich foam panels (rectangular panel: $210 \times 250 \times 15$ mm^3-foam core and two aluminum sheets 1.5 mm in thickness; long panel: $300 \times 35 \times 15$ mm^3-foam core; two aluminum sheets 1.5 mm in thickness) (**b**), and simple and in-situ reinforced foams (length: 150 and 200 mm; diameter: 25 mm) made of aluminum alloys (**c**).

The metallic foams prepared by powder metallurgy method have a more ductile behavior when compared to the direct foaming methods [32,33]. In addition to this, the broad spectrum of foam parts with a free design can be easily fabricated with a good surface finish (Figures 10 and 12c) as an outer surface dense (integral) skin around each component is created during its production. Results have demonstrated that the powder metallurgy is a near-net shape manufacturing technology that produces foam components close to the finished size and shape, minimizing the number of secondary finishing processes. Results have shown that this technology allows a reduction in the number of processing steps (no joining step), a significant reduction in the amount of waste material and an ability to easily produce complex and advanced geometries. Results have indicated that AlSi7 foam parts or components could be fabricated at 700 °C. Tables 2 and 3 give an overview of the main properties of some parts made of aluminum alloy foams, in which some of them are presented in Figures 10–12, such as the cubic and cylindrical integral-skin foams, in-situ foam filled tubes made of aluminum alloys and in-situ reinforced carbon steel reinforced foams. The compressive and bending strengths of these foams increase with the foam density.

Table 2. Compressive properties of the simple and composite structures made of aluminum alloy foams.

Foam Parts	Dimensions	Foam Density (g/cm^3)	Compressive Strength (MPa)
Integral-skin foams			
Cube foams	$20 \times 20 \times 20$ mm^3	0.4–0.6	5–8
Cylindrical foams	$\phi * = h * = 30$ mm	0.5–0.6	7–12
Cylindrical foams	$\phi * = 26$ mm; $h * = 23$ mm	0.6–0.8	8–21
In-situ foam filled tubes (FFTs)			
Cylindrical in-situ FFTs	$\phi_{outer} ** = 27$ mm; $\phi_{inner} ** = 26.4$ mm Tube wall thickness = 0.6 mm $h = 26$ mm	0.6–0.7 ****	19–23
Square in-situ FFTs	$25 \times 25 \times 25$ mm^3 Tube wall thickness = 1.5 mm	0.4–0.6 ****	38–48
Cylindrical in-situ FFTs	$\phi_{outer} ** = 30$ mm; $\phi_{inner} ** = 26$ mm Tube wall thickness = 2 mm $h = 23$ mm	0.5–0.6 ****	61–64
In-situ reinforced foams			
In-situ carbon steel bar *** reinforced FFTs	$\phi = 25$ mm; $h = 23$ mm	0.4–0.8 ****	19–42

* ϕ and h are the diameter and height of the specimen; ** Outer and inner diameters of the thin-walled tube; *** Carbon steel bars of S235JRG2C (0.20% C, 1.40% Mn, 0.045% P, 0.045% S); **** Foam core.

Table 3. Bending properties of the simple and composite structures made of aluminum alloy foams.

Foam Parts	Dimensions	Foam Density (g/cm^3)	Bending Strength (kN)
	Integral-skin foams		
Cylindrical foams	ϕ * = 25 mm; L * = 200 mm	0.6–0.7	0.8–1.8
Cylindrical foams	ϕ * = 25 mm; L ** = 150 mm	0.5–0.8	1.5–2.8
	In-situ foam filled tubes (FFTs)		
Cylindrical in-situ FFTs	ϕ_{outer} *** = 30 mm; ϕ_{inner} *** = 25 mm L = 150 mm	0.6–0.7 ****	9.9–10.6
	In-situ reinforced foams		
In-situ carbon steel bar **** reinforced FFTs	ϕ = 25 mm; L = 200 mm	0.5–0.8 ****	1.1–2.1
In-situ carbon steel bar **** reinforced FFTs	ϕ = 25 mm; L = 150 mm	0.5–0.8 ****	1.6–2.7

* ϕ and L are the diameter and the length of the specimen; ** Outer and inner diameters of the thin-walled tube; *** Carbon steel bars of S235JRG2C (0.20 %C, 1.40% Mn, 0.045% P, 0.045% S); **** Foam core.

The compressive strength of the foam components is higher than the bending strength. In fact, the closed-cell aluminum foams are used as crash energy absorbers.

Results have demonstrated that the larger the foam blocks, the higher the cellular architecture variation due to a non-uniform foam growth during the heating of the single precursor material [34,35]. For smaller foam blocks [34], a more regular distribution of the pores through the foam is obtained.

4. Conclusions

The automated continuous production line composed of a continuous furnace, a cooling sector and a robotic system is approximately 7 m long, 1.5 m high and 1 m wide. This foaming production line allows the production of larger quantities of foam materials when compared to those fabricated by the batch furnace. The new production line has several advantages in comparison to the conventional procedures as follows:

- The transition from a manual operation (piece by piece production) to an automated and continuous operation;
- The quality of the formed foam parts no longer depends on the experience and training of the worker since the process is automated;
- The safety of workers is ensured and no personnel protective equipment (e.g., heat protection clothing, gloves and safety glasses) is necessary. The workers only carry out tasks at the room temperature;
- No limitation in terms of weight, due to the support of the mechanical arm system;
- The automatic and controlled extraction operation of formed foams from the heating zone to the cooling sector prevents the collapse of the foams and reduces the number of rejected parts;
- The controlled cooling ensures uniform solidification of the foam in all directions and avoids the appearance of defects in formed parts, especially in larger pieces, avoiding inferior mechanical properties;
- The volume of production of foam parts per unit time increases.

Author Contributions: Conceptualization, I.D. and M.J.V.; methodology, I.D. and M.J.V.; validation, I.D., M.J.V. and M.V.; formal analysis, I.D. and M.V.; investigation, I.D. and M.V.; data curation, I.D. and M.V.; writing—original draft preparation, I.D. and M.V.; writing—review and editing, I.D. and M.V.; visualization, I.D.; supervision, I.D.

Acknowledgments: Supports given by M. J. Amaral-Equipamentos Industriais Lda (Vale de Cambra), the Portuguese Science Foundation (FCT), UID/EMS/00481/2019-FCT and CENTRO-01–0145-FEDER-022083 (Centro 2020 program-Portugal 2020).

Conflicts of Interest: The authors declare no conflict of interest.

References

1. García-Moreno, F. Commercial Applications of Metal Foams: Their Properties and Production. *Materials* **2016**, *9*, 85. [CrossRef]
2. Kim, S.; Lee, C.-W. A Review on Manufacturing and Application of Open-cell Metal Foam. *Procedia Mater. Sci.* **2014**, *4*, 305–309. [CrossRef]
3. Garcia-Avila, M.; Portanova, M.; Rabiei, A. Ballistic performance of composite metal foams. *Compos. Struct.* **2015**, *125*, 202–211. [CrossRef]
4. Vesenjak, M.; Borovinšek, M.; Ren, Z.; Irie, S.; Itoh, S. Behavior of Metallic Foam under Shock Wave Loading. *Metals* **2012**, *2*, 258–264. [CrossRef]
5. Shim, C.; Yun, N.; Robin Yu, R.; Byun, D. Mitigation of Blast Effects on Protective Structures by Aluminum Foam Panels. *Metals* **2012**, *2*, 170–177. [CrossRef]
6. Duarte, I.; Peixinho, N.; Andrade-Campos, A.; Valente, R. Special Issue on Cellular Materials. *Sci. Technol. Mater.* **2018**, *30*, 1–3. [CrossRef]
7. Obi, B. *Polymeric Foams Structure-Property-Performance: A Design Guide*, 1st ed.; Imprint William Andrew Elsevier Inc.: Oxford, UK, 2017.
8. Scheffler, M.; Colombo, P. *Cellular Ceramics, Structure, Manufacturing, Properties and Applications*; WILEY-VCH Verlag GmbH: Weinheim, Germany, 2005.
9. Duarte, I.; Banhart, J. A study of aluminium foam formation—Kinetics and microstructure. *Acta Mater.* **2000**, *48*, 2349–2362. [CrossRef]
10. Matijasevic-Lux, B.; Banhart, J.; Fiechter, S.; Görke, O.; Wanderk, N. Modification of titanium hydride for improved aluminium foam manufacture. *Acta Mater.* **2006**, *54*, 1887–1900. [CrossRef]
11. Stöbener, K.; Baumeister, J.; Rausch, G.; Rausch, M. Forming metal foams by simpler methods for cheaper solutions. *Met. Powder Rep.* **2005**, *60*, 12–14. [CrossRef]
12. Orbulov, I.N.; Szlancsik, A. On the Mechanical Properties of Aluminum Matrix Syntactic Foams. *Adv. Eng. Mater.* **2018**, *20*, 1700980. [CrossRef]
13. Duarte, I.; Vesenjak, M.; Krstulović-Opara, L.; Ren, Z. Crush performance of multifunctional hybrid foams based on an aluminium alloy open-cell foam skeleton. *Polym. Test.* **2018**, *67*, 246–256. [CrossRef]
14. Duarte, I.; Ferreira, J.M.F. 2D quantitative analysis of metal foaming kinetics by hot-stage microscopy. *Adv. Eng. Mater.* **2014**, *16*, 33–39. [CrossRef]
15. Gupta, N.; Rohatgi, P.K. *Metal Matrix Syntactic Foams: Processing, Microstructure, Properties and Applications*; DEStech Publications, Inc.: Lancaster, PA, USA, 2015.
16. Movahedi, N.; Murch, G.E.; Belova, I.V.; Fiedler, T. Functionally graded metal syntactic foam: Fabrication and mechanical properties. *Mater. Des.* **2019**, *168*, 107652. [CrossRef]
17. Duarte, I.; LovreKrstulović-Opara, L.; Dias-de-Oliveira, J.; Vesenjak, M. Axial crush performance of polymer-aluminium alloy hybrid foam filled tubes. *Thin-Walled Struct.* **2019**, *138*, 124–136. [CrossRef]
18. Duarte, I.; Krstulović-Opara, L.; Vesenjak, M. Axial crush behaviour of the aluminium alloy in situ foam filled tubes with very low wall thickness. *Compos. Struct.* **2018**, *192*, 184–192. [CrossRef]
19. Duarte, I.M.A.; Banhart, J.; Ferreira, A.J.M.; Santos, M.J.G. Foaming around fastening elements. *Mater. Sci. Forum* **2016**, *514–516*, 712–717. [CrossRef]
20. Duarte, I.; Vesenjak, M.; Krstulović-Opara, L.; Ren, Z. Static and dynamic axial crush performance of in situ foam-filled tubes. *Compos. Struct.* **2015**, *124*, 128–139. [CrossRef]
21. Banhart, J.; Seeliger, H.-W. Aluminium Foam Sandwich Panels: Manufacture, Metallurgy and Applications. *Adv. Eng. Mater.* **2008**, *10*, 793–802. [CrossRef]
22. Crupi, V.; Epasto, G.; Guglielmino, E. Impact Response of Aluminum Foam Sandwiches for Light-Weight Ship Structures. *Metals* **2011**, *1*, 98–112. [CrossRef]
23. Baumgärtner, F.; Duarte, I.; Banhart, J. Industrialization of powder compact foaming process. *Adv. Eng. Mater.* **2000**, *2*, 168–174. [CrossRef]
24. Paeplow, M.; García-Moreno, F.; Meagher, A.J.; Rack, A.; Banhart, J. Coalescence Avalanches in Liquid Aluminum Foams. *Metals* **2017**, *7*, 298. [CrossRef]
25. Mukherjee, M.; Garcia-Moreno, F.; Banhart, J. Solidification of metal foams. *Acta Mater.* **2010**, *58*, 6358–6370. [CrossRef]

26. Stanzick, H.; Duarte, I.; Banhart, J. Der schäumprozeß von aluminium. *Materialwissenschaft und Werkstofftechnik* **2000**, *31*, 409–411. [CrossRef]

27. Duarte, I.; Vesenjak, M.; Krstulović-Opara, L.; Vesenjak, M. Dynamic compressive behaviour of aluminium foams fabricated from rejected precursor materials. *Cienc. Tecnol. Mater.* **2016**, *28*, 19–22. [CrossRef]

28. Bangash, M.K. Graziano Ubertalli, Davide Di Saverio, Monica Ferraris and Niu Jitai Joining of Aluminium Alloy Sheets to Aluminium Alloy Foam Using Metal Glasses. *Metals* **2018**, *8*, 614. [CrossRef]

29. Chen, N.; Feng, Y.; Chen, J.; Li, B.; Chen, F.; Zhao, J. Vacuum Brazing Processes of Aluminum Foam. *Rare Met. Mater. Eng.* **2013**, *42*, 1118–1122.

30. Hangai, Y.; Kobayashi, R.; Suzuki, R.; Matsubara, M.; Yoshikawa, N. Aluminum Foam-Filled Steel Tube Fabricated from Aluminum Burrs of Die-Castings by Friction Stir Back Extrusion. *Metals* **2019**, *9*, 124. [CrossRef]

31. Duarte, I.; Krstulović-Opara, L.; Vesenjak, M. Analysis of performance of in situ carbon steel bar reinforced Al-alloy foams. *Compos. Struct.* **2016**, *152*, 432–443. [CrossRef]

32. Kuwahara, T.; Osaka, T.; Saito, M.; Suzuki, S. Compressive Properties of A2024 Alloy Foam Fabricated through a Melt Route and a Semi-Solid Route. *Metals* **2019**, *9*, 153. [CrossRef]

33. Rivera, N.M.T.; Torres, J.T.; Valdés, A.F. A-242 Aluminium Alloy Foams Manufacture from the Recycling of Beverage Cans. *Metals* **2019**, *9*, 92. [CrossRef]

34. Duarte, I.; Vesenjak, M.; Krstulović-Opara, L. Variation of quasi-static and dynamic compressive properties in a single aluminium foam block. *Mater. Sci. Eng. A* **2014**, *616*, 171–182. [CrossRef]

35. Ulbin, M.; Vesenjak, M.; Borovinšek, M.; Duarte, I.; Higa, Y.; Shimojima, K.; Ren, Z. Detailed Analysis of Closed-Cell Aluminum Alloy Foam Internal Structure Changes during Compressive Deformation. *Adv. Eng. Mater.* **2018**, *20*, 1800164. [CrossRef]

Article

Effect of Primary Crystals on Pore Morphology during Semi-Solid Foaming of A2024 Alloys

Takashi Kuwahara [1,*], Mizuki Saito [1], Taro Osaka [1] and Shinsuke Suzuki [1,2]**

1   Faculty of Science and Engineering, Waseda University, 3-4-1 Okubo, Shinjuku, Tokyo 169-8555, Japan;
    m.saitoh@akane.waseda.jp (M.S.); t-evolva@moegi.waseda.jp (T.O.); suzuki-s@waseda.jp (S.S.)
2   Kagami Memorial Research Institute of Materials Science and Technology, Waseda University,
    2-8-26 Nishi-Waseda, Shinjuku, Tokyo 169-0051, Japan
*   Correspondence: takuwahara@ruri.waseda.jp; Tel.: +81-(0)3-5286-8126

Received: 23 November 2018; Accepted: 9 January 2019; Published: 16 January 2019

Abstract: We investigated pore formation in aluminum foams by controlling primary crystal morphology using three master alloys. The first one was a direct chill cast A2024 (Al-Cu-Mg) alloy (DC-cast alloy). The others were A2024 alloys prepared to possess fine spherical primary crystals. The second alloy was made by applying compressive strain through a Strain-Induced Melt-Activated process alloy (SIMA alloy). The third one was a slope-cast A2024 alloy (slope-cast alloy). Each alloy was heated to either 635 °C (fraction of solid $f_s = 20\%$) or 630 °C ($f_s = 40\%$). TiH$_2$ powder was added to the alloys as a foaming agent upon heating them to a semi-solid state and they were stirred while being held in the furnace. Subsequently, A2024 alloy foams were obtained via water-cooling. The primary crystals of the DC-cast alloy were coarse and irregular before foaming. After foaming, the size of the primary crystals remained irregular, but also became spherical. The SIMA and slope-cast alloys possessed fine spherical primary crystals before and after foaming. In addition to average-sized pores (macro-pores), small pores were observed inside the cell walls (micro-pores) of each alloy. The formation of macro-pores did not depend on the formation of the primary crystals. Only in the DC-cast alloy did fine micro-pores exist within the primary crystals. The number of micro-pores in the SIMA and slope-cast alloys was one third of that in the DC-cast alloy.

Keywords: semi-solid; aluminum foam; primary crystals; SIMA process; slope casting; pore morphology

---

1. Introduction

New, lighter materials are required to reduce the environmental impact and cost of transportation equipment, including automobiles and aircrafts [1,2]. In addition, passenger safety should be maintained or enhanced by these new materials. Aluminum foam has attracted significant attention in recent years for meeting weight reduction and safety requirements. Aluminum foam has many pores and, owing to its ultralight and excellent shock-absorbing properties, could be applied to transportation equipment. However, the compressive properties of aluminum foam must be improved before it can be effectively used for this purpose. The compressive properties of foams are determined by their base metal and structure [3]. To this end, Fukui et al. [4,5] improved the compressive properties of aluminum foam using the super duralumin A2024 (Al-Cu-Mg) alloy, which is a well-known high-strength and lightweight material, as a base metal to improve the strength of the foam.

Foaming via the melt route is a common and efficient fabrication method for aluminum alloy foams [6]. With this method, the melt must be thickened to maintain pore stability. In general, thickening is achieved by adding Ca, which generates oxides that spread throughout the melt [7]. To fabricate the A2024 alloy foams, the oxide of the alloying element, Mg, is used as a thickener instead of Ca. However, decreases in base metal strength have been observed because the quantity of Mg as an alloying element also decreases.

To solve this problem, Hanafusa et al. [8] fabricated foams in a semi-solid state using primary crystals, instead of oxides, as a thickener. Sekido et al. [9] investigated the thickening effect of primary crystals in a semi-solid state. Our group demonstrated that it is possible to fabricate an A2024 alloy foam without adding a thickener in a semi-solid state. However, the effects of this fabrication on the properties of foam have been researched only experimentally, not systematically.

Also, the primary crystals acting as a thickener in the melt are larger in size than those of the oxides. Moreover, the primary crystals are generally larger than the preferred range of thickener sizes in the melt route [10]. Therefore, foams fabricated in a semi-solid state may have different stabilization mechanisms than those in the melt route. How these large thickeners affect the formation of pores in foams is not evident. However, compared to oxides, the sizes and shapes of primary crystals are easy to control and evaluate. Therefore, it may be possible to improve the foam fabrication process by aggressively controlling the primary crystals.

The objective of this study is to investigate the effects of the large primary crystals present inside the melt on pore formation in aluminum alloy foams in a semi-solid state. The effects of the diameter and shape of the crystals on pore morphology were examined. To observe the changes caused by the differences in the morphology of primary crystals, A2024 alloy foams were fabricated via the processing of an A2024 master alloy by controlling the formation of primary crystals to favor a fine spherical shape. The fabrication processes used were Strain-Induced Melt Activation (SIMA) [11] and slope casting [12]. Subsequently, we evaluated pore formation in the foam and the formation of primary crystals in the cell wall. In addition, we measured the Ti distribution to trace the path of TiH_2, which was the blowing agent used.

The innovations of this study are as follows. First, to systematically derive the effects of diameter, shape, and fraction of solid on primary crystals, the size and shape of the primary crystals were controlled by SIMA and slope casting. These controlling processes were difficult in oxide particle thickening. Second, we investigated fabrication through the stabilization of cell walls by controlled primary crystals.

2. Materials and Methods

2.1. Fabrication of Master Alloys

We fabricated three kinds of A2024 master alloys. Table 1 shows the composition of the A2024 alloys used in this study. Figure 1 shows a schematic of the fabrication method of the A2024 master alloys. The A2024 direct chill slab was sliced perpendicular to the pull-out direction of the DC casting. The A2024 plane was cut into several pieces. These pieces will be hereafter referred to as the DC-cast alloys. Then, some of the DC-cast alloys were processed via SIMA. The DC-cast alloys initially measured $20 \times 30 \times 30$ mm^3 and were compressed to a compressive strain $f_s = 10\%$, as done in the study presented by Sirong et al. [13]. Then, the pieces were annealed at 350 °C for 3 h in a muffle furnace. This master alloy will be hereafter referred to as the SIMA alloy. In addition, some of the DC-cast alloys were processed via slope casting. The A2024 pieces were heated at 649 °C and poured onto a cooling plate made of Cu. The degree of the slope, θ, and the contact length, l, were 60° and 200 mm, respectively, as in a previous study [14]. Afterwards, the alloy was cast; this master alloy will be referred to as the slope-cast alloy.

Table 1. Chemical composition of the A2024 alloys.

Element	Cu	Mg	Mn	Fe	Si	Cr	Zn	Ti	Al
mass%	4.44	1.60	0.67	0.15	0.08	<0.01	<0.01	<0.01	bal.

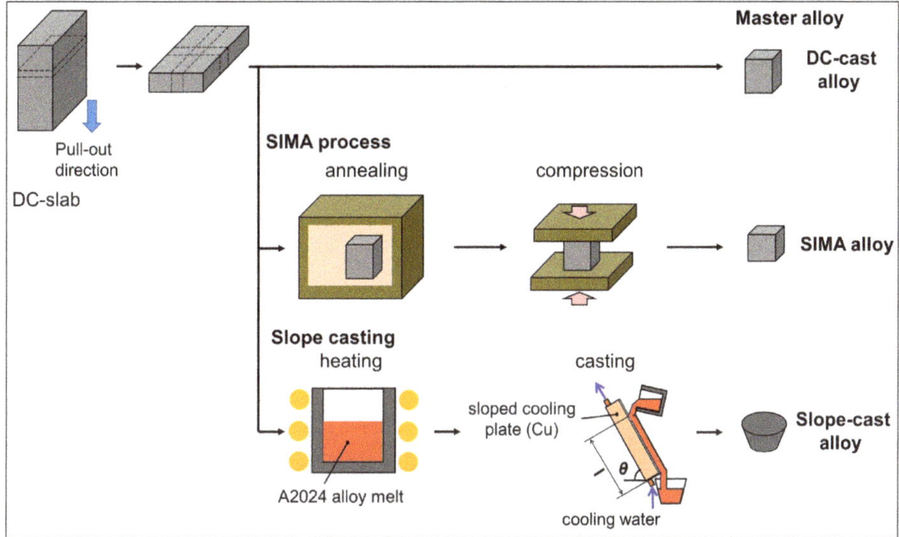

Figure 1. Schematic of the preparation of the master alloys.

2.2. Fabrication of A2024 Alloy Foam in a Semi-Solid State

Figure 2 shows the fabrication method of the A2024 alloy foam in a semi-solid state. First, 100 g of the DC-cast alloy, SIMA alloy, or slope-cast alloy was placed in a SUS304 crucible coated with Al_2O_3. Each alloy was heated to and held at 635 °C ± 0.3 °C (fraction of solid f_s = 20%) [15] or 630 ± 0.3 °C (f_s = 40%) [15] in an electric furnace, as shown in Figure 2a. The measured temperature was inside the error range. The temperature was measured using a K-type thermocouple covered with an Al_2O_3 protective tube. After the temperature stabilized, 1 mass% TiH_2 was added to the slurry after the thermocouple was taken out. Then, the slurry was stirred using an impeller coated with boron nitride at 900 rpm for 100 s, as shown in Figure 2b. After stirring, the slurry was held for 200 s and the crucible was removed from the furnace, as shown in Figure 2c. Subsequently, the foam was solidified and cooled using 3 L/min of water as shown in Figure 2d, and the A2024 alloy foam was obtained. Six foams were obtained in total, one for each combination of master alloy and fraction of solid. Additionally, to analyze the differences in the primary crystals before stirring and after foaming, metal samples held in a semi-solid state were obtained for each foam.

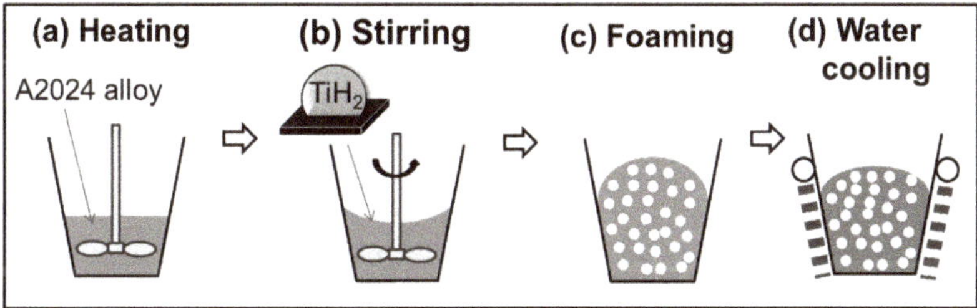

Figure 2. Fabrication method of the A2024 foam in a semi-solid state; (**a**) heating; (**b**) stirring; (**c**) foaming; (**d**) water cooling.

2.3. Evaluation of Pores and Primary Crystal Formation

To derive the porosity of the foam, we measured the volume of the foam via Archimedes' method using the mass of the foam and a spring scale. From Equation (1), we calculated the porosity p of the foam using the density of the A2024 alloy foam ρ_P, which in turn was calculated using the volume and mass of the foams and the density of the A2024 alloy, $\rho_{NP} = 2.77$ Mg/m^3. The pore morphology (average pore diameter d_p and average pore circularity e_p) of a cross-section from the middle of a sample was measured using the image-analysis software WinROOF^{TM} version 6.1 (Mitani Corporation, Fukui, Japan). Equations (2) and (3) describe the pore formations, where S is the area of the pores and L is their perimeter. To improve the accuracy of our measurements, pores smaller than $d = 0.2$ mm were not considered. The diameter of the primary crystals d_α and their circularity e_α before and after the foaming of each alloy were also measured using WinROOF^{TM} and calculated via Equations (2) and (3). Every measurement was taken once for every foam.

$$p = (1 - \rho_p/\rho_{NP}) \times 100\% \tag{1}$$

$$d_p = (4S/\pi)^{1/2} \tag{2}$$

$$e_\alpha = 4\pi S/L^2 \tag{3}$$

2.4. Analysis of the Pure Ti Distribution

We fabricated non-porous metal samples to estimate the distribution of the foaming agent and the foaming sites. These non-porous metal samples were fabricated using a similar procedure to the fabrication method for the A2024 foams, except that pure Ti was used instead of TiH$_2$. A Ti distribution analysis was performed via qualitative mapping using an electron probe micro-analyzer (EPMA, JEOL, JXA-8230, Tokyo, Japan; measurement elements: Ti, Cu; pressurization voltage: 15 kV; irradiation current value: 3×10^{-8} A; irradiation interval: 10 μm; acquisition time: 20 ms/point; measurement range: 5 mm × 3.75 mm). Next, mapping images of Ti and Cu were processed using the following method. First, the Ti mapping image was made monochrome and binarized using WinROOF^{TM} and Ti was colored green. Then, the parts that were not Ti were made transparent using the processing software paint.net Version 4.0.6. (dotPDN LLC, Washington, WA, USA). Finally, the Ti image and the monochrome Cu image were merged. One merged image was obtained for each DC-cast alloy and the SIMA alloy with a fraction of solid $f_s = 40\%$.

3. Results and Discussion

3.1. Pore Formation on the A2024 Alloy Foams

Figure 3 shows the microstructures of the (a) DC-cast, (b) SIMA, and (c) slope-cast A2024 alloys at $f_s = 40\%$. Figure 4 shows the average values of (a) the primary crystal diameter d_α and (b) the circularity e_α of the A2024 alloy foams. For Figure 4, approximately 100 primary crystals were measured for each sample, except for the after-foaming data of the DC-cast alloy, for which 27 primary crystals were measured. Dendritic primary crystals were present before melting and during holding at a semi-solid state in the DC-cast alloy, as shown in Figure 3a. Because some dendritic primary crystals grow while others do not, the diameters d_α of the primary crystals in the DC-cast alloy have high error values. In the SIMA alloy, shown in Figure 3b, dendritic primary crystals were present before melting, while fine and spherical primary crystals were formed during holding at a semi-solid state. In the slope-cast alloy, shown in Figure 3c, fine and spherical primary crystals were present before melting and during holding at a semi-solid state. Figures 3a and 4 show that the primary crystals were coarse and uneven in the DC-cast alloy before forming and remained coarse but spherical afterwards. The primary crystals in the SIMA and slope-cast alloys maintained their fine and spherical shape, as shown in Figure 3b,c and Figure 4. Therefore, the primary crystals of the DC-cast alloy spheroidized after foaming, whereas those of the SIMA and slope-cast alloys maintained their fine and spherical shape.

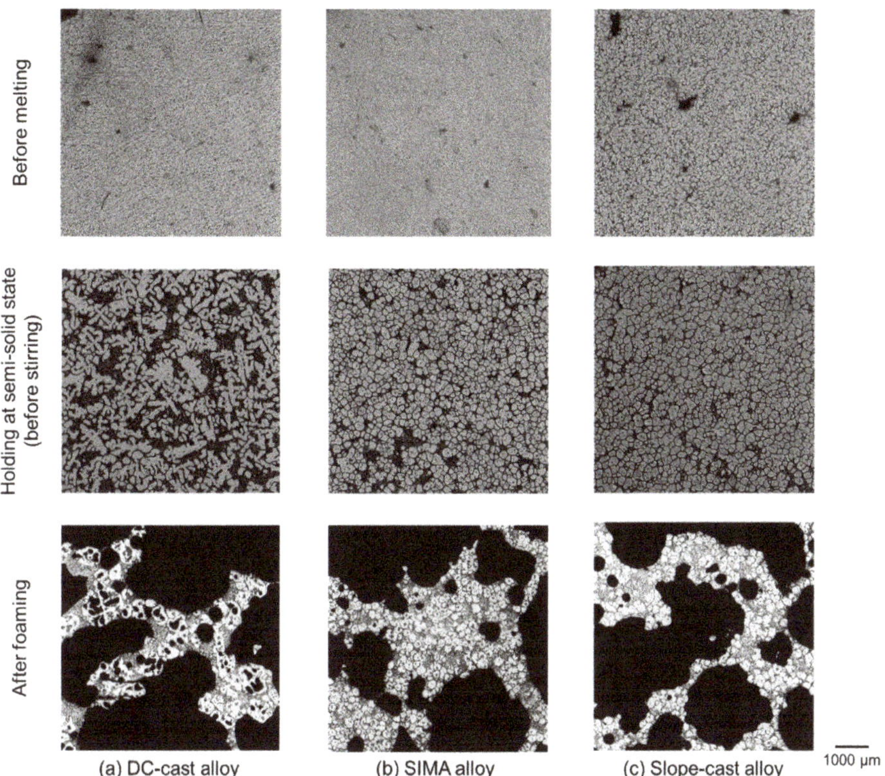

Figure 3. Images of the A2024 alloys (f_s = 40%); (**a**) direct chill cast (DC-cast); (**b**) Strain-Induced Melt-Activated (SIMA); and (**c**) slope-cast.

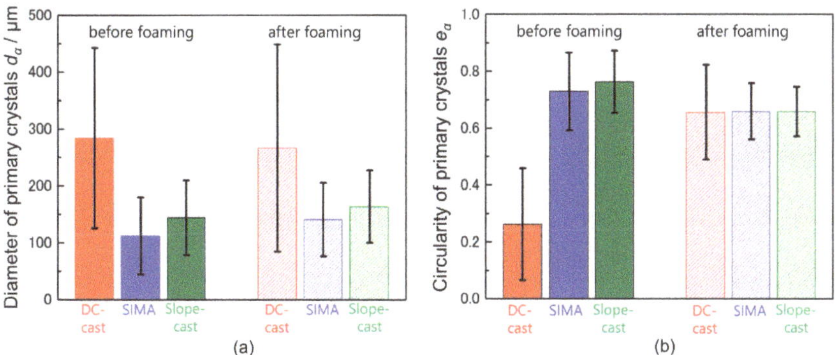

Figure 4. Average values of the primary crystals of the A2024 alloy foams fabricated via foaming in a semi-solid slurry (f_s = 40%); (**a**) diameter d_α; (**b**) circularity e_α. The error bars show the standard deviations.

3.2. Macro-Pores of the A2024 Alloys Fabricated in a Semi-Solid State

Figure 5 shows cross-sections of the A2024 alloy foams of the DC-cast, SIMA, and slope-cast alloys with an f_S of 20% and 40%. The pores were distributed homogeneously as shown in Figure 5. Although a non-porous part is conventionally found in the bottom part [4], the foams in this study did not show a clear difference of pore distribution at different heights. Therefore, the A2024 alloy foams of each alloy can be fabricated at f_s values of 20% and 40%. All of the foams except for the slope-cast alloy foam fabricated at $f_s = 40\%$ reached a porosity of 60%. Figure 6 shows the average pore (a) diameter d_p and (b) circularity e_p of the A2024 alloy foams with porosities of approximately 60%. In Figure 6, 1000–1900 pores were measured for each sample. The pores of the alloy foams fabricated at $f_s = 20\%$ were finer and more spherical than those of the foams fabricated at $f_s = 40\%$ (Figure 5; Figure 6). However, the differences between the primary crystals of the DC-cast and SIMA alloys had little effect on macro-pore morphology. Although the error value of the primary crystal size of the DC-cast was higher than that of the SIMA alloys, this difference did not affect macro-pore morphology. This result is due to the difference between the sizes of the primary crystals and the macro-pores. In each alloy, most primary crystals had a diameter d_α of less than 300 μm, whereas the diameter of the macro-pores d_p was larger than 1000 μm. Therefore, differences in the formation of finer primary crystals had little effect on the formation of macro-pores in terms of macro-pore diameter. Moreover, as shown in Figure 6, the diameter of the macro-pores d_p for a fraction of solid of 40% had high error values. These error values occur because a high fraction of solid makes it difficult for TiH_2 to be distributed uniformly. A schematic drawing of this mechanism is shown in Figure 7.

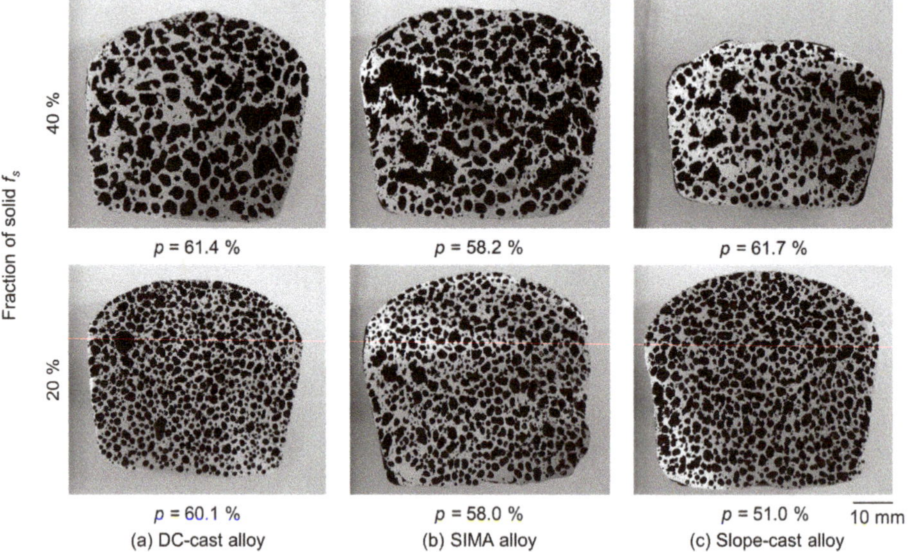

Figure 5. Cross-sections of foams: (**a**) direct chill cast (DC-cast); (**b**) Strain-Induced Melt-Activated (SIMA); and (**c**) slope-cast. *p*: porosity.

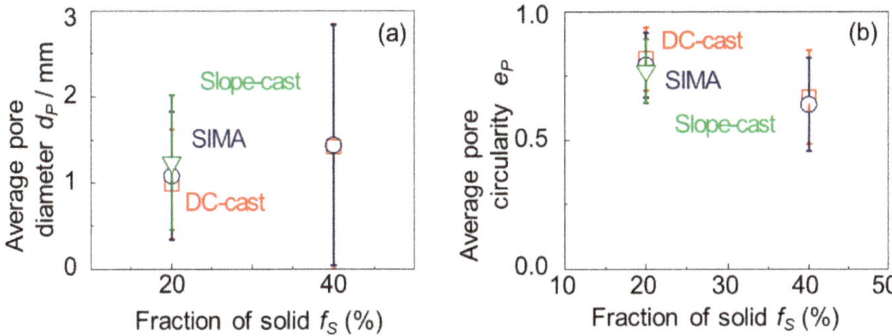

Figure 6. Average values of the pores of the A2024 alloy foams with a porosity of approximately 60%, fabricated via foaming in a semi-solid slurry: (**a**) pore diameter d_p, and (**b**) pore circularity e_p. The error bars show the standard deviations. DC-cast: direct chill cast, SIMA: Strain-Induced Melt-Activated.

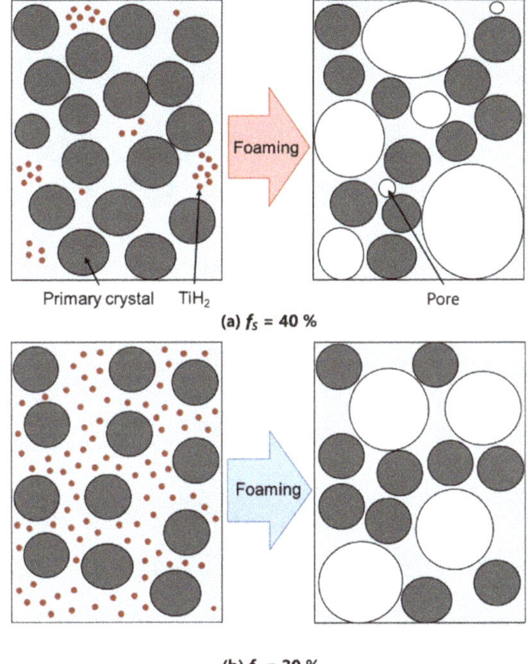

Figure 7. Schematic drawing of the TiH$_2$ distributed in the melt and pores after foaming. (**a**) f_s = 40%; (**b**) f_s = 20%.

3.3. Micro-Pores of the A2024 Alloys Fabricated in a Semi-Solid State

Figure 8 shows the location of the micro-pores in the A2024 alloy foams (f_s = 40%) for the (a) DC-cast, (b) SIMA, and (c) slope-cast alloys. Figure 9 shows the number of micro-pores per unit area in all the A2024 alloy foams. The f_S values of the foams were 20% and 40%. In this study, micro-pores are defined as pores in the cell wall with a diameter of less than 1000 μm, which is approximately the thickness of the cell walls. In addition, to improve image processing, pores with a diameter of less than 40 μm were ignored. Mukherjee et al. defined micro-pores as pores less than 350 μm in diameter [16]. Moreover, Ohgaki et al. reported that cracks initiate from pores with a diameter of 30 μm to 350 μm when an aluminum foam is compressed [17]. However, as shown in Figure 10, most of the micro-pores in this study were inside this range. Therefore, the micro-pores measured in this study corresponded to those in previous studies. In the DC-cast alloy, micro-pores were present inside the primary crystals, as shown in Figure 8a, whereas micro-pores were present between the primary crystals in the SIMA and slope-cast alloys, as shown in Figure 8b. As indicated in Figure 9, the number of micro-pores in the SIMA and slope-cast alloys was approximately 1/3 of that in the DC-cast alloy. As the number of micro-pores increased, the strength of the base metal decreased. However, micro-pores also suppressed densification under compression, as reported by Toda et al. [18].

In addition, Figure 5, Figure 6, Figure 9, and Figure 10 show that there was no major difference in the results of each master alloy except for the number of micro-pores. In contrast, the fraction of solid seems to be the major factor that determines each property in aluminum foam. Furthermore, aluminum foam was fabricated three times for each fraction of solid. Because this foam seems to be reproducible, it was considered unnecessary to repeat the experiment.

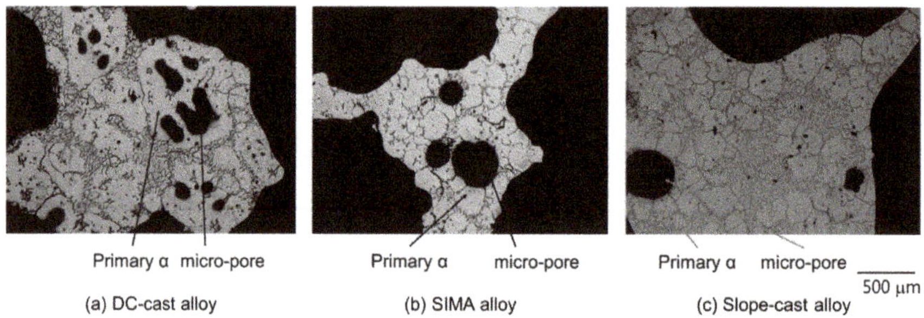

Figure 8. Locations of micro-pores (f_s = 40%). (**a**) direct chill cast (DC-cast) alloy; (**b**) Strain-Induced Melt-Activated (SIMA) alloy; and (**c**) slope-cast alloy.

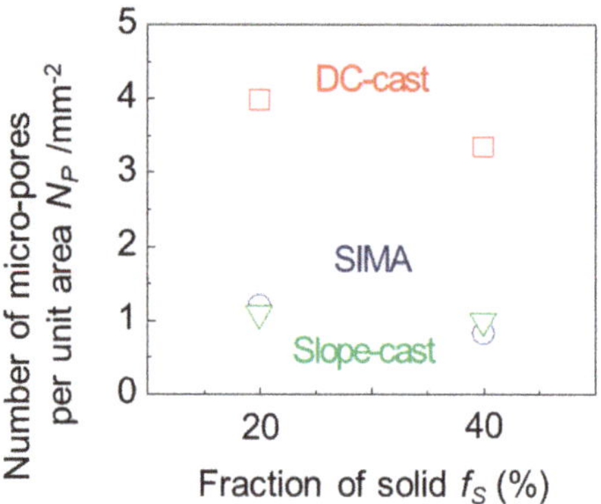

Figure 9. Number of micro-pores per unit area, N_p, in the A2024 alloy foams for the direct chill cast (DC-cast), Strain-Induced Melt-Activated (SIMA), and slope-cast alloys.

3.4. Effect of Primary Crystals on Micro-Pore Formation

As described in Section 3.2., the primary crystals were fine and spherical before and after stirring the SIMA and slope-cast alloys. Figure 11 shows the mappings for Ti and Cu obtained with the EPMA for the non-porous metal samples described in Section 2.4. In Figure 11, the darker parts with less Cu are considered to be primary crystals, which are mostly composed of Al. Moreover, Ti is not completely dispersed because the primary crystals exist as solids. For a better observation, Figure 12 shows the merged image made from the Ti and Cu images shown in Figure 11, as mentioned in Section 2.4. In the DC-cast alloy, the eutectic structure and Ti particles co-exist in one primary crystal, as shown in

Figure 12a. This eutectic structure was considered to have been liquid before solidification. In contrast, the liquid phase and the Ti particle exist between the primary crystals in the SIMA alloy, as shown in Figure 12b. From these results, we deduced a reason for the high quantity of micro-pores inside the DC-cast alloy. Figure 13 shows a schematic drawing of the mechanism behind micro-pore formation. As explained in Section 3.1, the uneven primary crystals in the DC-cast alloy spheroidize after foaming. Based on a previous study by Chen et al. [19], primary crystals have been shown to bend via stirring and spheroidize. Therefore, the liquid phase and TiH_2 were likely trapped in spheroidized primary crystals, similar to pure Ti, as shown in Figure 12a; for this reason, micro-pores could be found inside the primary crystals in the DC-cast alloy, as shown in Figure 13a.

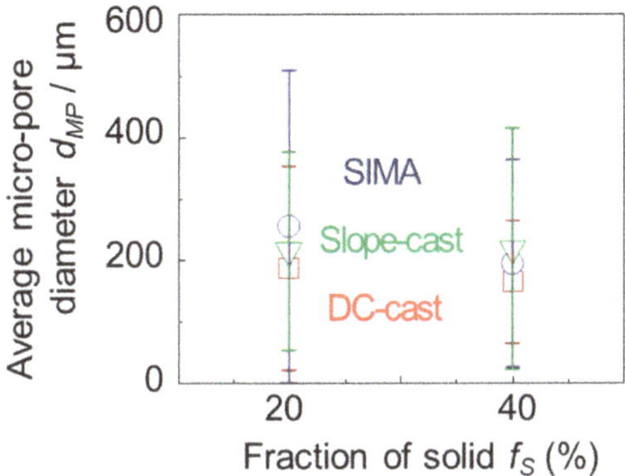

Figure 10. Average diameter of the micro-pores of the A2024 alloy foams fabricated via foaming in a semi-solid slurry for the direct chill cast (DC-cast), Strain-Induced Melt-Activated (SIMA), and slope-cast alloys.

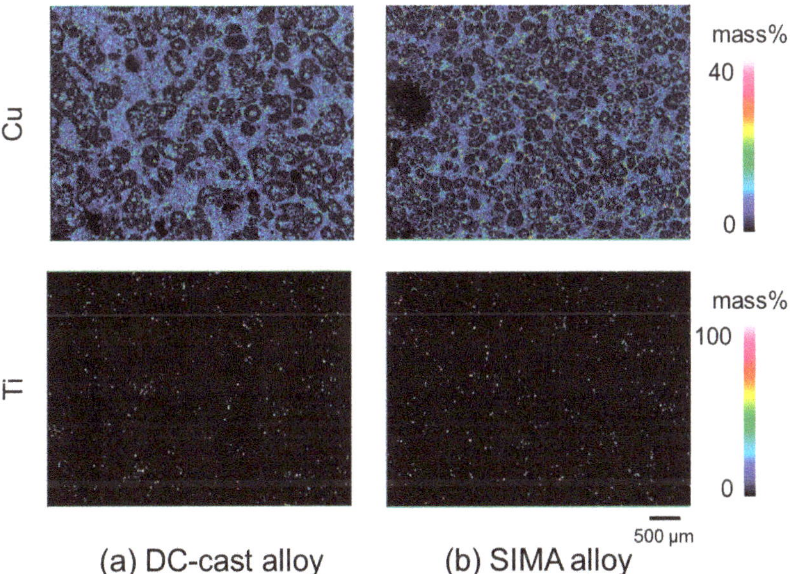

Figure 11. Mapping images of Ti and Cu obtained using the electron probe micro-analyzer (EPMA) (f_s = 40%). (**a**) direct chill cast (DC-cast) alloy; (**b**) Strain-Induced Melt-Activated (SIMA) alloy.

However, when the primary crystals were spherical before foaming, TiH_2 rarely got trapped in the primary crystals while stirring, similar to pure Ti as shown in Figure 12b; this would happen in the SIMA and slope-cast alloys. Therefore, micro-pores were present between the primary crystals, as shown in Figure 13b. In addition, micro-pores between the primary crystals easily coalesce with other micro- or macro-pores because they are not completely wrapped by primary crystals. Consequently, the number of micro-pores in the SIMA and slope-cast alloys was 1/3 of that in the DC-cast alloy.

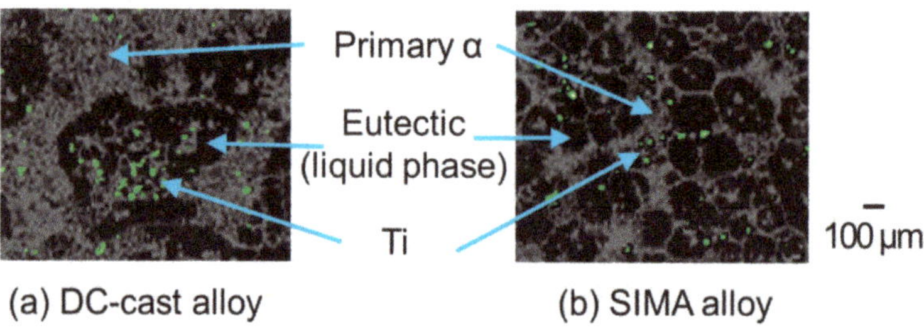

Figure 12. Merged images of Ti and Cu obtained using the electron probe micro-analyzer (EPMA) (f_s = 40%). (**a**) direct chill cast (DC-cast) alloy; (**b**) Strain-Induced Melt-Activated (SIMA) alloy.

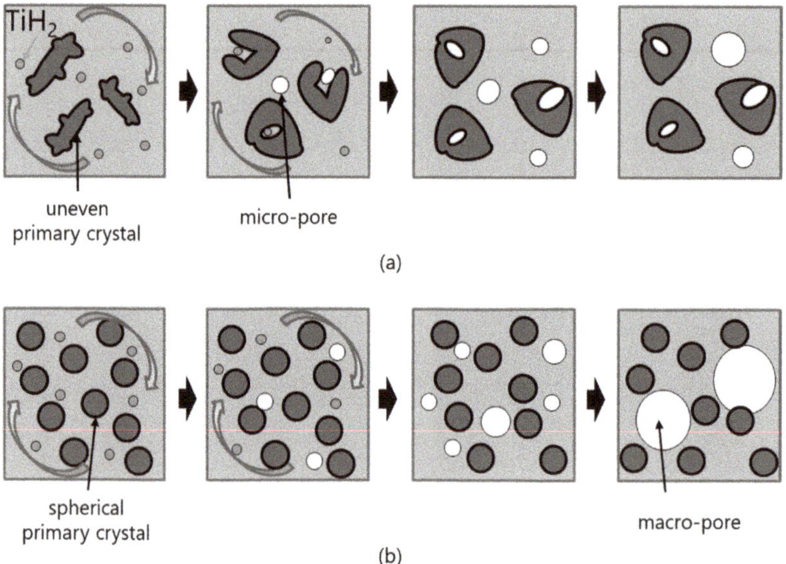

Figure 13. Schematic of the mechanism behind micro-pore formation in the (**a**) direct chill cast (DC-cast) and the (**b**) Strain-Induced Melt-Activated (SIMA) and slope-cast alloys.

4. Conclusions

A2024 alloy foams were fabricated in a semi-solid state using three different preparation methods for the master alloys. The most important finding of this study is that the morphology of the primary crystals influences the quantity of micro-pores inside the cell walls, but not on the quantity of macro-pores. The results can be summarized as follows.

1.	In the DC-cast alloy, coarse, dendritic, and uneven primary crystals were present before stirring. They were subsequently bent and spheroidized via stirring. On the other hand, the primary crystals in the SIMA and slope-cast alloys maintained their fine and spherical shape before and after stirring.

2. The differences between the uneven primary crystals and the fine/spherical primary crystals before stirring had little effect on macro-pore morphology. The size of the primary crystals was significantly smaller than that of the macro-pores.

3. In the DC-cast alloy, micro-pores existed inside the primary crystals after foaming. On the other hand, micro-pores were present between the primary crystals in the SIMA and slope-cast alloys. The number of micro-pores in the SIMA and slope-cast alloys was 1/3 of that in the DC-cast alloy.

Author Contributions: Writing–reviewing and editing, T.K. and S.S.; conceptualization, M.S. and S.S.; methodology, M.S. and S.S.; validation, M.S. and S.S.; formal analysis, T.K., M.S., and S.S.; investigation, M.S.; data curation, M.S. and S.S.; writing–original draft preparation, T.O.; visualization, M.S.; supervision, S.S.; project administration, S.S.

Funding: This study was supported by the Grant-in-Aid program of The Light Metal Educational Foundation.

Acknowledgments: The authors thank N. Sakaguchi from UACJ Corporation for suppling the A2024 ingots used in this study.

Conflicts of Interest: The authors declare no conflict of interest.

References

1. Banhart, J. Manufacture, characterisation and application of cellular metals and metal foams. *Prog. Mater Sci.* **2001**, *46*, 559–632. [CrossRef]

2. García-Moreno, F. Commercial Applications of Metal Foams: Their Properties and Production. *Materials* **2016**, *9*, 85. [CrossRef] [PubMed]

3. Gibson, I.J.; Ashby, M.F. *Cellular Solids Structure and Properties*, 2nd ed.; Pergamon Press: Oxford, UK, 1997; pp. 163–197.

4. Fukui, T.; Saito, M.; Nonaka, Y.; Suzuki, S. Fabrication of foams of A2024 and A7075 and influence of porosity and heat treatment on their mechanical properties. *Trans. Jpn. Soc. Mech. Eng.* **2014**, *80*, SMM0248. (In Japanese) [CrossRef]

5. Fukui, T.; Nonaka, Y.; Suzuki, S. Fabrication of Al-Cu-Mg Alloy Foams Using Mg as Thickener through Melt Route and Reinforcement of Cell Walls. *Proc. Mat. Sci.* **2014**, *4*, 33–37. [CrossRef]

6. Miyoshi, T.; Itoh, M.; Akiyama, S.; Kitahara, A. ALPORAS Aluminum Foam: Production Process, Properties, and Applications. *Adv. Eng. Mater.* **2000**, *2*, 179–183. [CrossRef]

7. Ueno, H.; Akiyama, S. Effects of calcium addition on the foamability of molten aluminum. *J. Jpn. Inst. Light Met.* **1987**, *37*, 42–47. [CrossRef]

8. Hanafusa, T.; Ohishi, K. Making of Porous Metallic Material by the Semi-Solid Aluminum Alloy. *Rpt. Hiroshima Prefect. Technol. Res. Inst. West Reg. Ind. Res. Center* **2010**, *23*, 28–29. (In Japanese)

9. Sekido, K.; Kitazono, K. Effect of Foaming Condition on Cell Morphology of Superplastic Zn-22Al Alloy Foams Manufactured through Melt Foaming Process. *J. Jpn. Inst. Met. Mater.* **2013**, *77*, 497–502. [CrossRef]

10. Heim, K.; García-Moreno, F.; Banhart, J. Particle size and fraction required to stabilise aluminum alloy foams created by gas injection. *Scr. Mater.* **2018**, *153*, 54–58. [CrossRef]

11. Young, K.P.; Kyonka, C.P.; Courtois, F. Fine Grained Metal Composition. U.S. Patent US4415374A, 15 November 1983.

12. Haga, T.; Kapranos, P. Billetless simple thixoforming process. *J. Mater. Process. Technol.* **2002**, *130–131*, 581–586. [CrossRef]

13. Sirong, Y.; Dongcheng, L.; Kim, N. Microstructure evolution of SIMA processed Al2024. *Mater. Sci. Eng. A* **2006**, *420*, 165–170. [CrossRef]

14. Kattamis, T.Z.; Piccone, T.J. Rheology of Semisolid Al-4.5%Cu-1.5%Mg Alloy. *Mater. Sci. Eng. A* **1991**, *131*, 265–272. [CrossRef]

15. Kim, W.Y.; Kang, C.G.; Lee, S.M. Effect of viscosity on microstructure characteristic in rheological behavior of wrought aluminum alloys by compression and stirring process. *Mater. Sci. Technol.* **2010**, *26*, 20–30. [CrossRef]

16. Mukherjee, M.; García-Moreno, F.; Jiménez, C.; Rack, A.; Banhart, J. Microporosity in aluminum foams. *Acta Mater.* **2018**, *131*, 156–168. [CrossRef]

17. Ohgaki, T.; Toda, H.; Kobayashi, M.; Uesugi, K.; Niinomi, M.; Akahori, T.; Kobayashi, T.; Makii, K.; Aruga, Y. *In situ* observations of compressive behavior of aluminum foams by local tomography using high-resolution X-rays. *Philos. Mag.* **2006**, *86*, 4417–4438. [CrossRef]
18. Toda, H.; Ohgaki, T.; Uesugi, K.; Kobayashi, M.; Kuroda, N.; Kobayashi, T.; Niinomi, M.; Akahori, T.; Makii, K.; Aruga, Y. Quantitative Assessment of Microstructure and its Effects on Compression Behavior of Aluminum Foams via High-Resolution Synchrotron X-Ray Tomography. *Metall. Mater. Trans. A* **2006**, *37*, 1211–1219. [CrossRef]
19. Chen, H.I.; Chen, J.C.; Liao, J.J. The influence of shearing conditions on the rheology of semi-solid magnesium alloy. *Mater. Sci. Eng. A* **2008**, *487*, 114–119. [CrossRef]

Article

Compressive Properties of A2024 Alloy Foam Fabricated through a Melt Route and a Semi-Solid Route

Takashi Kuwahara [1,*], Taro Osaka [1], Mizuki Saito [1] and Shinsuke Suzuki [1,2]

[1] Faculty of Science and Engineering, Waseda University, 3-4-1 Okubo, Shinjuku, Tokyo 169-8555, Japan; t-evolva@moegi.waseda.jp (T.O.); m.saitoh@akane.waseda.jp (M.S.); suzuki-s@waseda.jp (S.S.)

[2] Kagami Memorial Research Institute of Materials Science and Technology, Waseda University,2-8-26 Nishi-Waseda, Shinjuku, Tokyo 169-0051, Japan

* Correspondence: takuwahara@ruri.waseda.jp; Tel.: +81-03-5286-8126

Received: 14 December 2018; Accepted: 22 January 2019; Published: 29 January 2019

Abstract: A2024 alloy foams were fabricated by two methods. In the first method, the melt was thickened by Mg, which acts as an alloying element (melt route). In the second method, the melt was thickened by using primary crystals at a semi-solid temperature with a solid fraction of 20% (semi-solid route). A2024 alloy foams fabricated through the semi-solid route had coarse and uneven pores. This led to slightly brittle fracture of the foams, which resulted in larger energy absorption efficiency than that of the foams fabricated through the melt route. Moreover, A2024 alloy foams fabricated through the semi-solid route had a coarser grain size because of the coarse primary crystals. However, by preventing the decrease in the alloying element Mg, the θ/θ' phase was suppressed. Additionally, by preventing the precipitation of the S' phase, the amount of Guinier-Preston-Bagaryatsky (GPB) zone increased. This resulted in a larger plateau stress.

Keywords: semi-solid; aluminum foam; primary crystals; compression test; pore morphology; precipitation phase; age hardening

1. Introduction

Improvements in fuel consumption and safety are now required in transportation equipment, including automobiles and aircraft. To improve the fuel economy, it is necessary to reduce weight. However, shock absorption performance should be maintained in case of accidents. Aluminum foam has attracted attention as a material that satisfies such characteristics. Aluminum foams contain a large quantity of pores inside the material, which is thus ultralight and has excellent shock absorbing properties. Therefore, aluminum foam is expected to be applied in transportation as an ultralight shock absorbing material.

To manufacture aluminum foam, the method known as melt route or the Alporas method [1] can enlarge the foam at a relatively low cost. In this method, first, Ca as a thickener is added to the melt, and CaO is generated by reaction with the atmosphere [2]. Subsequently, CaO is spread into the melt, and the viscosity of the melt is increased. TiH_2 is then added to the melt as a blowing agent. Finally, the foam is obtained by water cooling. However, to apply aluminum foams as an ultralight shock absorbing material of transport equipment, an improvement in the compressive properties of foams is needed. The compressive properties of foams are known to be determined by the strength of the base metal and pore morphology [3]. Therefore, using super duralumin A2024 alloy (Al-Cu-Mg) which is known as high strength lightweight material for base metal to increase strength and controlling the pore morphology to improve compressive properties has been researched. Fukui et al. generated an oxide of Mg [4,5], which is contained in the A2024 alloy as a thickener instead of adding Ca.

However, the decrease in the base metal strength has been an issue of concern, because of the decrease in Mg, which acts as an alloying element. The decrease in Mg is known to decrease the plateau stress of the aluminum foam [6]. To solve this problem, foams were fabricated in a semi-solid state. In this study, this method is called the semi-solid route. In this method, primary crystals are used as a thickener [7] instead of an oxide. In our group, we found it is possible to fabricate the A2024 alloy foam without adding any thickener and confirmed the thickening effect of the primary crystal in a semi-solid state [8]. However, the size of the primary crystal is larger than that of the oxide. Therefore, the primary crystal existing inside the melt may affect the pore morphology of the aluminum alloy foam in a semi-solid state. Hence, the objective of this research is to investigate the effect of the cell wall structure and pore morphology on the compressive properties of A2024 alloy foam fabricated through the melt route and the semi-solid route.

We fabricated an A2024 alloy foam through melt and semi-solid routes and researched the effect of the cell wall structure and pore morphology on compressive properties (plateau stress and energy absorption efficiency). Nano-sized and micro-sized precipitates of the cell wall were analyzed. The hardness in cell walls of two kinds of foams was measured. Pore morphology of foams was evaluated by the pore diameter and pore circularity in the cross section of foams.

2. Materials and Methods

2.1. Sample Preparation

Table 1 shows the alloy composition of the A2024 alloy used in this study. Figure 1 shows the fabrication methods of the A2024 alloy foam through (a) the melt route and (b) the semi-solid route.

Table 1. A2024 alloy composition (mass %).

Cu	Mg	Mn	Fe	Si	Others	Al
4.59	1.66	0.66	0.15	0.12	0.08	Bal

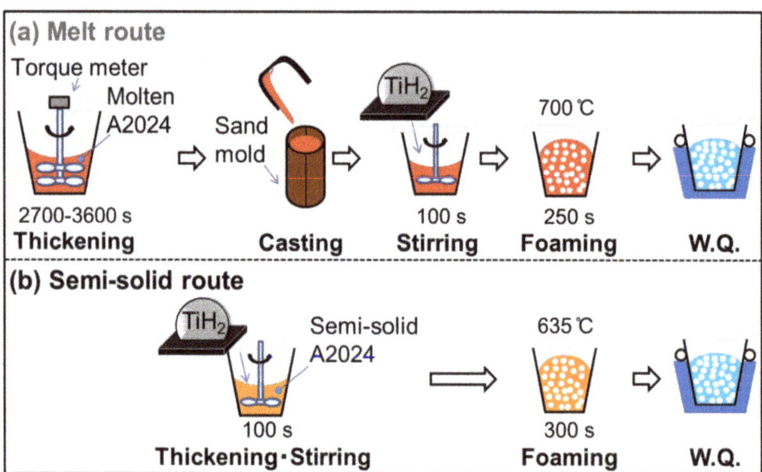

Figure 1. Fabrication methods of A2024 foam: (**a**) melt route; (**b**) semi-solid route.

In the melt route (Figure 1a), 700 g of the A2024 alloy was set in a graphite crucible. The alloy was heated and held at 700 °C in an electric furnace. After melting, the melt was thickened by using a stirring impeller coated with BN at 500 rpm until the torque became stable (2700–3600 s). This long stirring generates MgO, which acts as a thickener in the melt by reaction with the atmosphere. The melt was then poured into a mold, and 100 g of the cast alloy was cut and placed in an SUS304 crucible coated with Al_2O_3. The alloy was heated to, and held at, 700 °C in an electric furnace. After the temperature became stable, 1 g of the TiH_2 powder wrapped in an aluminum foil, which corresponds to 1 mass % of the alloy, was added to the melt as a blowing agent. The TiH_2 powder was previously dry-treated at 150 °C for 24 h. The melt was then stirred by an impeller coated with BN at 900 rpm for 100 s. After stirring, the melt was retained for 250 s so that TiH_2 decomposes. Subsequently, the crucible was taken out of the furnace. The foam was then solidified and cooled with 3 L/min of water.

In the semi-solid route (Figure 1b), 100 g of the alloy was cut and added to a crucible. The alloy was heated and held at 635 °C (solid fraction f_s = 20% [9]) in an electric furnace. The following procedures to obtain the foam were the same as those of the melt route, except that the retaining time was 300 s for the semi-solid route. The retaining time was set so as to obtain the same porosity as with the melt route.

In this study, A2024 alloy foams fabricated through the melt and semi-solid routes are referred as Melt-A2024 and Semi-A2024, respectively. After the foams were obtained, the T4 treatment was applied. The foams were solid-solutionized at 480 °C for 48 h in a muffle furnace. After that, the foams were aged at room temperature for over 96 h. Finally, the hardness in the cell walls of the samples was measured by a micro-Vickers hardness tester (HM-115, Akashi Co., Ltd., Kanagawa, Japan). The test load was 2.9 N, and the size of the impression was less than 1/3 of the cell wall thickness.

2.2. Evaluation of Pore Morphology

To derive the porosity of the foam, its volume was measured via Archimedes' method using the mass of the foam and a spring scale. In Equation (1), the porosity p of the foam was calculated on the basis of the density of the A2024 alloy foam ρ_P, which was calculated from the volume and mass of the foams, and the density of the A2024 alloy, ρ_{NP} = 2.77 Mg/m^3 [10]. The pore size and morphology (pore diameter d_p and pore circularity e_p) in a cross section in the middle of the sample were measured by using image analysis software (WinROOF^{TM}, Mitani Corp., Fukui, Japan). The pore morphology was calculated by Equations (2) and (3). The symbols S and L are the area of the pore and the perimeter of the pore, respectively. Because pores smaller than 0.2 mm in diameter cannot be distinguished from noises, these small pores were excluded from analysis.

$$p = (1 - \rho_P/\rho_{NP}) \times 100\% \tag{1}$$

$$d_p = (4S/\pi)^{1/2} \tag{2}$$

$$e_p = 4\pi S/L^2. \tag{3}$$

2.3. Measurement and Analysis of the Cell Wall

The difference in the cell wall structure between the two fabrication methods was evaluated. First, the average grain size $\bar{d}$ was measured by electron back scatter diffraction (EBSD) using the TSL-OIM4 system (TSL Solutions, Kanagawa, Japan). The dimension step was 6.0 μm. The average value $\bar{d}$ was calculated by Equation (4), which is the weighting average of the area for the size d_i of the i-th grain (d_i was calculated in the same way as Equation (2)). Second, electron probe micro analysis (EPMA) using the JXA-8230 electroprobe (JEOL, Kanagawa, Japan) was carried out at 15 kV. The micro-sized precipitates were analyzed by scanning electron microscope energy dispersive X-ray spectroscopy (SEM-EDS) on a JSM-6500F system (JEOL, Kanagawa, Japan) at 15 kV. The nano-sized precipitates were analyzed by transmission electron microscope energy dispersive X-ray spectroscopy (TEM-EDS), a differential scanning calorimeter (DSC), and X-ray diffraction (XRD). The elemental mapping of Mg, Mn and Cu was obtained by TEM-EDS on a JEM-2100F system (JEOL, Kanagawa, Japan) at 15 kV. Exothermic and endothermic peaks were analyzed by DSC on a DSC8500 apparatus (Perkin Elmer, Waltham, MA, USA). The weight of samples was 10 mg, the temperature zone was 0–500 °C, and the heating rate was 45 °C/min. Precipitates were analyzed by XRD on a SmartLab diffractometer (Rigaku, Tokyo, Japan).

$$\overline{d} \;=\; d_i^3 / \textstyle\sum_{i\,=\,1}^{N} d_i^2. \tag{4}$$

2.4. Evaluation of Compressive Properties

The fabricated foams were cut in dimensions of 20 × 20 × 20 mm³ by wire electrical discharge machining. After that, a compression test was performed according to ISO 13314 [11] by a universal testing machine (Autograph, AG-250lNI, Shimadzu Corp., Kyoto, Japan). In the test, the strain rate de/dt was 0.1 min⁻¹. The compressive properties, plateau stress σ_{pl}, and energy absorption efficiency η_V were calculated. The calculation method of each value is shown below. The plateau stress σ_{pl} was calculated by the average compressive stress in the range from 20 to 30% of compressive strain. The energy absorption efficiency η_V was calculated by the energy absorption amount E_V and ideal energy absorption amount E_{Videal} as in Equation (7). Further, the energy absorption amount E_V and ideal energy absorption amount E_{Videal} were calculated by Equations (5) and (6) using the maximum compressive stress σ_{max}. Moreover, as the porosity will change after cutting to cubic, the porosity of the cubic foam sample was derived before the compression test. The volume of the cubic foam sample was calculated by measuring each side of the cube by digital calipers. The porosity of the cubic foam sample was then calculated in the same way as above by Equation (1).

$$E_V \;=\; 1/100 \times \int_0^{50} \sigma(e)\,de \; \text{MJ/m}^3 \tag{5}$$

$$E_{Videal} = 1/100 \times \sigma_{max(e\,=\,0\sim50\%)} \times e \; \text{MJ/m}^3 \tag{6}$$

$$\eta_V = (E_V / E_{Videal}) \times 100\%. \tag{7}$$

3. Results and Discussion

3.1. Results of the Compression Test

Figure 2 shows the stress–strain curves obtained by compression tests of Melt-A2024 and Semi-A2024 with almost the same porosity of the cubic foam sample (about 60%). From Figure 2, the plateau stress of Semi-A2024 was σ_{pl} = 67 MPa, which was larger than that of Melt-A2024 (σ_{pl} = 57 MPa). In addition, Semi-A2024 shows a sudden decrease in compressive stress. Figure 3 shows the energy absorption efficiency η_V. The energy absorption efficiency of Semi-A2024 was between 86 and 92%, which was larger than that of Melt-A2024 (η_V = 70–85%). The sudden decrease in compressive stress may have prevented the increase in the maximum compressive stress σ_{max}. Therefore, Semi-A2024 may show larger energy absorption efficiency than that of Melt-A2024.

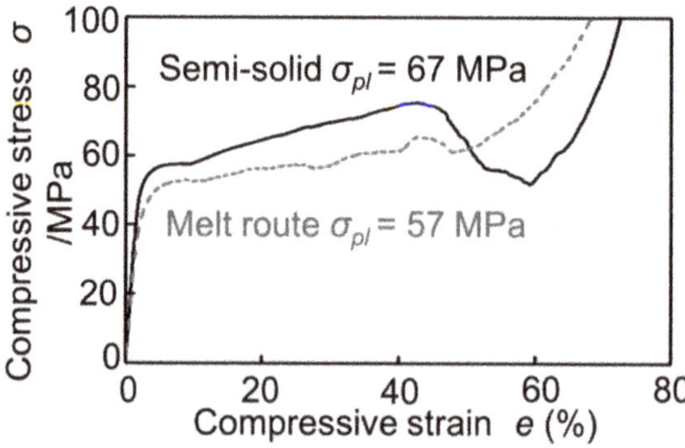

Figure 2. Stress–strain curves (porosity about p = 60%) of Melt-A2024 and Semi-A2024.

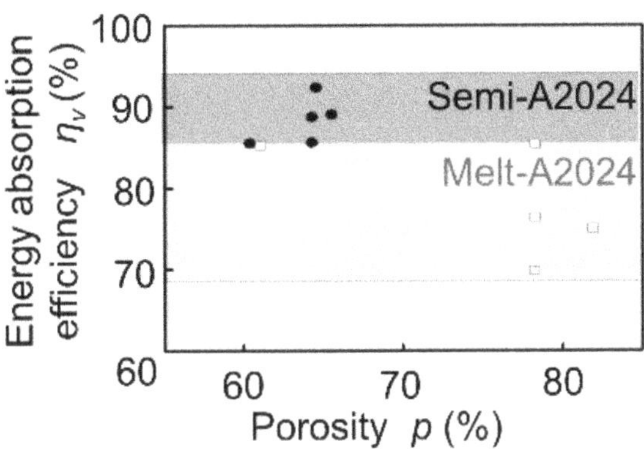

Figure 3. Energy absorption efficiency η_V of Melt-A2024 and Semi-A2024.

3.2. Effect of Pore Morphology on Compressive Properties

3.2.1. Pore Morphology of Melt-A2024 and Semi-A2024

Figure 4 shows cross sections of (a) Melt-A2024 and (b) Semi-A2024 with a porosity of foams of $p = 61\%$ and $p = 59\%$, respectively. Figure 4 shows that foams of the semi-solid route have coarse and uneven pores. Furthermore, Figure 5 shows image analysis results of the pore morphology of foams shown in Figure 4. Figure 5 shows that Semi-A2024 has coarser and more uneven pores than those of Melt-A2024. These tendencies of the Melt-A2024 and Semi-A2024 samples were observed in almost all other samples. Moreover, by increasing the retaining time, the pores will expand and become coarser, and the porosity will also increase until the end of TiH_2 decomposition [5]. Moreover, it is thought that, with a change in foaming temperature, there will be no significant difference in the pores of Melt-A2024, unless the temperature falls below the liquidus line. However, for Semi-A2024, the change in temperature will generate a difference in pore formation because the fraction of solid changes [8].

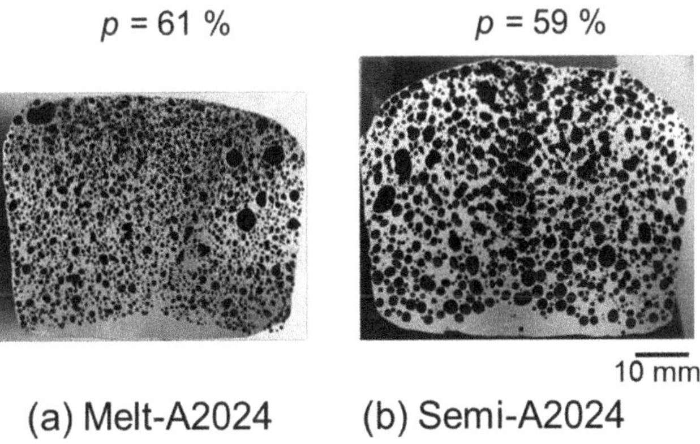

Figure 4. Cross sections of samples: (**a**) Melt-A2024; (**b**) Semi-A2024.

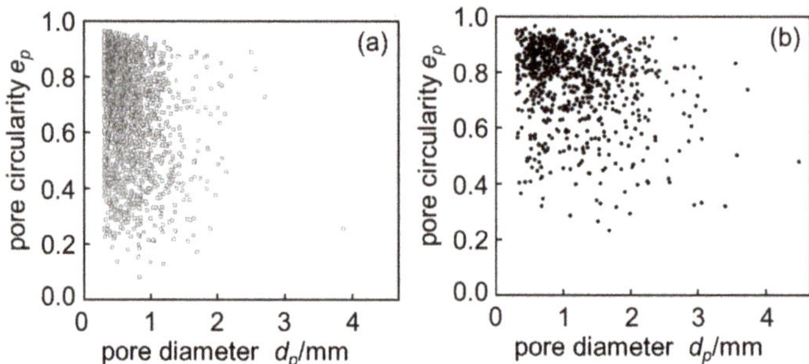

Figure 5. Distribution map of pore circularity: (**a**) Melt-A2024 ($p = 61\%$); (**b**) Semi-A2024 ($p = 59\%$).

3.2.2. The Compressive Behavior of A2024 Alloy Foams

Figure 6 shows the compressive behavior of the samples during the compression test. The porosities shown in Figure 6 correspond to the cubic foam sample porosity. From Figure 6a, before the compression test ($e = 0\%$), Melt-A2024 had finer pores. Moreover, during the compression test ($e = 50\%$), a sudden decrease in compressive stress was not observed. Therefore, the material showed a ductile fracture. At this moment, the energy absorption efficiency of the samples was $\eta_V = 70\%$. The sample shown in Figure 6b is Semi-A2024. This sample had finer pores, so it is different from that previously mentioned in Section 3.2.1. This is because Semi-A2024 showed variations in pore diameter and circularity (Figure 5b); when the sample was cut into a cube, it included much finer pores. Further, during the compression test ($e = 50\%$), a sudden decrease in compressive stress was not observed. Therefore, the material showed a ductile fracture. At this moment, the energy absorption efficiency of the sample was $\eta_V = 86\%$. Figure 6c shows the results of Semi-A2024. Before the compression test ($e = 0\%$), the sample had coarse and uneven pores, as mentioned in Section 3.2.1. During the compression test ($e = 50\%$), there was a sudden decrease in compressive stress after a compressive strain of approximately 40%. Therefore, the material showed a slightly brittle fracture. At this moment, the energy absorption efficiency of the sample was $\eta_V = 89\%$. Thus, its energy absorption efficiency was larger than that of the other two samples. This is because the coarse and uneven pores were crushed preferentially [12], and the plateau region became flat. Therefore, Semi-A2024, which had coarse and uneven pores, showed larger energy absorption efficiency.

Figure 6. Compressive behavior of the sample during compression test.

3.3. *The Effect of Cell Wall Structure on Compressive Properties*

3.3.1. Cell Wall Hardness

Figure 7 shows the results of micro-Vickers hardness tests of (a) T4 treated A2024, (b) cell walls of Melt-A2024, and (c) cell walls of Semi-A2024. From Figure 7, the cell wall hardness of Semi-A2024 (HV = 127) was close to that of the T4-treated A2024 (HV = 133) and larger than that of Melt-A2024 (HV = 108).

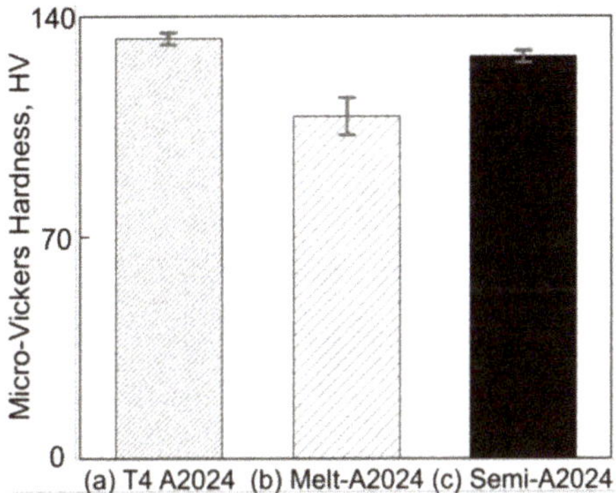

Figure 7. Micro-Vickers hardness. (**a**) T4-treated A2024; (**b**) cell walls of Melt-A2024; (**c**) cell walls of Semi-A2024. The error bars show the standard deviations.

3.3.2. The Grain Size of Cell Wall

Figure 8 shows the inverse pole figure map of (a) Melt-A2024 and (b) Semi-A2024 after T4 treatment. Semi-A2024 has coarser grains because of the existence of coarse primary crystals (Figure 8b). Further, Figure 9 shows the average grain size $\bar{d}$ of (a) Melt-A2024 and (b) Semi-A2024. From Figure 9, these foams had an $\bar{d}$ of 84 μm and 239 μm, respectively. Therefore, the average grain size of Semi-A2024 is larger than that of Melt-A2024.

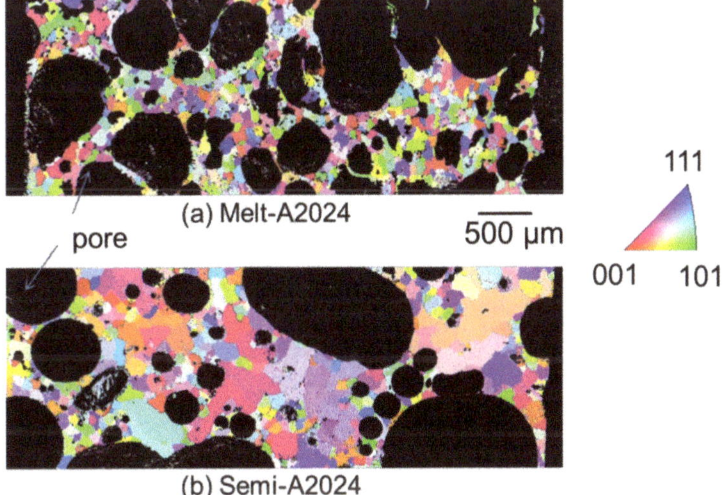

Figure 8. Inverse pole figure map. (**a**) Melt-A2024; (**b**) Semi-A2024.

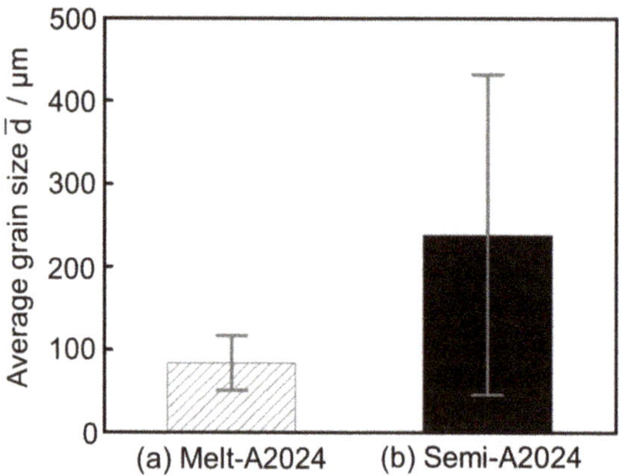

Figure 9. Average grain size $\bar{d}$; (**a**) Melt-A2024; (**b**) Semi-A2024. The error bars show the standard deviations.

3.3.3. MgO Amounts of Cell Wall

For Melt-A2024, the decrease in the base metal strength has been an issue of concern because of the decrease in Mg, which acts as an alloying element for strengthening. To estimate the amount of MgO in the cell walls of foams, an EPMA was performed. Figure 10 shows the qualitative mapping of Mg and O by EPMA. The part with high concentration of both Mg and O is considered to be MgO. As shown in the red square in Figure 10a, the concentration of MgO was high in the Melt-A2024 foams. This is due to the long period of stirring performed to thicken the melt by generating MgO. Meanwhile, parts with high concentrations of both Mg and O were rarely observed in the Semi-A2024 foams because the long stirring for thickening was excluded. Therefore, the amount of MgO in the cell walls is expected to be larger in the Melt-A2024 foams, so the hardness in the cell walls is expected to be lower.

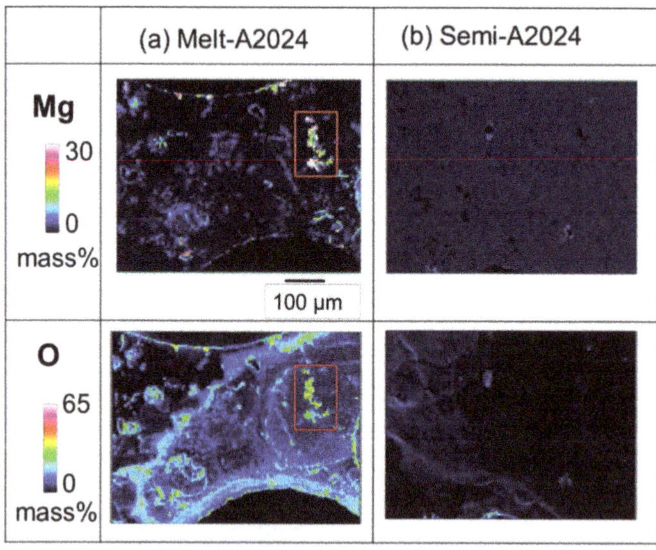

Figure 10. X-ray elemental mapping of Mg and O in cell wall obtained by electron probe micro analysis (EPMA): (**a**) Melt-A2024; (**b**) Semi-2024.

3.3.4. Identification of Micro-Sized Precipitates

Figures 11 and 12 show the images of intermetallic compounds and EDS analysis by SEM for Melt-A2024 and Semi-A2024, respectively. First, Al–Cu compounds were observed. These particles are considered to be Al_2Cu (θ phase), which is a typical precipitate of A2024 [13] (Figures 11a and 12a). Al–Cu–Fe–Mn compounds were also observed [13] (Figures 11b and 12b). Furthermore, Ti was detected (Figures 11c and 12c). This is considered to be the foaming agent, TiH_2, which did not decompose and remained in the matrix. Further, an Al–Ti compound was detected (Figures 11d and 12d). The titanium decomposed from TiH_2 could have reacted with the aluminum matrix. No differences in the amounts of these precipitates were seen. Therefore, the effect of the difference in these micro-sized precipitates on compressive properties is considered to be small.

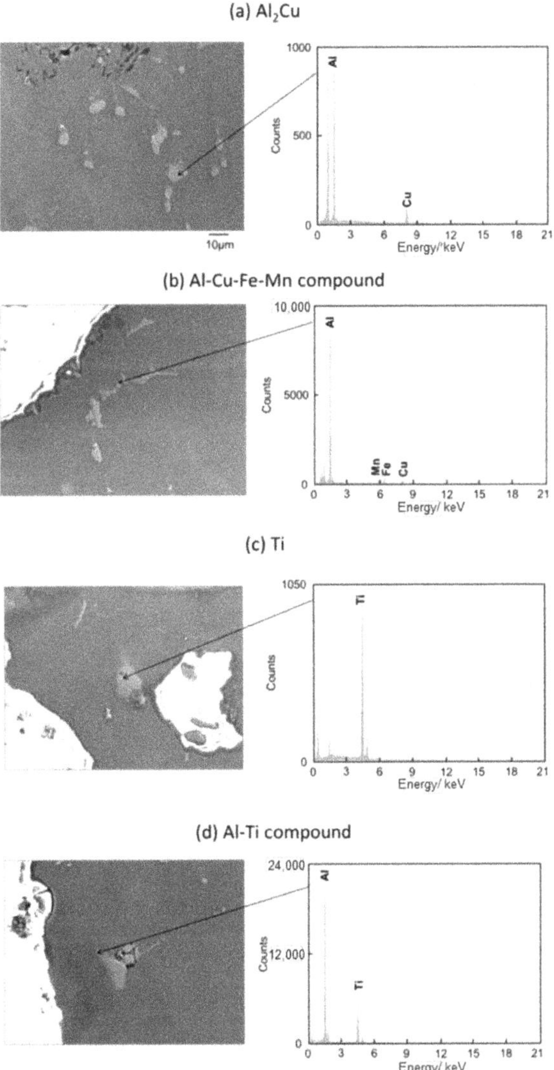

Figure 11. Scanning electron microscope (SEM) images of intermetallic compounds and energy dispersive X-ray spectroscopy (EDS) analysis in the cell wall of Melt-A2024.

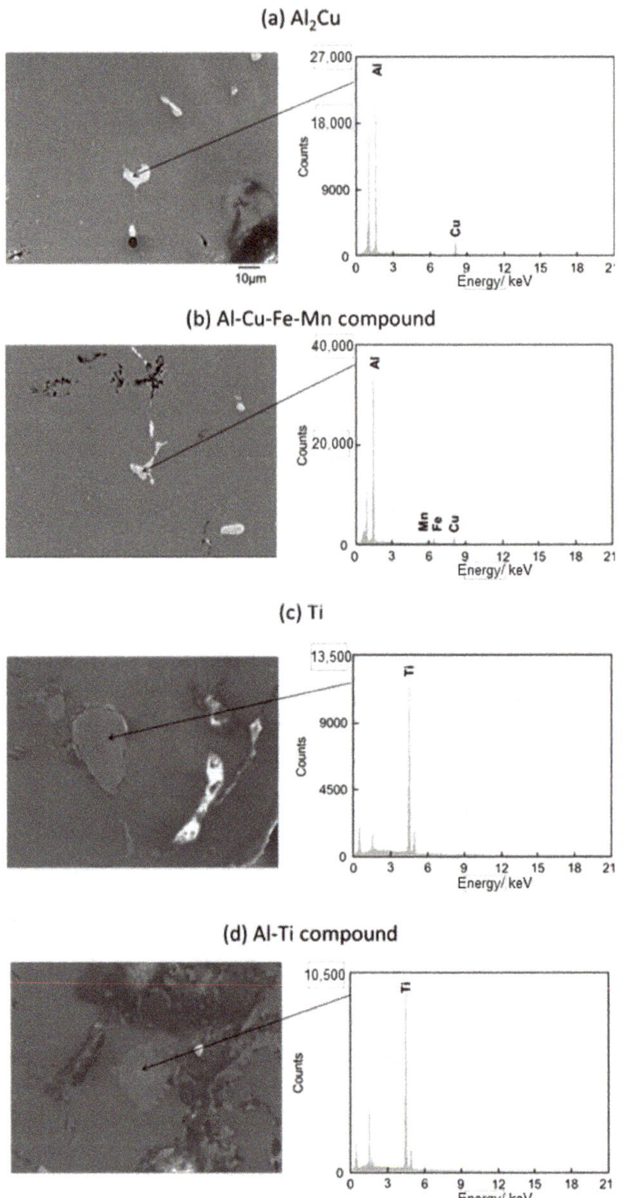

Figure 12. SEM images of intermetallic compounds and EDS analysis of the cell wall of Semi-A2024.

3.3.5. Identification of Nano-Sized Precipitates

Differential Scanning Calorimeter

Figure 13 shows the DSC curves of (a) Melt-A2024 and (b) Semi-A2024. In Figure 13, endothermic Peak I and the exothermic Peak II are exhibited in both alloys. For the endothermic process, Peak I at approximately 240 °C could be ascribed to the dissolution of the GPB zone [14]. For the exothermic process, Peak II at approximately 280 °C could be ascribed to the precipitation of

S′/S phases [15]. Therefore, the Guinier-Preston-Bagaryatsky (GPB) zone and the S′ phase, which are strengthening phases of the A2024 alloy, were observed in both alloys.

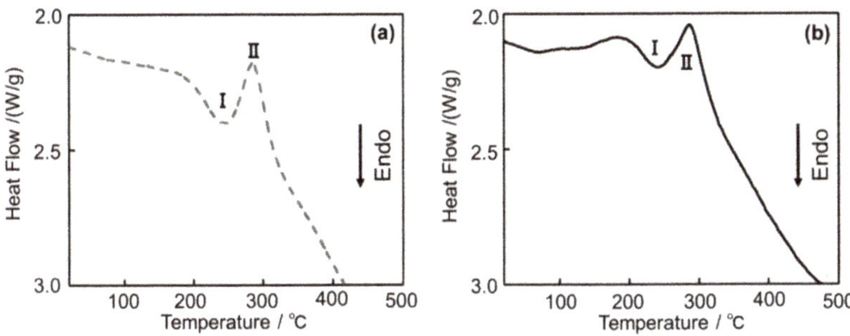

Figure 13. Differential Scanning Calorimeter (DSC) curves of the A2024 alloy foams: (**a**) Melt-A2024; (**b**) Semi-A2024.

X-ray Diffraction

Given that the S and S′ phases and θ and θ′ phases cannot be distinguished by XRD, they are expressed as the S/S′ phase and θ/θ′ phase, respectively. Figure 14 shows the XRD patterns of (a) Melt-A2024 and (b) Semi-A2024. In Figure 14a, S/S′ and θ/θ′ phases are indicated in Melt-A2024. The precipitation of the θ phase is known to decrease the degree of age hardening [13]. Only the S/S′ phase is indicated in Semi-A2024 (Figure 14b). The θ/θ′ phase was found in Melt-A2024 and not in Semi-A2024 because the low Mg ratio in the Al–Cu alloy accelerates the precipitation of the θ/θ′ phase [16].

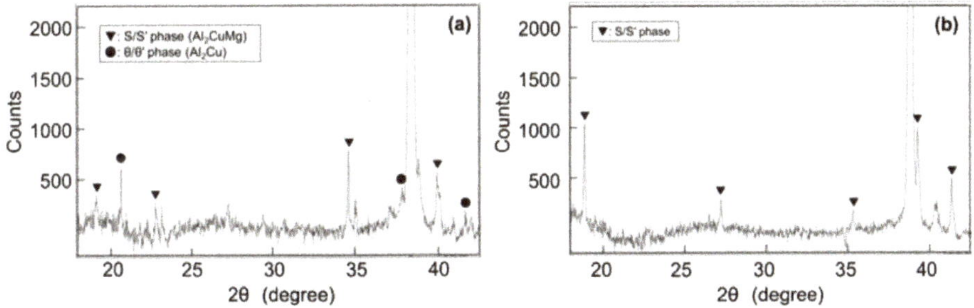

Figure 14. X-ray Diffraction (XRD) patterns; (**a**) Melt-A2024; (**b**) Semi-A2024.

Transmission Electron Microscope Energy Dispersive X-ray Spectroscopy

Finally, TEM-EDS was carried out to identify nano-sized precipitates. Figure 15 shows the X-ray mapping image of (a) Melt-A2024 and (b) Semi-A2024 by TEM-EDS. The upper figures show the mapping image of Mg. The middle figure shows the mapping image of Mn piled on a bright field (BF) image. The lower figures show the mapping image of Cu. According to Figure 15a, Mg was distributed completely over both Melt-A2024 and Semi-A2024. Moreover, in Melt-A2024, both Mn and Cu concentration parts were observed and are circled in red. This needle- or round-shaped precipitates are considered to be the T phase ($Al_{20}Mn_2Cu_3$) [17]. Concentration parts of only Cu were also observed and are circled in green. These round-shaped precipitates are considered to be the θ/θ′ phase. In Semi-A2024, only Mn and Cu concentration parts were observed, which are circled in red (Figure 15b). These needle-shaped precipitates are considered to be the T phase. Figure 16 shows the number of T phases in the cell wall per unit area, N_T, of (a) Melt-A2024 and (b) Semi-A2024. The numbers of T phases of these alloys were 3.3 and 5.6, respectively. Therefore, the number of T phases in the cell walls of Semi-A2024 is larger than that of Melt-A2024. From the above, in Semi-A2024, the precipitation of the θ/θ′ phase is suppressed, and this results in an increase in the precipitation of the T phase.

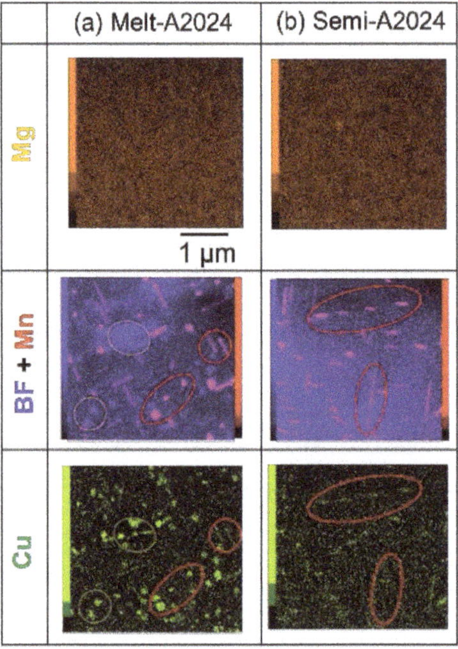

Figure 15. X-ray mapping images by TEM-EDS: (**a**) Melt-A2024; (**b**) Semi-A2024.

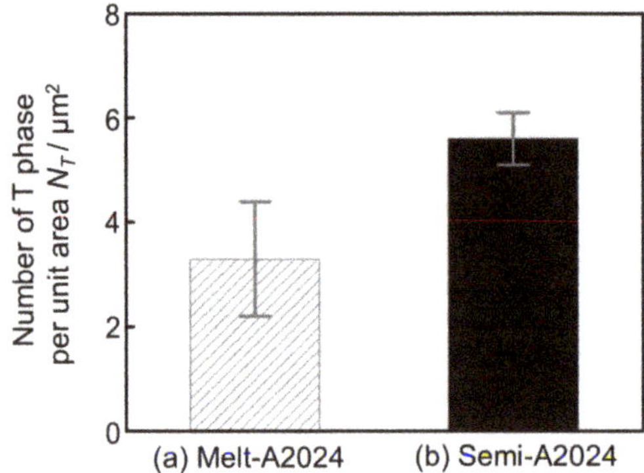

Figure 16. Number of T phases per unit area N_T: (**a**) Melt-A2024; (**b**) Semi-A2024. The error bars show the standard deviations.

3.3.6. The Effect of the Difference of Nano-Sized Precipitates on Compressive Properties

There were some differences in nano-sized precipitate between Melt-A2024 and Semi-A2024. In this section, the entire effect of the nano-sized precipitate on compressive properties is summarized. First, the θ phase, which decreases the degree of age hardening, was found only in Melt-A2024, as explained in Section 3.3.5. (X-ray Diffraction). The number of T phases existing in the cell wall of Semi-A2024 was larger than that of Melt-A2024. As the S′ phase precipitates on the T phase, the total number of S′ phases that did not precipitate on the T phases decreased in Semi-A2024. This will cause the number of GPB zones, which contribute most to the strengthening, to increase owing to the decrease in the total number of S′ phases precipitating [18]. Therefore, in Semi-A2024, the total number of S′ phases is considered to decrease, so the total number of GPB zones is considered to increase. From the above, the suppression of θ/θ′ phase precipitation and the increase in GPB zone resulted in larger value of hardness in the cell walls of Semi-A2024. Therefore, the plateau stress of Semi-A2024 was larger than that of Melt-A2024. Moreover, Mg will diffuse in the S/S′ phase.

3.3.7. The Effect of the Difference in Pore Morphology and the Cell Wall Structure on Compressive Properties

In this section, the whole effect of the difference in pore morphology and cell wall structure on the compressive property tendencies is explained. Figure 17 shows schematic drawings of the pore morphology and cell wall structure of (a) Melt-A2024 and (b) Semi-A2024. The fine pores of Melt-A2024 resulted in the ductile fracture of samples. Semi-A2024 includes coarser and more uneven pores than those of Melt-A2024. This resulted in a slightly brittle fracture of samples [12]. Therefore, the plateau area became flat, and the energy absorption efficiency became larger. However, Melt-A2024 has finer grains, whereas Semi-A2024 has coarser grains. Therefore, the effect of grain boundary strengthening is weaker than that in Melt-A2024. However, the θ/θ′ phase, which decreases the degree of age hardening, precipitates in Melt-A2024. In Melt-A2024, the total number of S′ phases will increase, causing the GPB zone to decrease. The θ/θ′ phase was suppressed in Semi-A2024. Thus, the S′ phase decreases and the GPB zone will increase. Therefore, the degree of age hardening of Semi-A2024 is larger than that of Melt-A2024. In the age hardening-type aluminum alloys, the precipitation effect will contribute more to the strength of cell walls than the effect of grain boundary strengthening [19]. Therefore, the hardness of cell walls of the Semi-A2024 alloy is larger than that of Melt-A2024.

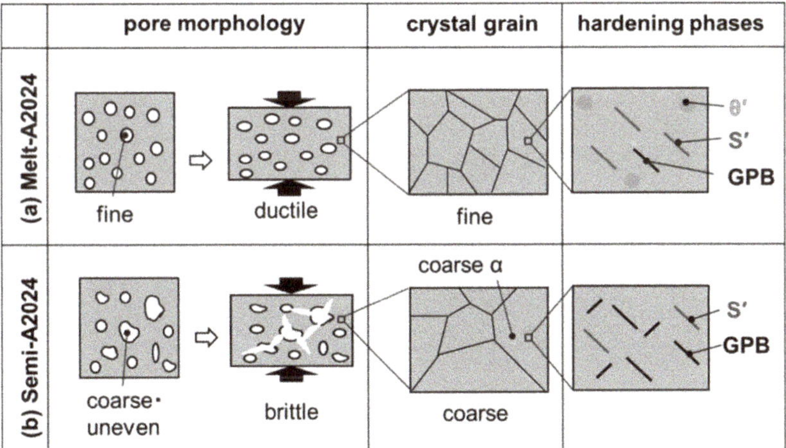

Figure 17. Schematic drawing of trend pore morphology and cell wall structure: (a) Melt-A2024; (b) Semi-A2024.

4. Conclusions

A2024 alloy foams were fabricated through the melt route at above the liquidus temperature (648 °C [9]) and through the semi-solid route in the range between the liquidus and solidus temperature (510 °C [9]). The effect of the differences in pore morphology and the cell wall structure on compressive properties was then clarified. The most important finding of this study is that A2024 alloy foams fabricated through the semi-solid route prevent the precipitation of the θ/θ′ phase and increase the participation amount of the GPB zone, leading to larger plateau stress σ_{pl}. The results can be summarized as follows.

1. A2024 alloy foams fabricated through the semi-solid route have coarse and uneven pores. The pores led to a slightly brittle fracture of the foams and may have resulted in larger energy absorption efficiency η_V.
2. A2024 alloy foams fabricated through the semi-solid route have a coarser grain size because of the coarse primary crystals. However, the prevention of the decrease in the alloying element Mg by excluding the long stirring for thickening suppressed the θ/θ' phase and increased the number of GPB zones by preventing the precipitation amount of the S$'$ phase. This led to a larger plateau stress σ_{pl}.

Author Contributions: Writing—review and editing, T.K. and S.S.; conceptualization, T.O. and S.S.; methodology, T.O. and S.S.; validation, T.O. and S.S.; formal analysis, T.K., T.O., and S.S.; investigation, T.O.; data curation, T.O. and S.S.; writing—original draft preparation, T.K. and T.O.; visualization, T.O.; supervision, M.S. and S.S.; project administration, M.S. and S.S.

Funding: This study was supported by the Grant-in-Aid the Light Metal Educational Foundation.

Acknowledgments: The authors thank N. Sakaguchi from UACJ Corporation for supplying the A2024 ingots used in this study.

Conflicts of Interest: The authors declare no conflict of interest.

References

1. Miyoshi, T.; Itoh, M.; Akiyama, S. ALPORAS Aluminum foam: Production process, properties, and applications. *Adv. Eng. Mater.* **2000**, *2*, 179–183. [CrossRef]
2. Ueno, H.; Akiyama, S. Effects of calcium addition on the foamability of molten aluminum. *J. Jpn. Inst. Light Met.* **1987**, *37*, 42–47. [CrossRef]
3. Hamada, T.; Nishi, S.; Miyoshi, T.; Kanetake, N. Effect of foaming condition in the melt on cell structure and compression strength of porous aluminum. *J. Jpn. Inst. Met.* **2008**, *72*, 825–831. [CrossRef]
4. Fukui, T.; Saito, M.; Nonaka, Y.; Suzuki, S. Fabrication of foams of A2024 and A7075 and influence of porosity and heat treatment on their mechanical properties. *Trans. Jpn. Soc. Mech. Eng.* **2014**, *80*, 1–9. (In Japanese)
5. Fukui, T.; Nonaka, Y.; Suzuki, S. Fabrication of Al-Cu-Mg alloy foams using Mg as thickener through melt route and reinforcement of cell walls. *Procedia Mater. Sci.* **2014**, *4*, 33–37. [CrossRef]
6. Hamada, T.; Nishi, S.; Takagi, T.; Miyoshi, T.; Kanetake, N. Effect of alloy contents on the compression performances of porous aluminum by melt route. *J. Jpn. Inst. Met.* **2009**, *2*, 88–94. [CrossRef]
7. Hanafusa, T.; Ohishi, K. *Making of Porous Metallic Material by the Semi-Solid Aluminum Alloy*; Report of Hiroshima Prefectural Technology Research Institute West Region Industrial Research Center; Hiroshima Prefectural Technology Research Institute West ern Region Industrial Research Center: Hiroshima, Japan, 2010; Volume 23, pp. 28–29. (In Japanese)
8. Kuwahara, T.; Saito, M.; Osaka, T.; Suzuki, S. Effect of primary crystals on pore morphology during semi-solid foaming of A2024 alloys. *Metals* **2019**, *9*, 88. [CrossRef]
9. Kim, W.Y.; Kang, C.G.; Lee, S.M. Effect of viscosity on microstructure characteristic in rheological behavior of wrought aluminum alloys by compression and stirring process. *Mater. Sci. Technol.* **2010**, *26*, 20–30. [CrossRef]
10. Japan Aluminum Association. (Ed.). Physical property. In *Aluminum Hand Book*, 7th ed.; Japan Aluminum Association: Tokyo, Japan, 2007; p. 32.
11. International Organization for Standardization (ISO). *Mechanical Testing of Metals-Ductility Testing-Compression Test for Porous and Cellular Metals*; ISO13314; ISO: Geneva, Switzerland, 2011; pp. 1–7.
12. Kadar, M.A.; Islam, M.A.; Saadatfar, M.; Hazell, P.J.; Brown, A.D.; Ahmed, A.; Escobedo, J.P. Macro and micro collapse mechanisms of closed-cell aluminum foams during quasi-static compression. *Mater. Des.* **2017**, *118*, 11–21. [CrossRef]
13. Zhang, X.; Hashimoto, T.; Lindsay, J.; Zhou, X. Investigation of the de-alloying behavior of θ-phase (Al$_2$Cu) in AA2024-T351 aluminum alloy. *Corros. Sci.* **2016**, *108*, 85–93. [CrossRef]

14. Kumaran, S.M. Identification of high temperature precipitation reactions in 2024 Al-Cu-Mg alloy through ultrasonic parameters. *J. Alloys Compd.* **2012**, *539*, 179–183. [CrossRef]

15. Son, S.K.; Takeda, M.; Mitome, M.; Bando, Y.; Endo, T. Precipitation behavior of an Al-Cu alloy during isothermal aging at low temperatures. *Mater. Lett.* **2005**, *59*, 629–632. [CrossRef]

16. Silcock, J.M. The Structural Ageing Characteristics of Al-Cu-Mg Alloys with Copper: Magnesium Weight Ratio of 7:1 and 2.2:1. *J. Inst. Met.* **1961**, *89*, 203–210.

17. Chen, Y.Q.; Pan, S.P.; Liu, W.H.; Liu, X.; Tang, C.P. Morphologies, orientation relationships, and evolution of the T-phase in an Al-Cu-Mg-Mn alloy during homogenisation. *J. Alloys Compd.* **2017**, *709*, 213–226. [CrossRef]

18. Kanno, M.; Suzuki, H.; Itoi, K. The effects of zirconium addition on aging phenomena of Al-Cu-Mg alloy. *J. Jpn. Inst. Light Met.* **1986**, *36*, 616–621. [CrossRef]

19. Ono, N.; Gouya, J.; Miura, S. Hall-Petch Relation and strengthening mechanisms in Al-Zn-Mg-Cu alloy polycrystals. *Trans. Jpn. Soc. Mech. Eng.* **2002**, *68*, 138–144. (In Japanese) [CrossRef]

Article

Stabilization Mechanism of Semi-Solid Film Simulating the Cell Wall during Fabrication of Aluminum Foam

Takashi Kuwahara [1,*], Akira Kaya [1], Taro Osaka [1], Satomi Takamatsu [1] and Shinsuke Suzuki [1,2]

[1] Faculty of Science and Engineering Faculty of Science and Engineering, Waseda University, 3-4-1 Okubo, Shinjuku, Tokyo 169-8555, Japan; s49y5wrgw-sl@suou.waseda.jp (A.K.); t-evolva@moegi.waseda.jp (T.O.); s-takamatsu@asagi.waseda.jp (S.T.); suzuki-s@waseda.jp (S.S.)

[2] Kagami Memorial Research Institute of Materials Science and Technology, Waseda University, 2-8-26 Nishi-Waseda, Shinjuku, Tokyo 169-0051, Japan

* Correspondence: takuwahara@ruri.waseda.jp; Tel.: +81-3-5286-8126

Received: 23 January 2020; Accepted: 26 February 2020; Published: 2 March 2020

Abstract: Semi-solid route is a fabrication method of aluminum foam where the melt is thickened by primary crystals. In this study, semi-solid aluminum alloy films were made to observe and evaluate the stabilization mechanism of cell walls in Semi-solid route. Each film was held at different solid fractions and holding times. In lower solid fractions, as the holding time increases, the remaining melt in the films lessens and this could be explained by Poiseuille flow. However, the decreasing tendency of the remaining melt in the films lessens as the solid fraction increases. Moreover, when the solid fraction is high, decreasing tendency was not observed. These are because at a certain moment, clogging of primary crystals occurs under the thinnest part of the film and drainage is largely suppressed. Moreover, clogging is occurring in solid fraction of 20–45% under the thinnest part of the film. Moreover, the time to occur clogging becomes earlier as the solid fraction increases.

Keywords: porous metal; semi-solid; aluminum foam; primary crystals; drainage; clogging

---

1. Introduction

Recent days, improved fuel consumption is required for transportation equipment such as automobiles. Furthermore, passenger safety including the crash safety must be maintained. For this purpose, aluminum foams with many pores dispersed inside are actively investigated. They are ultralight materials and exhibit shock absorbing characteristics called plateau phenomenon when compressed. Additionally, aluminum foam can absorb shocks isotropically [1]. Therefore, it is expected to be applied for transportation equipment [2]. Moreover, they have excellent property in sound insulation equivalent to glass wool [3,4]. Furthermore, they also have heat insulation properties [5]. Furthermore, it can be applied for architecture material [6].

However, there is a problem of drainage during fabricating aluminum foam which coarsens pores and thus, deteriorates the mechanical properties. Usually, aluminum foams are obtained by solidifying the aluminum alloy melt with pores generated inside. During holding of the foam before solidification, liquid flows downwards in a cell wall (a film between pores) due to the gravity. This phenomenon is called drainage. Due to the remarkable collapse of cell walls, pore morphology becomes uneven. Moreover, unfoamed parts would appear at lower part of the foam. However, by suppressing the drainage by thickening the melt, foam with uniformed pore morphology could be obtained [7]. Fabrication of foams with uniform pores becomes possible by stabilizing the cell walls through thickening.

There are several routes to fabricate aluminum foams. Melt-route, shown in Figure 1 is the most known route to fabricate aluminum foam. First, the melt is thickened by oxide particles. Then, the melt is stirred with adding blowing agent (TiH$_2$) included. Finally, an aluminum foam can be obtained by foaming and water cooling. However, our group is fabricating foams in the Semi-solid route shown in Figure 1. The fabrication method is almost the same as the Melt route except for thickening the melt by primary crystals existing in semi-solid state [8]. In the Semi-solid route, the adding of thickener, such as ceramic particles and Ca, can be excluded. Therefore, foam with less impurities can be obtained. Moreover, foams obtained by the Semi-solid route are thought to have higher compressive yield stress than that obtained by the Melt route [9].

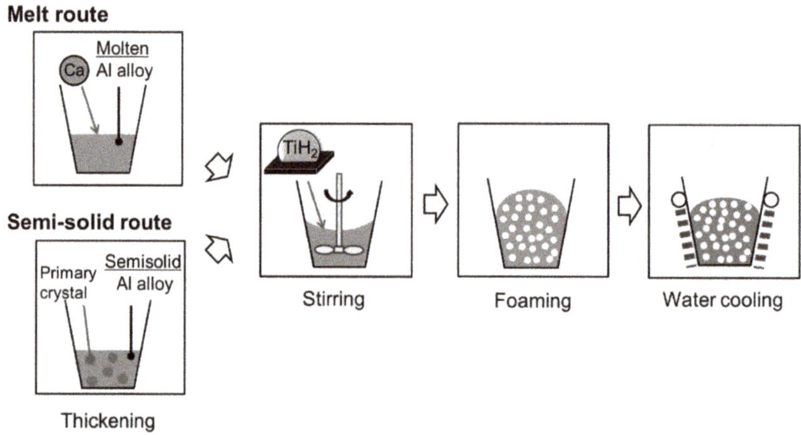

Figure 1. Fabrication methods of aluminum foam (Melt route and Semi-solid route).

Jin et al. evaluated the Melt route with SiC particles as a thickener and elucidated that the parameters to fabricate stabilized foams (foams with stabilized cell walls) are particle size and particle fraction. Similarly, they expressed the parameter map for fabricating of stabilized foams [10]. Moreover, Heim et al. fabricated foams with various types of thickener including primary crystal and classified them as foamable, partially foamable, or unfoamable. They also expressed that the parameter map proposed by Jin et al. could be adapted to various types of thickener [11].

Moreover, in the Melt route, Heim et al. expressed that some thickener particles are covered by oxide skin of cell wall and interact and builds bridge with other particles resulting to prevent drainage [12]. This is thought to be the stabilization mechanism of cell wall in the Melt route. However, the stabilization mechanism of cell wall in the Semi-solid route has not been evaluated. Furthermore, size of primary crystals is quite large and out of the stabilization region in the parameter map proposed by Jin et al. [10]. Likewise, primary crystals were classified as partially foamable or unfoamable by Heim et al. [11]. Despite, our group found it possible to fabricate stabilized foams in the Semi-solid route out of the stabilization region in the map. Therefore, cell wall of Semi-solid route is thought to have different stabilization mechanism from Melt route. Therefore, elucidation of stabilization mechanism of cell wall of Semi-solid route is required.

However, it is difficult to evaluate the cell walls since the size and shape of cell walls are uneven. As a solution, Heim et al. simulated the ideal shape of the cell wall using aluminum alloy film [13]. In this method, we can evaluate the cell wall in higher comparability. Therefore, the objective of this study is to elucidate the stabilization mechanism of cell wall of the Semi-solid route using aluminum alloy film.

2. Materials and Methods

2.1. Sample Material

Sample material used in this study was Al-6.4mass%Si. Since the phase diagram of Al-Si is simple and primary crystals of hypoeutectic alloy are easy to observe, this material is suitable for evaluation of stabilization mechanism. The Al-6.4mass%Si ingots were prepared by mixing pure aluminum with Al-25mass%Si. The compositions of the pure aluminum and the Al-25mass%Si ingots used in this study are shown in Tables 1 and 2, respectively.

Table 1. Composition of pure aluminum.

Element	Si	Fe	Cu	Al
mass%	0.004	0.003	0.001	bal.

Table 2. Composition of Al-25mass%Si.

Element	Si	Fe	Cu	Al
mass%	25.1	0.17	0.00	bal.

2.2. Forming of Aluminum Alloy Film

Aluminum alloy films which are simulating the cell walls of aluminum alloy foam were formed to evaluate the stabilization mechanism of cell wall in higher comparability. In the Semi-solid route, the melt was thickened in semi-solid state as shown in Figure 1. Therefore, aluminum alloy films were also formed in the semi-solid state to observe the cell walls in the Semi-solid route. The solid fractions of the melt of Al-6.4mass%Si were obtained accurately using Thermo-Calc 2019a (Itochu Techno-Solutions Corporation, Tokyo, Japan) as shown in Figure 2.

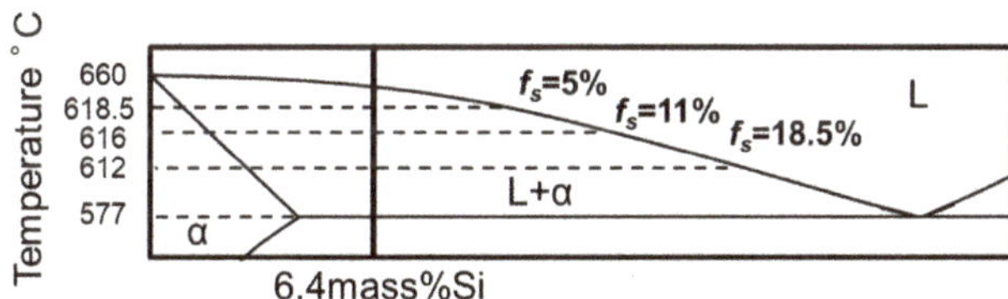

Figure 2. Al-Si binary phase diagram. The values of solid fraction were calculated by Thermo-Calc.

All of processes to form aluminum alloy films were done inside a chamber under pressure condition of 5000 Pa of air to suppress heat dissipation from the film during the retaining process described later. To obtain this condition, pressure was adjusted by vacuuming continuously while controlling the leaking rate with valve. To form the aluminum alloy film, Al-6.4mass%Si was completely molten in the crucible. Then, the melt was slowly cooled to the semi-solid temperature. The temperature was measured by K-type thermocouple coated with heat resistant inorganic adhesive (Toagosei Co. Ltd., Tokyo, Japan) soaked in the melt. Then, the melt was stirred for 60 s at 900 rpm so as to round the shape of primary crystals [14]. Before this process, thermocouple was taken out of melt to make the stirring possible. Since the stirring generates the temperature of melt to change, the thermocouple was soaked again in the melt and the temperature was adjusted again to semi-solid temperature.

After the semi-solid temperature reached the setting temperature, a wire shown in Figure 3a was dipped into the melt. This wire was shaped so as to form a liquid film between the rings when pulled up from the melt. The design of this wire originates from a previous research [13]. In this study, it was adopted to study the semi-solid densification phenomena. The material of the wire was stainless steel SUS304 and the drainage channel was provided for the melt to flow down from the film and reproduce the drainage. The ratio of distance between the rings versus its diameter was one versus three following the previous study to form a film with a continuous curvature [13]. After pulling up the film from the melt, the film was retained over the melt with drainage channels contacted with the melt to reproduce the drainage [13]. This process simulates the progression of drainage which causes collapse of cell walls in aluminum form. The holding time was changed in each film. Finally, the aluminum alloy film was obtained by pulling up the film at 5 mm/s and rapid cooling by dipping into a melt pool of low melting point alloy U78 (Osaka Asahi Co. Ltd., Osaka, Japan) held at 150–200 °C in the chamber. The schematic of the experimental procedure is shown in Figure 4.

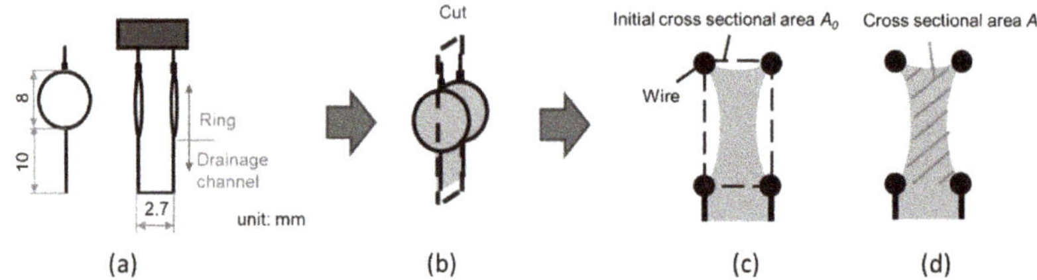

Figure 3. Schematic of wire and film: (**a**) size of wire; (**b**) cutting of film; (**c**) measurement of initial cross-sectional area; and (**d**) measurement of cross-sectional area.

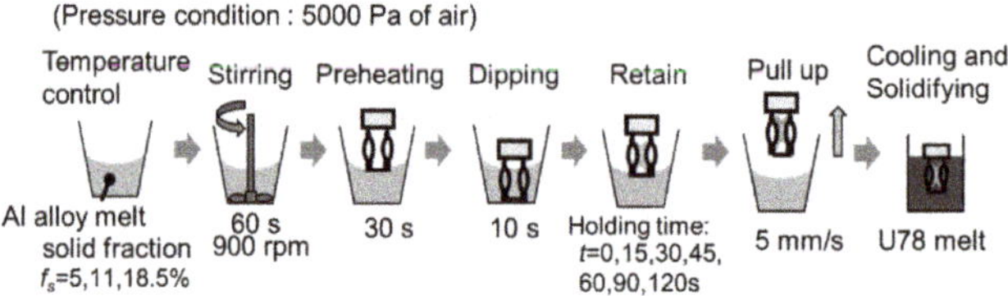

Figure 4. Schematic of forming liquid film of aluminum alloy.

Holding times of the aluminum alloy films were 0, 30, 60, 90, and 120 s. However, since it took a little time to move film to cool down, actual time of drainage was longer than the time designated. The solid fractions of melt were 5%, 11%, and 18.5%. Aluminum alloy film was formed for each combination of holding time and solid fraction. For solid fraction of 5%, holding time of 15 s and 45 s were also obtained for further consideration.

2.3. Observation of Aluminum Alloy Film

The formed film was first rinsed inside with the boiled water to remove U78 used for rapid cooling. Then, formed films were cut in the middle as shown in Figure 3b and the cross sections were observed by optical microscope. Primary crystals with a diameter larger than 200 μm, which seemed to have existed in a semi-solid state, were colored on micrographs with using an image processing software GIMP 2.10.4 (The GIMP Development Team, Berkeley, CA, USA). Then, the cross-sectional area of the films and the area of colored primary crystals were measured with using an image processing software WinROOF™ 6.4.0 (Mitani corporation, Fukui, Japan).

The cross-sectional area was measured to evaluate the change of melt quantity during the holding time. Furthermore, comparison of cross-sectional area was done by normalized cross-sectional area, as a parameter of the area of the remained melt compared with the initial one. To obtain the normalized cross-sectional area, initial cross-sectional area A_0 is defined as area surrounded with wires (Figure 3c). Then the normalized cross-sectional area was obtained through dividing the cross-sectional area A (Figure 3d) by the initial cross-sectional area. However, small deformation occurred on wires because of the flow stress of the semi-solid state. Therefore, the wires of the actual results were not aligned as schematic in Figure 3a. Likewise, this tendency is remarkable in the lengths between the rings while the lengths from the diameter of rings show relatively close value. Therefore, the initial cross-sectional area was obtained by calculating the area of trapezoid. The upper and lower length between the rings were used as upper and lower bases of the trapezoid. The average diameter of rings was used as its height.

3. Results

3.1. Temporal Change of Cross Sectional Area

The plots in Figure 5 show the temporal change of the normalized cross-sectional area obtained from the experimental results. From Figure 5, the normalized cross-sectional area decreases as the holding time increases at lower solid fractions. Additionally, decreasing tendency of the normalized cross-sectional area lessens as the solid fraction increases. Moreover, at high solid fraction this decreasing tendency was not observed.

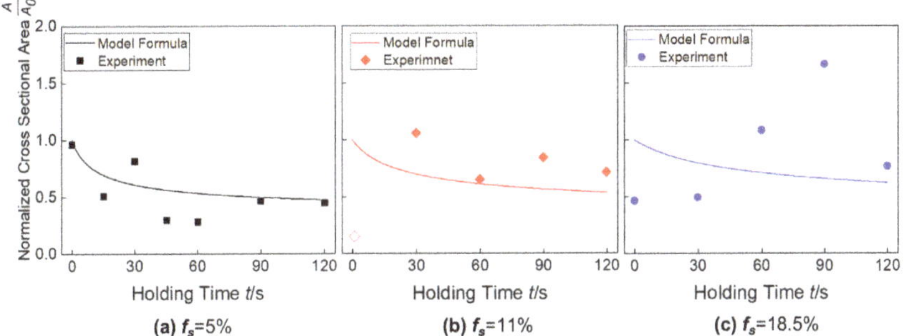

Figure 5. Temporal change of normalized cross-sectional area: (**a**) solid fraction $f_s = 5\%$; (**b**) $f_s = 11\%$; and (**c**) $f_s = 18.5\%$.

3.2. Observation of Primary Crystals

Figure 6b shows the cross-sectional photomicrograph of an aluminum alloy film shown in Figure 6a. In Figure 6c, primary crystals that seem to have been existing in the semi-solid state are colored as expressed in Section 2.3. Most of them are observed in the lower part of the film in the micrograph and less primary crystals were observed in the drainage channel than in the ring. Therefore, primary crystals seem to have clogged in the lower part of the film. However, only for the film for $f_s = 11\%$, $t = 0$ s, primary crystals that seem to have been existing from semi-solid state were not observed (Figure 7). This is an exception. In this experiment, the part with a small solid fraction might have been pulled up for some reason, such as inhomogeneous solid fraction in the crucible.

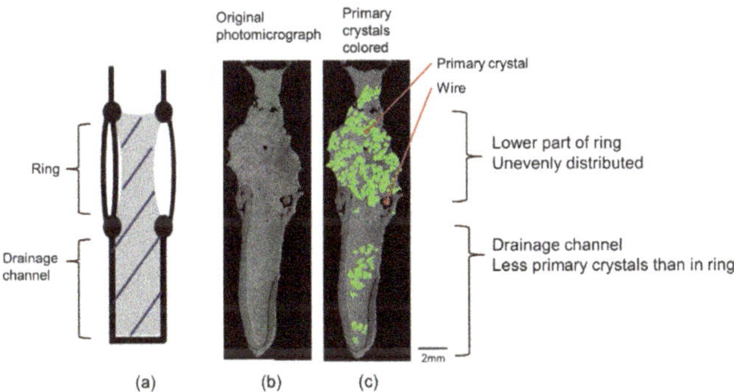

Figure 6. Cross section of aluminum alloy film (solid fraction $f_s = 18.5\%$, holding time $t = 120$ s): (**a**) cross section in film; (**b**) photomicrograph; and (**c**) primary crystals colored on photomicrograph.

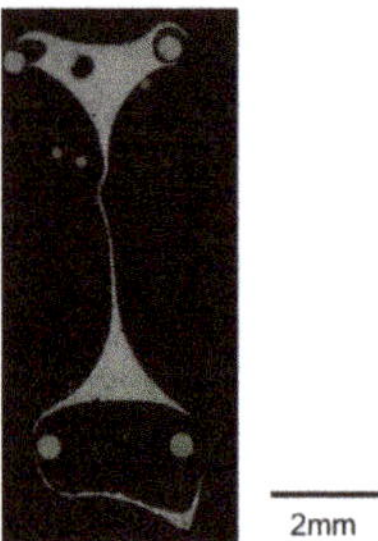

Figure 7. Photomicrograph of cross section of aluminum alloy film (solid fraction f_s = 11%, holding time t = 0 s).

4. Discussion

4.1. Suppression of Drainage by Thickening

As descripted in Introduction, thickening of melt stabilizes the cell walls and could homogenize the pore morphology. Increase of viscosity by thickening suppresses the liquid flow in aluminum alloy film. This is thought to be the stabilization mechanism of cell wall. For confirmation, we derived a model formula of melt flow and compared it with the experimental results. The model formula describes relationship between the holding time and the normalized cross-sectional area of the film.

4.1.1. Derivation of Model Formula

Figure 8 shows the change in geometry of the aluminum alloy film. Heim et. al. approximated the cross-sectional shape of the pulled-up film as a rectangle [12]. Then, the film will start to curve as the drainage progresses. The curvature radius of the film is decided by approximation to an arc defined by three points; the top and bottom of the wire and the middle of the film. The middle point of the film should shrink as the drainage progresses. Therefore, the curvature radius of the film should change continuously. The depth of the film L would not change in this model. This is because from observation of films including Figure 6, change in depth of film is small compared to width of film x_0. This may be because it is difficult to make curvature radius smaller when the original curvature radius is small. In this situation, the relationship between the cross-sectional area A of film and width of film x_t is shown in Equation (1). (The discussion is shown in Appendix A).

$$A = hx_0 - \frac{1}{4\Delta x}\left\{\frac{\pi}{2\Delta x}\left(h^2 + \Delta x^2\right)^2 \frac{\theta}{360°} - \left(h^2 + \Delta x^2\right)h\right\} \tag{1}$$

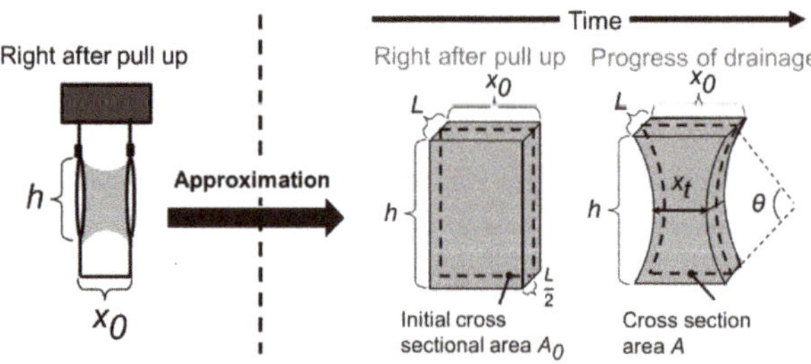

Figure 8. Geometry model of aluminum alloy film.

Here, θ is curvature radius of the film, h is height of rectangle, x_0 is width of rectangle, and x_t is width of middle point of rectangle which would change continuously. Furthermore, Δx is defined as Equation (2).

$$\Delta x = x_0 - x_t \tag{2}$$

Moreover, the curvature radius of the film θ can be shown in Equation (3).

$$\theta = \mathrm{Arcsin}\frac{4h(x_0 - x_t)\{h^2 - (x_0 - x_t)^2\}}{\{h^2 + (x_0 - x_t)^2\}^2} \tag{3}$$

However, since the changing parameter is width of film at middle x_t, Equation (1) could not explain the temporal change of the cross-sectional area of film A. Therefore, a fluid dynamic model of film is considered. Brady et. al. considered that liquid (semi-solid) flow in a film by drainage can be described as the Poiseuille flow under condition that the ends of both sides are fixed [15]. The situations are, one way flow, steady flow, and gravity as the only external force, as shown in Figure 9a. Following the Poiseuille flow, width of film at middle x_t can be shown as Equation (4).

$$x_t = x_0 \sqrt{\frac{6\mu h}{\rho g t x_0^2 + 6\mu h}} \tag{4}$$

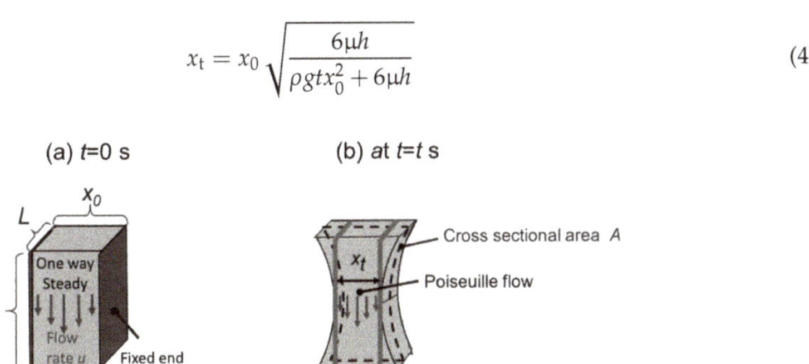

Figure 9. Schematics of Poiseuille flow and deformation of the film: (**a**) at $t = 0$ s and (**b**) at t.

Here, μ is the viscosity of melt, ρ is density of melt and g is the gravitational acceleration. By Equation (4), relationship between the width of film at middle x_t and holding time t could be obtained. By substituting the width of film at middle x_t obtained in Equation (4) into Equation (2), relationship between the cross-sectional area of film A and holding time t could be obtained by Equations (1) and (3). Therefore, a model formula to compare with the result is obtained. Figure 9b shows a schematic model of the phenomenon expressed by this model formula. Correspondingly, in this model, since the mainly considered area is middle of the film, which is assumed that drainage occurs mainly, three-dimensional deformation and drainage from ends of film is not considered.

4.1.2. Comparing Cross Sectional Areas Obtained from the Model with Experimental Results

To compare the model and the experimental results, the model formula was transformed into a dimensionless formula shown in Equation (5). A_0 in Equation (5) is the initial cross-sectional area and can be shown in Equation (6).

$$\frac{A}{A_0} = 1 - \frac{1}{4hx_0\Delta x}\left\{\frac{\pi}{2\Delta x}\left(h^2 + \Delta x^2\right)^2\frac{\theta}{360°} - \left(h^2 + \Delta x^2\right)h\right\} \tag{5}$$

$$A_0 = hx_0 \tag{6}$$

Figure 5 shows the cross-sectional areas obtained from the model with experimental results. According to the experimental data of Al-6.5mass%Si obtained by Moon et. al. using a Searle-type viscometer [16], the viscosity values of 22, 45, and 100 mPa·s were used for 5%, 11%, and 18.5% of solid fraction, respectively. Furthermore, the density value used was $2.4×10^3$ kg/m^3 for every solid fraction. This value was assumed from study done by Kudoh et. al. which measured the density of liquid phase in semi-solid state of Al-2.4mass%Si [17]. However, change of density between $2.3–2.7×10^3$ kg/m^3 only produces about 3% change in the calculated value of the model formula and has no effect on discussion. At lower solid fractions, experimental results seem to fit model formula. However, fitting becomes worse as the solid fraction increases. Furthermore, for high solid fraction, fitting of model formula to experimental results was not good. Therefore, it is difficult to conclude that only the increase in fluid viscosity is the mechanism to stabilize the film. Therefore, consideration of other factors to suppress the drainage is required.

4.2. Suppression of Drainage by Clogging of Primary Crystals

As expressed in Section 3.2, primary crystals seem to have been clogging in the lower part of the film. To confirm and to find the clogging part more finely, the distribution of primary crystals in the films was measured in the height direction from the lower wires (Figure 10). In Figure 10, the primary crystals that seem to have been existing from semi-solid state are colored as expressed in Section 2.3. As the primary crystals were found more in the lower part than in the upper part and drainage channel, clogging should have occurred in the lower part under the thinnest part of the film. Thus, this factor effects to suppress drainage more strongly than thickening. Likewise, at solid fraction of 5%, this tendency was seen remarkably in films with long holding time.

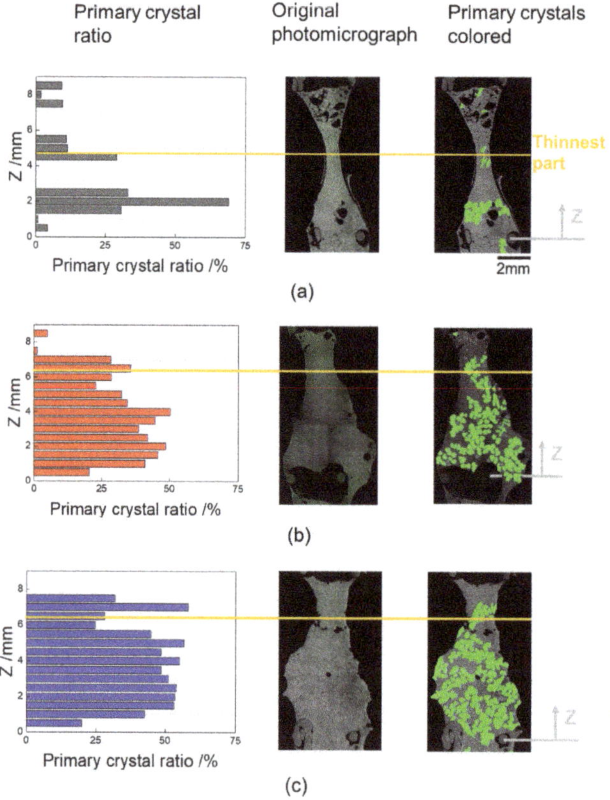

Figure 10. Primary crystal ratio: (**a**) solid fraction $f_s = 5\%$, holding time $t = 120$ s; (**b**) $f_s = 11\%$, $t = 90$ s; and (**c**) $f_s = 18.5\%$, $t = 120$ s.

To derive this clogging mechanism more quantitatively, the critical solid fraction f_{s_cr} for clogging was evaluated. Areas of primary crystals A_p and film A_f were measured in the region between the thinnest part of the film and the lower wires (Figure 11). In Figure 11, primary crystals that seem to have been existing in semi-solid state are colored in the same way as expressed in Section 2.3. The area of aluminum alloy film is also colored in the same way as expressed in Section 2.3. By dividing the area of primary crystals A_p by area of aluminum alloy film A_f, the critical solid fraction for clogging was obtained. Figure 12 shows the measured results of the critical solid fraction for clogging. In Figure 12, the critical solid fraction for clogging seems not to change in increase of holding time. Similarly, critical solid fraction for clogging is in the range from 20% to 45%. Therefore, clogging seems to be occurring in this region of solid fraction.

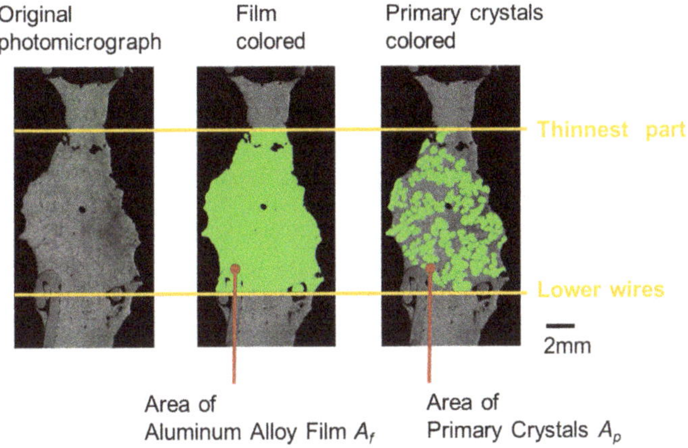

Figure 11. Evaluations of area of aluminum alloy film and primary crystals.

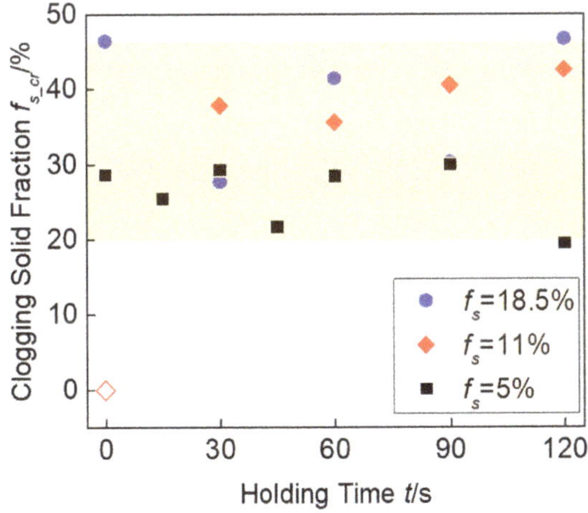

Figure 12. Relationship between critical solid fraction for clogging f_{s_cr} and holding time for initial solid fraction f_s of 5%, 11%, and 18.5%.

Clogging of thickener has also been reported for the Melt route [13]. However, bridge formation of thickener attributed to oxide film is thought to be the main stabilization mechanism in Melt route [12]. For Semi-solid route, since the size of thickener is large, clogging is possible to occur below the thinnest part of film and it is assumed that film is not required to be thin for the clogging to occur.

4.3. Fluid Flow and Clogging of Primary Crystals

According to the results in Figure 5 clogging occurs at a certain timing, which becomes earlier with increasing solid fraction. For solid fraction of 18.5%, clogging seems to have occurred right after the pull-up. However, for the lower solid fractions, drainage following the model formula occurs in a pulled-up film until the clogging. Figure 13 which is a qualitative representation based on Figure 5 summarizes the schematic and graph of this phenomenon. The solid lines show the actual change of normalized cross-sectional area A/A_0. The broken lines and curved solid lines show the values obtained from model formula.

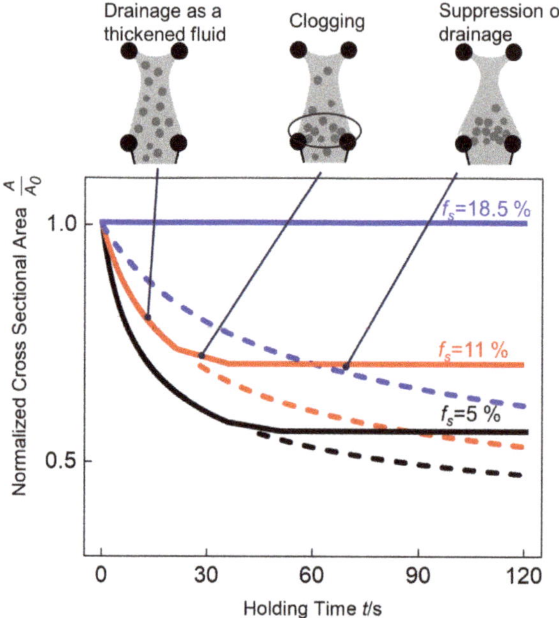

Figure 13. Clogging time in model formula and schematic of clogging mechanism. The solid lines show the actual change of normalized cross-sectional area A/A_0. The broken lines and curved solid lines show the values obtained from model formula.

5. Conclusions

The stabilization mechanism of cell wall in Semi-solid route was evaluated using aluminum alloy films. The results can be summarized as follows.

1. Pulled-up film drainages as a thickened melt. Since this could be explained as Poiseuille flow, drainage rate becomes slower with increasing solid fraction.
2. At a certain timing clogging occurs and drainage is largely suppressed. Clogging is occurring in the range of solid fraction of 20–45% under the thinnest part of the film.
3. The clogging occurs earlier as the solid fraction increases. For solid fraction f_s = 18.5% clogging seems to be already occurring in holding time t = 0 s.

Author Contributions: Conceptualization, T.K. and S.S.; methodology, T.K., A.K., T.O. and S.S.; validation, T.K., A.K., T.O. and S.S.; formal analysis, T.K.; investigation, T.K., A.K., T.O. and S.S.; data curation, T.K. and S.S; writing—original draft preparation, T.K.; writing—review and editing, T.K., S.T. and S.S.; visualization, T.K.; supervision, S.S.; project administration, S.S.; funding acquisition, T.K. and S.S. All authors have read and agreed to the published version of the manuscript.

Funding: This study was supported by the Grant-in-Aid the Light Metal Educational Foundation.

Acknowledgments: The authors thank The Light Metal Educational Foundation for suppling the pure Aluminum ingots used in this study.

Appendix A

Here, derivation of geometry model is expressed. As expressed in Section 4.1.1 and Figure 8, a pulled-up film is approximated as a rectangle. Therefore, the cross-sectional area of the film at this point can be expressed as Equation (A1).

$$A_0 = hx_0 \tag{A1}$$

Here, A_0 is the initial cross-sectional area, h is height of rectangle, and x_0 is width of rectangle. Then, the side wall of the film will start to become a curved surface as the drainage progresses. At this point cross section of the film can be shown as Figure A1. Here, θ is the curvature radius of the film, x_t is the width of middle point of rectangle which would change continuously.

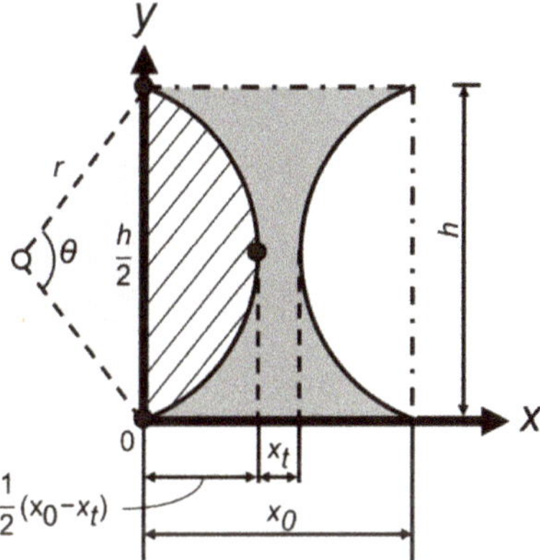

Figure A1. Coordinate axes of cross section of aluminum alloy film during progress of drainage.

Figure A1 also shows the coordinate axes of aluminum alloy film. Rectangle inside chain line shows the initial cross section. To obtain the cross-sectional area A as a formula, area of shaded part S in Figure A1 is obtained as follows. The equation of a circle with this curvature can be shown as Equation (A2) since the circle is known to pass through three dark points. Therefore, the radius of the circle can be shown as Equation (A3). Middle point of circle is expressed as light point in Figure A1.

$$\left\{ x - \frac{(x_0 - x_t)^2 - h^2}{4(x_0 - x_t)} \right\}^2 + \left(y - \frac{h}{2} \right)^2 = \frac{1}{16(x_0 - x_t)^2} \left\{ (x_0 - x_t)^2 + h^2 \right\}^2 \tag{A2}$$

$$r = \frac{1}{4(x_0 - x_t)} \left\{ (x_0 - x_t)^2 + h^2 \right\} \tag{A3}$$

Then the area of the light gray triangle S' in Figure A1 can be obtained as Equation (A4) from Heron's formula using three sides of triangle.

$$S' = \frac{h}{8(x_0 - x_t)} \left\{ h^2 - (x_0 - x_t)^2 \right\} \tag{A4}$$

Since the area of triangle can also be shown as Equation (A5), the curvature radius θ can be obtained as Equation (A6) from Equations (A4) and (A5).

$$S' = \frac{1}{2}r^2 \sin\theta \tag{A5}$$

$$\theta = \mathrm{Arcsin}\frac{4h(x_0 - x_t)\{h^2 - (x_0 - x_t)^2\}}{\{h^2 + (x_0 - x_t)^2\}^2} \tag{A6}$$

From the curvature radius, the area of the sector composed of the shaded part and the light gray triangle can be obtained as Equation (A7). Then, by subtracting Equation (A4) from Equation (A7), the area of the shaded part can be obtained as Equation (A8). In the equations, S is the shaded part which shows the area of outer part the of curved film.

$$(S + S') = \frac{\pi}{16(x_0 - x_t)^2}\{h^2 + (x_0 - x_t)^2\}^2 \frac{\theta}{360°} \tag{A7}$$

$$S = \frac{1}{8(x_0 - x_t)}\left[\frac{\pi}{2(x_0 - x_t)}\{h^2 + (x_0 - x_t)^2\}^2 \frac{\theta}{360°} - \{h^2 - (x_0 - x_t)^2\}h\right] \tag{A8}$$

Since there are curved parts at both side of film, the total area of the outer part of the curved film can be shown as Equation (A9).

$$2S = \frac{1}{4(x_0 - x_t)}\left[\frac{\pi}{2(x_0 - x_t)}\{h^2 + (x_0 - x_t)^2\}^2 \frac{\theta}{360°} - \{h^2 - (x_0 - x_t)^2\}h\right] \tag{A9}$$

Finally, by subtracting Equation (A9) from Equation (A1), the cross-sectional area of the film with a curvature can be shown as Equation (A10) with Δx defined as Equation (A11).

$$A = A_0 - 2S = hx_0 - \frac{1}{4\Delta x}\left\{\frac{\pi}{2\Delta x}(h^2 + \Delta x^2)^2 \frac{\theta}{360°} - (h^2 + \Delta x^2)h\right\} \tag{A10}$$

$$\Delta x = x_0 - x_t \tag{A11}$$

References

1. Baumeister, J.; Banhart, J.; Weber, M. Aluminium foams for transport industry. *Mater. Des.* **1997**, *18*, 217–220. [CrossRef]
2. Banhart, J. Manufacture, characterisation and application of cellular metals and metal foams. *Prog. Mater Sci.* **2001**, *46*, 559–632. [CrossRef]
3. Miyoshi, T.; Itoh, M.; Akiyama, S.; Kitahara, A. ALPORAS aluminum foam: production process, properties, and applications. *Adv. Eng. Mater.* **2000**, *2*, 179–183. [CrossRef]
4. Lu, T.J.; Hess, A.; Ashby, M.F. Sound absorption in metallic foams. *J. Appl. Phys.* **1999**, *85*, 7528–7539. [CrossRef]
5. Kitazono, K. Superplastic forming and foaming of cellular aluminum components. *Mater. Trans.* **2006**, *47*, 2223–2228. [CrossRef]
6. Banhart, J.; Vinod-Kumar, G.S.; Kamm, P.H.; Neu, T.R.; García-Moreno, F. Light metal foams—Some recent developments. *Cienc. Tecnol. Mater.* **2015**, *27*, 1–80. [CrossRef]
7. Kuwahara, T.; Saito, M.; Osaka, T.; Suzuki, S. Effect of primary crystals on pore morphology during semi-solid foaming of A2024 alloys. *Metals* **2019**, *9*, 88. [CrossRef]
8. Hanafusa, T.; Ohishi, K. Making of Porous Metallic Material by the Semi-Solid Aluminum Alloy. Available online: https://www.pref.hiroshima.lg.jp/uploaded/life/206737_384148_misc.pdf (accessed on 23 January 2020).
9. Kuwahara, T.; Osaka, T.; Saito, M.; Suzuki, S. Compressive properties of A2024 alloy foam fabricated through a Melt route and a Semi-solid route. *Metals* **2019**, *9*, 153. [CrossRef]

10. Jin, I.; Kenny, L.D.; Sang, H. Method of producing lightweight foamed metal. U.S. Patent 4,973,358, 27 November 1990.

11. Heim, K.; García-Moreno, F.; Banhart, J. Particle size and fraction required to stabilize aluminum alloy foams created by gas injection. *Script. Mater.* **2018**, *153*, 54–58. [CrossRef]

12. Heim, K.; Vinod Kumar, G.S.; García-Moreno, F.; Rack, A.; Banhart, J. Stabilization of aluminium foams and films by the joint action of dispersed particles and oxide films. *Acta Mater.* **2015**, *99*, 313–324. [CrossRef]

13. Heim, K.; Vinod Kumar, G.S.; García-Moreno, F.; Manke, I.; Banhart, J. Drainage of particle-stabilised aluminium composites through single films and Plateau borders. *Colloids Surf. A* **2013**, *438*, 85–92. [CrossRef]

14. Chen, H.I.; Chen, J.C.; Liao, J.J. The influence of shearing conditions on the rheology of semi-solid magnesium alloy. *Mater. Sci. Eng.* **2008**, *A487*, 114–119. [CrossRef]

15. Brady, A.P.; Ross, S. The measurement of foam stability. *J. Am. Chem. Soc.* **1944**, *66*, 1348–1356. [CrossRef]

16. Moon, H.K. Rheological behavior and microstructure of ceramic particulate/aluminum alloy composites. Ph.D. Thesis, Massachusetts Institute of Technology, Cambridge, MA, USA, 1990.

17. Kudoh, M.; Takahashi, T. Estimation of the fluidity of the liquid in the solid-liquid coexisting zone of an Al-Si alloy. Available online: https://eprints.lib.hokudai.ac.jp/dspace/bitstream/2115/41753/1/110_69-80.pdf. (accessed on 23 January 2020).

 *metals*

Article

A-242 Aluminium Alloy Foams Manufacture from the Recycling of Beverage Cans

Nallely Montserrat Trejo Rivera *, Jesús Torres Torres and Alfredo Flores Valdés

Centro de Investigación y de Estudios Avanzados del Instituto Politécnico Nacional, Unidad Saltillo, Avenida Industria Metalúrgica 1062, Parque Industrial Saltillo-Ramos Arizpe, 25900 Ramos Arizpe, Coahuila, Mexico; jesus.torres@cinvestav.edu.mx (J.T.T.); alfredo.flores@cinvestav.edu.mx (A.F.V.)

* Correspondence: nallely.trejo@cinvestav.edu.mx; Tel.: +52-844-438-9600

Received: 5 December 2018; Accepted: 11 January 2019; Published: 16 January 2019

Abstract: This paper presents and discusses a methodology implemented to study the process of the preparation of aluminium alloy foams using the alloy A-242, beginning from the recycling of secondary aluminium obtained from beverage cans. The foams are prepared by a melting process by adding 0.50 wt.% calcium to the A-242 aluminium alloy with the aim to change its viscosity in the molten state. To obtain the foam, titanium hydride is added in different concentrations (0.50 wt.%, 0.75 wt.%, and 1.00 wt.%) and at different temperatures (923, 948 K, and 973 K) while the foaming time is kept constant at 30 s. For a set of experimental parameter values, aluminium alloy foams with the average relative density of 0.12 were obtained and had an 88.22% average porosity. In this way, it is possible to state that the preparation of aluminium alloy foams A-242 processed from the recycling of cans is possible, with characteristics and properties similar to those obtained using commercial-purity metals.

Keywords: aluminium alloy foam; recycling; beverage cans; direct foaming method; A-242 alloy

1. Introduction

Aluminium and its alloys are highly recyclable metals that are easily used for the preparation of several specific alloys, and there are claims that the practice of recycling aluminium allows to reduce pollution and contributes to saving electrical energy as compared to the primary aluminium obtaining process. The recycling process is suitable as aluminium cans from discarded beverages are composed mainly of aluminium alloys, so any separation methods for other materials are not required [1].

Recycling is desirable when the environmental and economic implications of their reintegration do not exceed the limits of their primary production [2]. Aluminium is used in present times like no other metal, together with steel, so its production is continuously growing, with an average growth of 3.7% per annum [3]. Its physical properties make it an ideal candidate for a range of applications in industries such as packaging, transportation, construction, and aerospace, among others [4]. However, secondary aluminium production faces problems with quality losses (when the purity of the aluminium produced is lower than the input material, e.g., by adding alloying elements during re-melting) and dilution losses (addition of primary aluminium during melting to dilute the concentration of residual elements that cannot be refined), accumulation of impurities, and unlimited options for purification of the molten mass [5–8]. Nevertheless, several studies have addressed the fact that aluminium recycling needs no more than 5% of the energy needed for its primary production, so it presents a real opportunity to reduce environmental impacts if managed in a sustainable way [9,10].

On the other hand, it is well known that for the 2XX series of alloys, copper is the main alloying element, so these types of alloys are thermally treatable. Some of the properties that they present are good ductility, high tensile strength (275 MPa) and compressive yield strength (235 MPa). During solidification, the main alloying elements form intermetallic phases which precipitate in different morphologies, giving rise to various properties. Copper is slightly soluble in aluminium at room temperature (0.2%); however, increasing the temperature (821 K) can dissolve up to 5.65% copper. During the solidification process of the A-242 aluminium alloy, the intermetallic compound Al_2Cu is formed as a result of a eutectic reaction, improving resistance and hardness in the as-cast condition. With high copper content (4 to 6%), the aluminium alloy responds strongly to heat treatments. On the other hand, nickel forms the intermetallic Al_3Ni with aluminium on solidification, which improves the properties at high temperatures, also reducing the coefficient of thermal expansion. This alloy is used in the manufacture of cylinder heads for motorcycles and pistons [11]. Table 1 shows the chemical composition of the A-242 alloy [12].

Table 1. Chemical composition of A-242 aluminium alloy (wt.%).

Al	Si	Fe	Cu	Mn	Mg	Cr	Ni	Zn	Ti	Others
Balance	0.70	1.00	3.5–4.5	0.35	1.2–1.8	0.25	1.7–2.3	0.35	0.25	0.15

The automotive industry is constantly searching for innovation in its products, giving the opportunity to develop and innovate new materials which allow a better fuel efficiency and passenger safety [13]. Aluminium foams are an attractive alternative for this purpose because their properties are ideal for this purpose as its lightness and ability to absorb impact energy.

Metal foams are light materials in which a gas dispersed in a metal matrix occupies between 50 and 90% of the total volume, obtaining low density values ($0.3–0.8\,g/cm^3$). They are characterized by a combination of mechanical and physical properties resulting from both the porous structure and the characteristics of the metal from which they were manufactured [14,15]. Depending on the production method, the foam structure is more or less homogeneous and comprises different characteristic features that determine its properties and, therefore, the fields of application [16,17].

However, less common and familiar are applications and products based on metal foams. The reason is that they are still not widespread, although they have a very high potential, and a large number of applications already exist on the market. Some reviews about the applications of metallic foams are available in the literature [16–22]; however, in recent years, new application fields have emerged, and not all are considered commercially relevant.

The main applications of metal foams can be grouped into structural and functional applications, and are based on several excellent properties of the material [17]. Structural applications take advantage of the light-weight and specific mechanical properties of metal foams; functional applications are based on a special functionality, i.e., a large open area in combination with very good thermal or electrical conductivity for heat dissipation or as electrode for batteries, respectively [16].

There are a large number of manufacturing methods for metal foams. The closed porosity metal foams are commonly fabricated by direct and indirect foaming methods, such as the melting (ML) and powder metallurgical (PM) methods [16,17], respectively. These methods are already described in the literature [16,17,23–26].

Shinko Wire provided the first commercial production of metal foams in the late 1980s. In this process, calcium is first added to an aluminium melt and stirred in air to produce oxides and raise its viscosity. Subsequently, the powder of the blowing agent (TiH_2) is dispersed quickly into the melt by stirring, decomposing into gaseous hydrogen and titanium at the melt temperature. The melt starts then foaming inside the crucible or inside a mold. Finally, it is cooled down to a big block and usually sliced into plates of the desired thickness. This type of metallic foam and the corresponding process is called Alporas, which was patented in America in 1987 [16,27].

Another process to manufacture aluminium alloy foams melt was invented in the early 1990s by Alcan International Limited in Montreal (QC, Canada) [28], and Norsk Hydro (Oslo, Norway). In this process, the melt needs to be prepared with ceramic particles, such as silicon carbide or aluminium oxides, ranging from 5 to 20 vol.% to increase the viscosity of the melt and to stabilize the liquid cell walls. Subsequently, air bubbles are injected and, at the same time, dispersed into the melt using rotating impellers. The bubbles rise to the top of the melt, where they are collected in a liquid foam and start solidifying after leaving the furnace, where the foam can be continuously drawn off using a conveyor belt. This method is used nowadays by Cymat (Mississauga, ON, Canada) to produce foam panels or to fill molds with foams that do not need further processing [16].

In the 1950s, Allen et al. patented a method for foaming metal [29]. This route consists of an indirect foaming of solid precursors by heating. The precursor is produced by mixing aluminium powders with the corresponding alloying elements and a blowing agent, typically 0.5 to 1.0 wt.% of TiH_2. Once the powder mixture is prepared, it is consolidated and sintered by extrusion, uniaxial compaction or rolling to yield a foamable precursor. By the heating of the precursors, the matrix starts melting, and the gas of the blowing agent nucleates. In the course of the temperature increasing, hydrogen production increases, and the gas diffuses to the nucleated pores, letting them grow into big bubbles and expanding the foam. The resident oxides in the metal powders (usually 0.5 to 1%) provide the stability of the foam during the holding time in the liquid state [30]. After several minutes, the foam development is fulfilled, and the foamed metal structure can be conserved by temperature reduction, leading to foam solidification [16].

Applications of metal foams are strongly linked to the properties that such kinds of materials can offer and especially to those that are excellent or even unique. Some of the properties are obviously mainly related to those of the matrix metal itself, e.g., elasticity, temperature or corrosion resistance, etc., while others appear only in combination with the cellular structure, e.g., low density, large surface area or damping [16].

The mechanical properties of metal foams are of course correlated to the ones of the corresponding bulk metal, but in a specific manner. The dominating factors here are the density and the structure itself. The foam structure is obviously the characteristic feature of a foam. Mechanical properties depend mainly on the density but are also influenced by the quality of the cellular structure in the sense of cell connectivity, cell roundness and diameter distribution, fraction of the solid contained in the cell nodes, edges or the cell faces [16,26,31–36].

The main objective of this work was the preparation A-242 aluminium alloy foams, obtained by adjusting the chemical composition of secondary aluminium from the recycling of beverage cans. On the other hand, taking into account that during solidification of the aluminium alloy foam, the precipitation of various intermetallic compounds occurs (i.e., Al_2Cu, Al_3Ni, Al_9FeNi, Al_2CuMg, Ti, Al_3Ti), it was interesting to study the effects of the processing temperature and the content of titanium hydride used on the formation in the aluminium alloy foams and mechanical behavior during the compression test.

2. Materials and Methods

2.1. Preparation of Aluminium Alloy Foam

The cans are composed mainly of three different alloys, i.e., the body corresponds to the A-3004 alloy, the lid to the A-5182 alloy, and the seal to the A-5082 alloy [37]. Once melted, the composition of the obtained alloy is similar to that described in Table 2. Compacted cans were melted in a gas-fired furnace containing a silicon carbide crucible with a capacity of 60 kg. Once the temperature of 1023 K was attained, a flux was added to the molten bath to remove impurities from the alloy.

Once the beverage cans were melted, the chemical composition was adjusted with additions of pure copper (4 wt.%) and electrolytic nickel (2 wt.%) at a temperature of 1293 K to bring the mixture to the A-242 alloy specification. The amounts of alloying elements to be added were calculated using the following equation:

$$X = \frac{(Pc)(C)}{100\%},$$

(1)

where X is the amount, in grams, of the alloying element to be added (Cu, Ni); Pc is the weight of the load as melted (g); and C is the difference between the weight percentage necessary to attain the composition and the initial one. Table 3 shows the final chemical composition of the A-242 aluminium alloy obtained from the adjustment of the chemical composition after the melting step.

Table 2. Chemical compositions of the alloys contained in the aluminium cans and the final composition of the fused alloy (base alloy) in wt.%.

Alloy	Al	Si	Fe	Cu	Mn	Mg	Ni	Ti	Cr	Zn
Body A-3004	Balance	0.30	0.70	0.25	1.00–1.50	0.80–1.30	-	-	-	0.25
Lid A-5182	Balance	0.20	0.35	0.15	0.20–0.50	4.00–5.00	-	0.10	0.10	0.25
Seal A-5082	Balance	0.20	0.35	0.15	0.15	4.00–5.00	-	0.10	0.15	0.25
Base alloy	Balance	0.27	0.72	0.16	0.89	0.86	0.01	0.01	0.02	0.09

Table 3. Chemical composition of the A-242 aluminium alloy obtained after the melting and composition adjustment steps.

Alloy	Al	Si	Fe	Cu	Mn	Mg	Ni	Ti	Cr	Zn
A-242	Balance	0.25	0.66	3.68	0.81	0.77	1.94	-	0.02	0.09

The chemical compositions of the secondary aluminium and of the A-242 aluminium alloy were determined using a SpectroLAB spark emission spectrometer (Spectro Inc., Kleve, Germany).

To determine the melting temperature range of the A-242 alloy during heating, a Perkin Elmer Differential Thermal Analyzer 7 (Seiko Instruments Inc, Chiba, Japan) was used. Three tests were carried out at a heating speed of 10 °C/min under an argon atmosphere. Figure 1 shows a differential thermal analysis (DTA) pattern for the obtained A-242 aluminium alloy. This shows an endothermic event in the range from 869 to 931 K, related to the starting temperatures of the melting of the A-242 aluminium alloy.

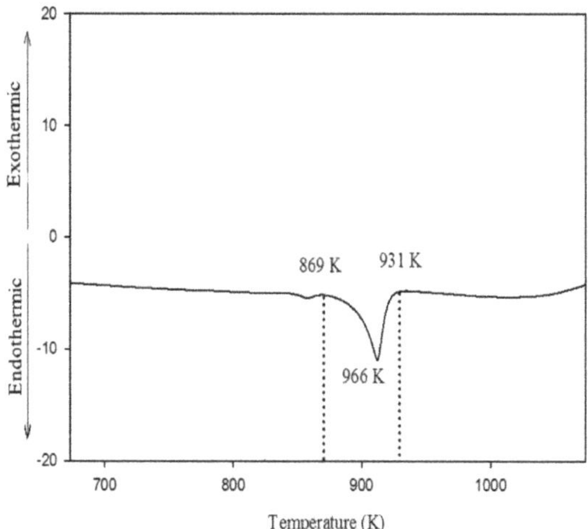

Figure 1. DTA pattern for the A-242 aluminium alloy.

The next step was the preparation of A-242 aluminium alloy foams. At this step, the effects of the content of the foaming agent and the foaming temperature on the mechanical properties and microstructure of the aluminium alloy foams were investigated. Table 4 presents the parameters and their values studied.

Table 4. Values selected for the preparation of A-242 aluminium alloy foams (foaming time is constant for 30 s).

Sample	% TiH$_2$	Foaming Temperature (K)
A1	0.50	923
A2	0.75	923
A3	1.00	923
A4	0.50	948
A5	0.75	948
A6	1.00	948
A7	0.50	973
A8	0.75	973
A9	1.00	973

The A-242 aluminium alloy foams were prepared in an electrical resistance furnace equipped with a mechanical agitator using a bipartite steel mold. Once the constant temperature of 923 K was attained, 0.50 wt.% Ca was added, and the mixture was shaken for 5 min at a speed of 1500 rpm in order to modify the viscosity of the aluminium alloy. It is worth mentioning that the inclined plane technique was used to determine the viscosity value attained after this addition. We found that, under the conditions imposed, the aluminium alloy developed a viscosity of 0.196 Pa·s—enough to carry out the foaming process. The foaming agent was added according to the conditions depicted in Table 4, where the TiH$_2$ thermally decomposes, releasing hydrogen [38]. The release of this gas gives rise to the formation of bubbles inside the metal held in a semisolid state. The TiH$_2$ was allowed to react for 30 s with constant agitation at 3000 rpm. After the foaming time had elapsed, the mold was removed from the furnace, allowing the prepared aluminium alloy foam to solidify to room temperature (298 K). When the aluminium alloy solidifies, the bubbles are trapped in the metal, giving rise to the formation of pores inside the alloy. The aluminium alloy foams obtained are cylindrical in shape, an average of 118.06 mm in height, and with a diameter of 75 mm.

2.2. Morphological Characterization

The expansion of the aluminium alloy only takes place in the direction of the height. Therefore, Equation (2) [39] is used to find the linear expansion of the aluminium alloy as a function of the heights of the aluminium alloy foam and of the molten metal in the mold:

$$\alpha_{LE} = \frac{h_1 - h_2}{h_2} \times 100\%, \tag{2}$$

where α_{LE} is the linear expansion of the foam, h_1 is the height of the foam, and h_2 is the height of the molten metal in the mold.

The density of the aluminium alloy foams was determined using Equation (3), where the samples used were in the form of cubes made by cutting the aluminium alloy foam. Each sample was weighed using a digital apparatus obtaining the mass of the specimen (m) expressed in grams. The dimensions of the samples were also measured in order to calculate their volume (V).

$$\rho = \frac{m}{V} \tag{3}$$

The relative density (ρ^*) of the samples was estimated using Equation (4), where ρ corresponds to the density of the aluminium alloy foams obtained from Equation (3), and ρ_s for which the density of A-242 aluminium alloys (2.823 g/cm^3) [11] was used.

$$\rho^* = \frac{\rho}{\rho_s} \tag{4}$$

The percentage of porosity was determined using Equation (5) [39]:

$$P = \left(1 - \frac{\rho}{\rho_s}\right) \times 100 \tag{5}$$

where P is the percentage of porosity of the aluminium alloy foams.

To find the pore diameter, the photomicrographs obtained from the stereographic microscope were used. Image Pro Plus software (4.1, Media Cybernetics Inc., Rockville, MD, USA) was used to trace two perpendicular lines to each other within each pore; these lines give the length between two points, so 30 measurements were made per sample to find the average diameter. Figure 2 shows the image used, illustrating the pore diameter of some of the measurements.

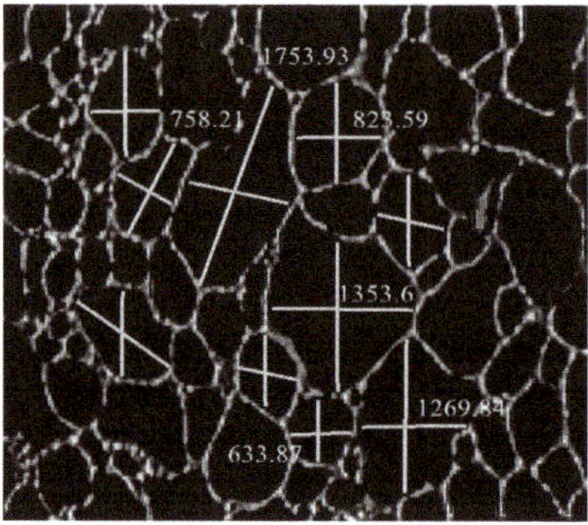

Figure 2. Measurement of the pore diameter using Image Pro Plus software.

The wall thickness measurements were performed using a scanning electron microscope (SEM) (Royal Philips Inc., Amsterdam, The Netherlands), as is shown in Figure 3 by a micrograph of the foam, indicating the measurements performed. To find the average wall thickness, 10 measurements per sample were performed.

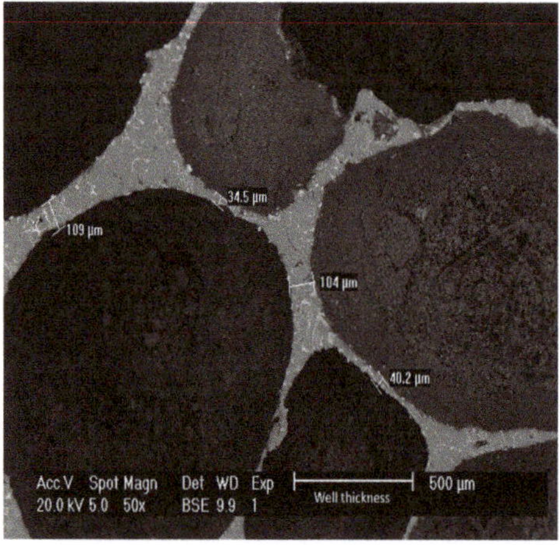

Figure 3. Scanning electron microscope (SEM) micrograph of A-242 aluminium alloy foam where the wall thickness measurements are illustrated.

The microstructure of the A-242 aluminium alloy foams was observed using an XL30 ESEM scanning electron microscope (Royal Philips, Amsterdam, The Netherlands) equipped with a GENESIS 400 EDS (energy dispersion spectroscopy, Hi-Tech Instruments, Las Pinas, Philippines) microanalysis system to identify the intermetallic compounds present in the samples.

2.3. Mechanical Characterization

Finally, in order to evaluate the mechanical strength of the aluminium alloy foams under compression loads, uniaxial compression testing was carried out according to the ASTM E9 Standard procedure at room temperature. The tests were carried out using a Qtest Elite 100 model MTS electromechanical universal testing machine (MTS Inc., Berlin, Germany) with a capacity of 100 KN, equipped with TestWork software (Version 4, MTS Systems Corporation, Berlin, Germany). Cubic samples ($25 \times 25 \times 25$ mm^3) were tested as shown in Figure 4. The compression tests were performed at cross-head rates of 3 mm/min. The force and the displacement were recorded during the compression tests. The engineering stress-strain data were determined through load-displacement measurements taking into account the initial dimensions of specimens.

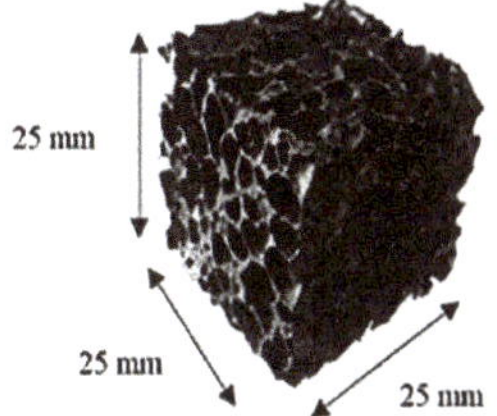

Figure 4. Image of A-242 aluminium alloy foam, showing the closed porosity attained with 0.75% TiH$_2$ at 948 K (A5).

3. Results and Discussion

3.1. Foam Morphology

Figure 4 shows an image of A-242 aluminium alloy foam obtained using the process described in this work: a structure of closed porosity having an average relative density of 0.12.

Figure 5 shows a photomicrograph of a A-242 aluminium alloy foam prepared with 0.75% TiH$_2$ at 948 K (A5) using a stereographic microscope, where it can be observed that the pores of the sample are not connected to each other, thus A-242 aluminium alloy foams prepared using the technique described in this work presented a structure with closed porosity.

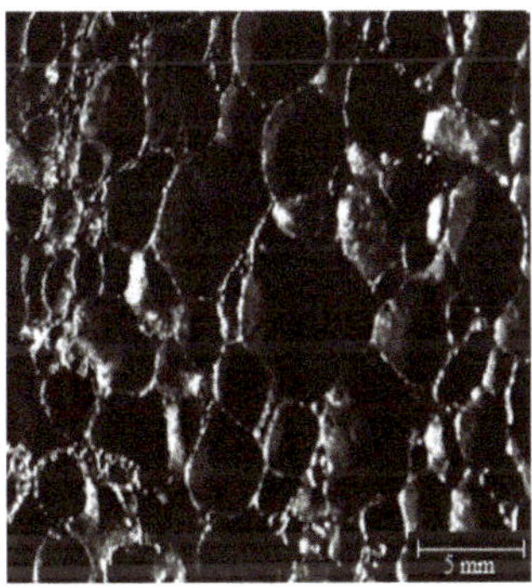

Figure 5. Stereographic photomicrography of A-242 aluminium alloy froam, obtained with 0.75% TiH$_2$ at 948 K (A5).

The samples obtained have an average of 88.22% porosity, average pore size of 1.29 mm (0.50 mm standard deviation) and an average wall thickness of 114.67 μm (32.61 μm standard deviation). Table 5 presents the results obtained from the linear expansion, relative density, porosity, pore diameter, and wall thickness values for A-242 aluminium alloy foams obtained in the experiments indicated. We analyzed the effect of foaming temperature and TiH_2 content on the properties of A-242 aluminium alloy foams.

Table 5. Linear expansion, relative density, porosity, pore diameter and wall thickness of the indicated A-242 aluminium alloy foams (σ = standard deviation).

Sample	Linear Expansion (%)	Relative Density	Porosity (%)	Pore Diameter (μm)	σ Pore Diameter (μm)	Wall Thickness (μm)	σ Wall Thickness (μm)
A1	79.50	0.1293	87.07	633.87	279.74	148.11	66.84
A2	75.01	0.1512	84.88	864.40	362.54	90.07	56.54
A3	79.21	0.1451	85.49	1220.96	742.55	124.74	65.61
A4	77.20	0.1006	89.94	1753.93	848.31	87.56	29.32
A5	71.56	0.1327	86.73	1071.93	486.78	89.44	45.04
A6	77.63	0.0883	91.17	2149.79	979.64	181.69	75.13
A7	77.35	0.1410	85.90	823.51	268.20	97.48	38.31
A8	80.21	0.1118	88.82	1388.27	867.70	92.55	55.82
A9	76.98	0.0604	93.96	1710.11	699.65	120.43	54.8

As can observed from the values reported in Table 5, the foaming temperature does not greatly affect the linear expansion of the aluminium alloy foams obtained or the percentage of TiH_2. Duarte et al. [40] report the effects of foaming temperature for 6061 alloy foams where they found that, when the foaming temperature is close to the solidus temperature, only a slight expansion occurs. If the foaming temperature is in the solid-liquid range, a greater expansion of the foam can be observed. However, increasing the temperature above the solid-liquid region reduces the viscosity of the alloy and promotes the production of more gas (H_2) so that a greater expansion of the metal can be observed. Therefore, it is evident that the foaming process is sensitive to the foaming temperature chosen as well as the TiH_2 content.

The relative density of A-242 aluminium alloy foams is highly sensitive to foaming temperature and TiH_2 content. An increase in some of these parameters causes the relative density of the foam to decrease. The porosity of the foam is related to the relative density; therefore, this property is affected by both the foaming temperature and the content of TiH_2.

The pore diameter and wall thickness of the pores is highly affected by the foaming temperature and the content of the foaming agent. An increase in foaming temperature produces a thinning of the pore walls effect of the coalescence and drainage phenomena of the foam (drained is a flow of molten metal from the walls into the pores edges (driven by surface tension) and through pore edges downwards driven by gravity). The same is true when a high content of H_2 is released, giving rise to the phenomenon of coalescence (coalescence occurs whenever two pores merge to form a larger one) [40].

3.2. Compression Behavior

During the compression behavior of the foams, three characteristic zones must be evaluated: quasi-elastic, plateau, and densification [17,39].

The first area represents the quasi-elastic deformation behavior attained at smaller values of compression. A more complete analysis revealed that the deformation is partly reversible and a certain process of irreversible deformation of the foam structure occurs during the first load (depending on the density gradients, the structural composition, and microstructure of the foam). The second zone is that of constant stress or plateau, resulting in the abrupt and repeated failure of successive layers of pores (that in porous materials, the deformation takes place in low resistance regions, this being the one that presents the thinnest pore wall and the first contact zone during the compression test). If the foam is not perfect, then the stress in this zone shows ups and downs due to defects in the structure of the foam such as pore size distribution, low-density regions, very long pores, and walls of fractured pores. The third zone is that of densification; it begins when there are no longer enough walls of intact pores to withstand the load. The stress therefore increased quickly because the walls of the pores collided with each other, occupying the space left by the pores and causing the foam to densify, increasing its mechanical resistance [15,41]. Figure 6 shows the stress-strain curves of the A-242 aluminium alloy foams prepared to different processing parameters.

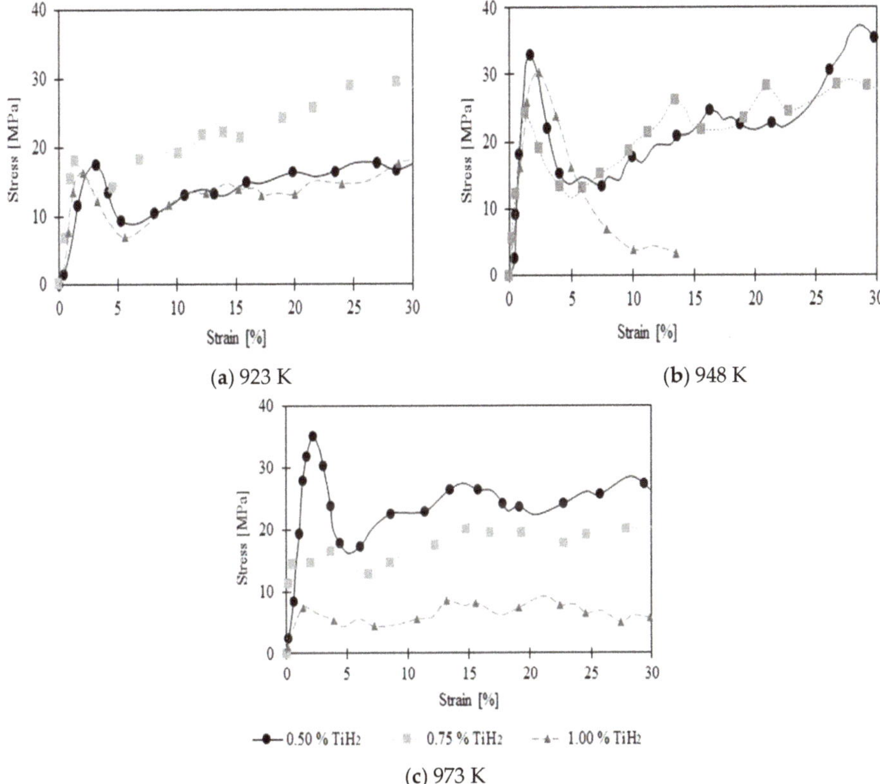

Figure 6. Effect of the percentage of TiH_2 and temperature on the mechanical behavior of A-242 aluminium alloy foams, (**a**) 923 K, (**b**) 948 K, (**c**) 973 K.

In general, an increase in the content of the foaming agent added in the foaming process affects the internal structure of the aluminium alloy foams, i.e., by increasing pore size and decreasing resistance to deformation. The content of H_2 retained in the aluminium alloy, which results from the thermal decomposition of the TiH_2, causes the presence of large pores (greater than 2 mm) and low relative density values (less than 0.1). Therefore, the presence of intermetallic compounds such as Al_3Fe and Al_2CuMg causes brittleness during the compression test.

The samples foamed at 923 K presented very similar mechanical behaviors, with the sample foamed with 0.75 wt.% of TiH_2 being the one that presented greater densification. The samples foamed at 948 K presented higher resistance compared with the samples obtained at 923 K. However, the foam with 1.00 wt.% TiH_2 did not densify; this effect is due to its structure presenting large pores and low density. The aluminium alloy foams manufactured at 973 K presented a lower mechanical resistance, mainly for contents of 0.75 wt.% and 1.00 wt.% TiH_2, the latter of which in general developed the lowest value of density. The exception to this was the aluminium alloy foam to which was added 0.50% TiH_2, since this foam is the one that presented the highest value of resistance to deformation.

The most important characteristic that affects the mechanical properties of the foams is the relative density (relation between the density the foam and that of the solid). The metal foams have relative density values of less than 0.3 [14]. An increase in the density value is related to an increase in resistance.

The literature [14] establishes that the pore size of most metal foams lies in the range from 2 to 10 mm. Although the mechanical properties of the foams are sensitive to the wall thickness ratio (metal layer that separates one pore from another), most do not depend on the absolute pore size. The shape of the pores of the metal foams varies from equiaxial to ellipsoidal, and this has an important effect on mechanical behavior. In addition, the curvature of the walls and the chemical composition of the same affects the properties of the foam. Additives and foaming agents used to manufacture metal foams often result in unconventional alloys. An example of this is the use of Ca as a modifier of viscosity of the molten metal and TiH_2 as a foaming agent, which introduces precipitates of Al, Ca, and Ti into the microstructure, weakening the walls of the pores [14]. The presence of calcium in the aluminium alloys forms intermetallic Al_4Ca which presents a polyhedral morphology, being a precursor for the fracture and collapse of the foam during the compression test. In this case, the resistance of the aluminium alloy foam is expected to decrease in comparison with materials that do not present the formation of such intermetallic compounds.

Prieto et al. [42] presented a comparison of the stress-strain curves for aluminium foams obtained from the recycling of cans for beverages and pure aluminium, where the alloying elements such as Fe, Mn, and Mg form intermetallic compounds in the pore wall. The foam obtained from recycling presents a higher resistance than does the pure aluminium foam, an effect which is attributed to the presence of alloying elements such as Ca, Ti, Fe, and Mg which form intermetallic compounds such as Al_4Ca, Al_3Ti, and $Al_6(Fe, Mn)$.

Figure 7 shows a comparison of the stress-strain curves of foams obtained with pure aluminium, recycled aluminium cans, and the A-242 aluminium alloy obtained from the recycling of cans. The data of pure aluminium and of recycled aluminium were obtained from the work of Prieto et al. [42]. The pure aluminium and secondary aluminium foams have a relative density of 0.46, porosity of 82%, and pore diameter of 3 mm.

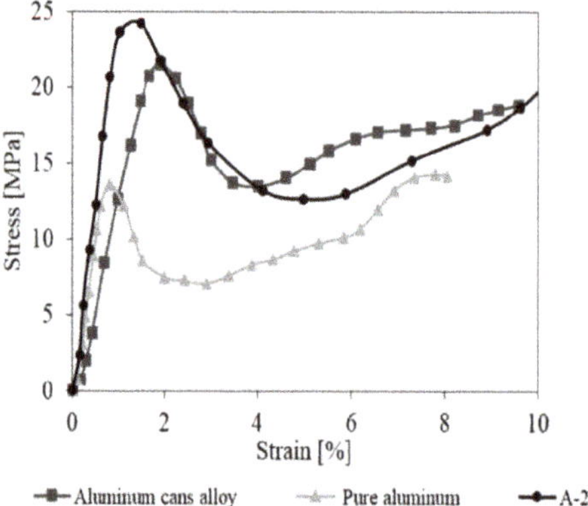

Figure 7. Stress–strain curves of a pure aluminium foam, an aluminium foam of an alloy prepared from beverage cans, and foams prepared with the A-242 aluminium alloy prepared in this work.

As can be observed in this figure, the foam of A-242 aluminium alloy prepared with 0.75 wt.% TiH_2 at 948 K presents the highest mechanical strength (24.13 MPa) and short plateau zone, as compared to foams prepared with pure aluminium (13 MPa) and secondary aluminium prepared from beverage cans (21 MPa). For the preparation of A-242 aluminium alloy foams, a lesser Ca content (0.50 wt.%) is required to stabilize the aluminium alloy foam, as well as a lesser content of TiH_2 to get low relative density values (0.13), a porosity of 82%, and a pore size of 1.5 mm. This could be an advantage for developing this kind of foam at an industrial level. In addition, due to the high Cu content in the A-242 aluminium alloy, the A-242 aluminium alloy foams can be thermally treated in order to improve their mechanical properties. The addition of TiH_2 during foaming seems to also affect grain refinement during solidification; the $TiAl_3$ particles formed improve the mechanical resistance of the aluminium alloy [43].

On the other hand, Ca particles seem to greatly affect the collapsing behavior of the foam during compression testing, as Ca forms new intermetallic compounds with Cu, weakening the foam wall and causing the fracture of the same before the application of the compression load [44]. Due to this fact, in this work we propose the use of a lower content of Ca to stabilize the aluminium alloy foam and avoid the formation of Cu-rich intermetallic compounds with calcium.

3.3. Microstructural Analysis

Figure 8 shows micrographs obtained via SEM where the pore walls of aluminium alloy foams with different contents of TiH_2 are evident. The aluminium alloy foams depicted were obtained as follows: a) 0.50 wt.% TiH_2, b) 0.75 wt.% TiH_2, c) 1.00 wt.% TiH_2 at 948 K. The intermetallic phases were identified using EDS in the SEM. The microstructure of the aluminium alloy foam cell wall is composed of aluminium dendrites, in addition to intermetallic compounds of Al_2Cu (alternating lamellae of α-Al + θ-Al_2Cu). Intermetallic compounds of the Al_9FeNi and Al_3Fe in their acicular forms and polymorphs of the Al_2CuMg intermetallic compound are evident. The presence of Fe- and Mg-rich intermetallics comes from the raw material (beverage cans), in contrast to those rich in Ni and Cu, which come from the alloying elements added to adjust the chemical composition of the alloy. The presence of Ti-rich particles is a product of the decomposition reaction of the foaming agent (TiH_2) during the process of frothing, where Al_3Ti is formed by the reaction between aluminium and titanium. The cooling rate and the chemical composition of the aluminium foam are the main variables that determine the morphology, size and distribution of the different intermetallic compounds.

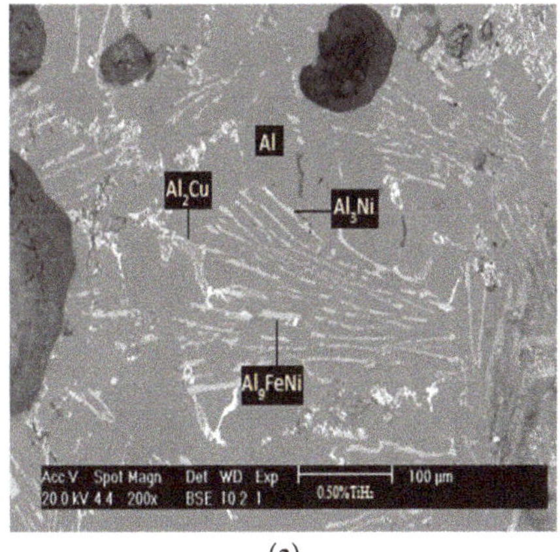

(a)

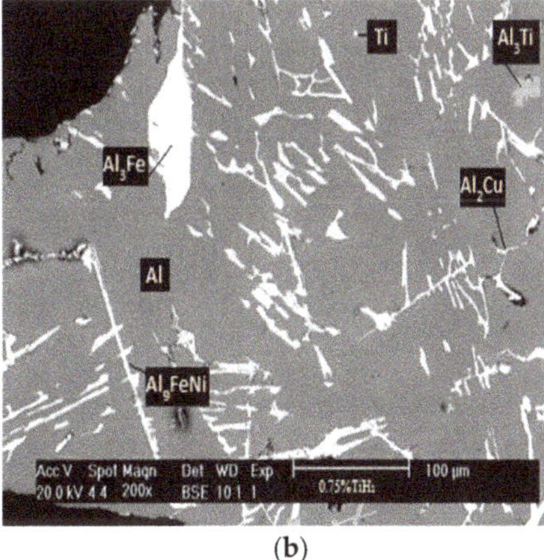

(b)

Figure 8. *Cont.*

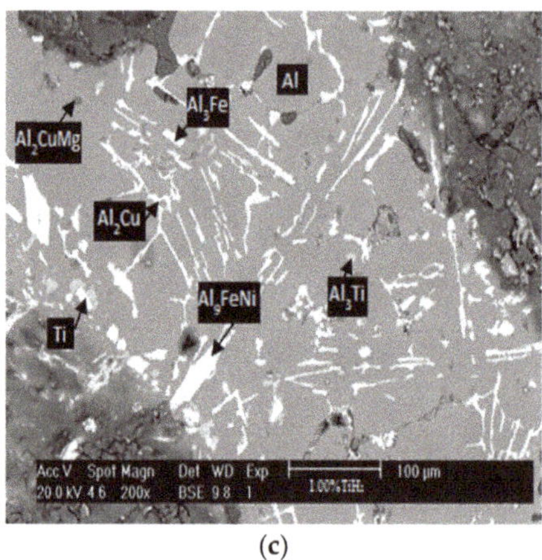

(**c**)

Figure 8. Micrographs of A-242 aluminium alloy foams: (**a**) 0.50 wt.% TiH$_2$; (**b**) 0.75 wt.% TiH$_2$; and (**c**) 1.00 wt.% TiH$_2$ at 948 K.

Figure 9 shows a micrograph obtained through SEM of a Ti-rich particle, surrounded by particles of the Al$_3$Ti intermetallic compound, in an A-242 aluminium alloy foam prepared with 1.00 wt.% TiH$_2$ at 948 K. The compound Al$_3$Ti appears when the processing temperature is low, and the Al$_3$Ti intermetallic gives the alloy greater mechanical strength [45]. The appearance of this intermetallic compound based on the equilibrium diagram is due to the preparation conditions used during the process (reaction time and foaming temperature). Figure 10 shows EDS patterns of the matrix, Ti-rich particle and Al$_3$Ti intermetallic.

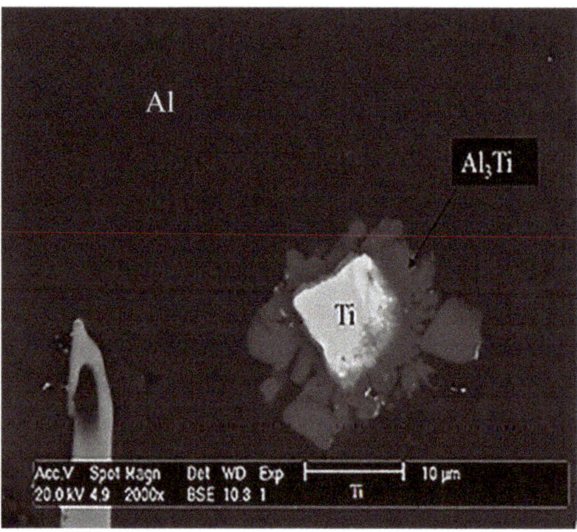

Figure 9. SEM micrograph of a Ti-rich particle, presumably the Al$_3$Ti intermetallic compound.

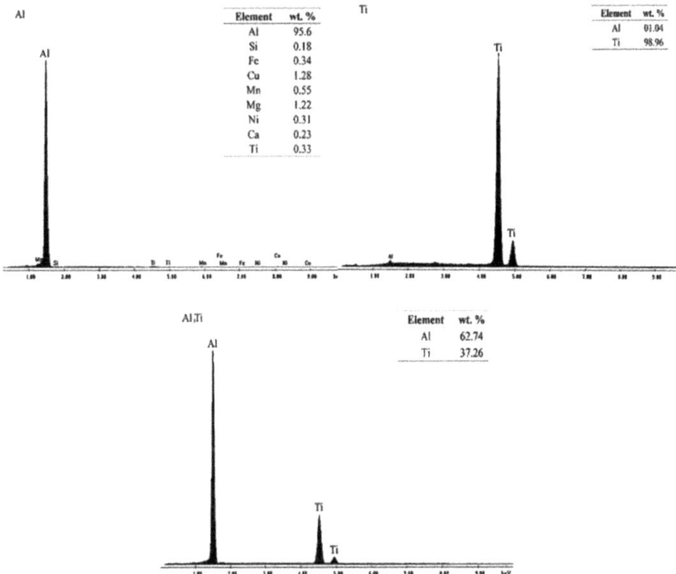

Figure 10. Energy dispersion spectroscopy (EDS) patterns for the Ti-rich particle.

4. Conclusions

From the results obtained in this work, it is important to emphasize the possibility of using secondary aluminium for the preparation of aluminium alloy foams with characteristics similar to those obtained by the use of commercial-purity elements.

Aluminium alloy foams were prepared with closed porosity in the order of 88.22% and with a relative density of 0.12. The pore size was 1.29 mm, and the wall thickness was 114.67 μm.

Various intermetallic compounds were identified scattered on the pore walls, such as Al_2Cu, Al_9FeNi, Al_3Fe, Al_2CuMg, Ti, and Al_3Ti. The aluminium alloy foams manufactured with 1.00 wt.% TiH_2 showed higher concentrations of intermetallics.

With low TiH_2 contents (0.50 wt.% and 0.75 wt.%), the samples presented high mechanical strength values; when analyzing the microstructure of this group of samples, it was observed that they present lower contents of intermetallic compounds distributed in the pore wall and the thickness of thinner wall.

The temperature of 948 K turned out to be the best temperature for foaming because the resulting aluminium alloy foams presented the highest values of compressive strength.

The aluminium alloy foams prepared from the A-242 aluminium alloy presented higher values of mechanical strength and lower relative density values when compared to foams prepared using commercial-purity elements.

Author Contributions: Investigation: N.M.T.R.; Supervision: J.T.T. and A.F.V.

Funding: This research was financed by the research stimulus program of CONACYT Mexico.

Acknowledgments: The authors thank the Consejo Nacional de Ciencia y Tecnología (CONACYT) and the Centro de Investigación y de Estudios Avanzados del Instituto Politécnico Nacional (CINVESTAV) Unidad Saltillo for their support in making the realization of this project possible.

References

1. Castillo, E.; Hector, E. Caracterización y Recuperación (Reciclado) de Chatarra de Aluminio (Latas). Bachelor's Thesis, Universidad Nacional Autónoma de México, Mexico City, Mexico, 1995.
2. Stotz, P.; Niero, M.; Bey, N.; Paraskevas, D. Environmental screening of novel technologies to increase material circularity: A case study on aluminium cans. *Resour. Conserv. Recycl.* **2017**, *127*, 96–106. [CrossRef]
3. Bauxite Index. Available online: https://thebauxiteindex.com/en/cbix/industry-101/alumina-101/production-trade/outlook#alumina_101 (accessed on 10 July 2018).
4. European Aluminium. Available online: https://www.european-aluminium.eu/about-aluminium/aluminium-in-use/ (accessed on 16 July 2018).
5. Paraskevas, D.; Kellens, K.; Dewulf, W.; Duflou, J.R. Environmetal modelling of aluminium recycling: A Life Cycle Assessment tool for sustainable metal management. *J. Clean. Prod.* **2015**, *105*, 357–370. [CrossRef]
6. United Nations Environment Programme. Available online: https://wedocs.unep.org/bitstream/handle/20.500.11822/8702/Recycling%20rates%20of%20metals:%20A20status%20report2011Recycling_Rates.pdf?amp%3BisAllowed=&sequence=3 (accessed on 24 April 2018).
7. United Nations Environment Programme. Available online: https://www.unenvironment.org/resources/report/metal-recycling-opportunities-limits-infrastructure (accessed on 24 April 2018).
8. Modaresi, R.; Müller, D.B. The role of automobiles for the future of aluminum recycling. *Environ. Sci. Technol.* **2012**, *46*, 8587–8594. [CrossRef]
9. Paraskevas, D.; Kellens, K.; Van de Voorde, A.; Dewulf, W.; Duflou, J.R. Environmetal impact analysis of primary aluminium production at country level. *Procedia CIRP* **2016**, *40*, 209–213. [CrossRef]
10. Comercio Exterior Bancomext. Available online: www.revistacomercioexterior.com/articulo.php?id=120&t=la-industria-del-aluminio-en-mexico (accessed on 16 July 2018).
11. American Society for Metals. *ASM Metals Handbook, Properties and Selection: Nonferrous Alloys and Special-Purpose Materials*; ASM International: Geauga County, OH, USA, 1990; Volume 2, pp. 583–590.
12. American Society for Metals. *ASM Metals Handbook, Casting*; ASM International: Geauga County, OH, USA, 2008; pp. 44–49.
13. Fuki, T.; Nonaka, Y.; Suzuki, S. Fabrication of Al-Cu-Mg Alloy Foams Using Mg as Thickener Through Melt Route and Reinforcement of Cell Walls by Heat Treatment. *Procedia Mater. Sci.* **2014**, *4*, 33–37. [CrossRef]
14. Gibson, L.J. Mechanical Behavior of Metallic Foams. *Ann. Rev. Mater. Sci.* **2002**, *30*, 191–227. [CrossRef]
15. Banhart, J.; Baumeister, J. Deformation characteristics of metal foams. *Mater. Sci.* **1998**, *33*, 1431–1440. [CrossRef]
16. García-Moreno, F. Commercial Application of Metal Foams: Their Properties and Production. *J. Mater. Sci.* **2016**, *9*, 85. [CrossRef] [PubMed]
17. Banhart, J. Manufacture, characterization and application of cellular metals and metal foam. *Prog. Mater. Sci.* **2001**, *46*, 559–632. [CrossRef]
18. Davies, G.J.; Zhen, S. Metallic foams: Their production, properties and applications. *J. Mater. Sci.* **1983**, *18*, 1899–1911. [CrossRef]
19. Nakajima, H. Fabrication, properties and application of porous metals with directional pores. *Prog. Mater. Sci.* **2007**, *52*, 1091–1173. [CrossRef]
20. Kranzlin, N.; Niederberger, M. Controlled fabrication of porous metals from the nanometer to the macroscopic scale. *Mater. Horiz.* **2015**, *2*, 359–377. [CrossRef]
21. Qin, J.; Chen, Q.; Yang, C.; Huang, Y. Research process on property andapplication of metal porous materials. *J. Alloys Compd.* **2016**, *654*, 39–44. [CrossRef]
22. Lefebvre, L.P.; Banhart, J.; Dunand, D.C. Porous metals and metallic foams: Current status and recent developments. *Adv. Eng. Mater.* **2008**, *10*, 775–787. [CrossRef]
23. Ashby, M.F.; Evans, A.G.; Fleck, N.A.; Gibson, L.J.; Hutchinson, J.W.; Wadley, H.N.G. *Metal Foams: A Design Guide*; Butterworth-Heinemann: Boston, MA, USA, 2000.
24. Baumeister, J.; Weise, J. Metallic foams. In *Ullmann's Encyclopedia of Industrial Chemistry*; Wiley-VCH Verlag GmbH & Co. KGaA: Weinheim, Germany, 2000.
25. Weaire, D.; Cox, S.J.; Banhart, J. Methods and models of metallic foam fabrication. In Proceedings of the 8th Annual International Conference on Composites Engineering, Canary Islands, Spain, 5–11 August 2001; pp. 977–978.

26. Gibson, L.; Ashby, M. *Cellular Solids: Structure and Properties*; Cambridge University Press: Cambridge, UK, 1997.
27. Akiyama, S.; Ueno, H.; Imagawa, K.; Kitahara, A.; Nagata, S.; Morimoto, K.; Nishikawa, T.; Itoh, M. Foamed Metal and Method of Producing Same. U.S. Patent 4,713,277, 15 December 1987.
28. Jin, I.; Kenny, L.D.; Sang, H. Method of Producing Lightweight Foamed Metal. U.S. Patent 4,973,358, 27 November 1990.
29. Allen, B.C.; Mote, M.W.; Sabroff, A.M. Method of Making Foamed Metal. U.S. Patent 3,087,807, 30 April 1963.
30. Asavavisithchai, S.; Kennedy, A.R. Effect of powder oxide content on the expansion and stability of PM-route Al foams. *J. Colloid Interface Sci.* **2006**, *297*, 715–723. [CrossRef]
31. Weaire, D.; Hutzler, S. *The Physics of Foams*; Oxford University Press: Oxford, UK, 1999.
32. Hall, I.W.; Guden, M.; Yu, C.-J. Crushing of aluminum closed cell foams: Density and strain rate effects. *Scr. Mater.* **2000**, *43*, 515–521. [CrossRef]
33. Olurin, O.B.; Fleck, N.A.; Ashby, M.F. Deformation and fracture of aluminium foams. *Mater. Sci. Eng. A* **2000**, *291*, 136–146. [CrossRef]
34. Kennedy, A.R. Aspects of the reproducibility of mechanical properties in Al based foams. *J. Mater. Sci.* **2004**, *39*, 3085–3088. [CrossRef]
35. Ramamurty, U.; Paul, A. Variability in mechanical properties of a metal foam. *Acta Mater.* **2004**, *52*, 869–876. [CrossRef]
36. Beals, J.T.; Thomson, M.S. Density gradient effects on aluminium foam compression behaviour. *J. Mater. Sci.* **1997**, *32*, 3595–3600. [CrossRef]
37. Antônio, A.; Oliveira, J.R.; Marques, R. Thixoforging of Al-3.8% Si alloy recycled from aluminum cans. *Mater. Sci. Eng. A* **2014**, *A607*, 219–225. [CrossRef]
38. Soltani, R.; Sarajan, Z.; Soltani, M. Foaming of pure aluminium by TiH_2. *Mater. Res. Innov.* **2014**, *18*, 401–406. [CrossRef]
39. Geramipour, T.; Oveisi, H. Effects of foaming parameters on microstructure and compressive properties of aluminum foams produced by powder metallurgy method. *Trans. Nonferrous Met. Soc. China* **2014**, *27*, 1569–1579. [CrossRef]
40. Duarte, I.; Banhart, J. A study of aluminium foam formation kinetics and microstructure. *Acta Mater.* **2000**, *48*, 2349–2362. [CrossRef]
41. Baumeister, J.; Banhart, J.; Weber, M. Aluminium Foam for Transport Industry. *Mater. Des.* **1997**, *18*, 217–220. [CrossRef]
42. Prieto, V.; Torres, J.; Flores, A. Recycling of aluminum beverage cans for metallic foams manufacturing. *J. Porous Mat.* **2017**, *24*, 707–712. [CrossRef]
43. Li, P.; Kandalova, E.G.; Nikitin, V.I. Grain refining performance of Al-Ti master alloys with different microstructures. *Mater. Lett.* **2005**, *59*, 723–727. [CrossRef]
44. Huang, L.; Wang, H.; Yang, D.H.; Ye, F.; Wang, S.Q.; Lu, Z.P. Effects of calcium on mechanical properties of celular Al-Cu foams. *Mater. Sci. Eng. A* **2014**, *A618*, 471–478. [CrossRef]
45. Mishin, Y.; Herzing, C. Diffusion in the Ti-Al system. *Acta Mater.* **2000**, *48*, 589–623. [CrossRef]

Article

Influence of MWCNT Coated Nickel on the Foaming Behavior of MWCNT Coated Nickel Reinforced AlMg4Si8 Foam by Powder Metallurgy Process

Ferdinandus Sarjanadi Damanik * and Günther Lange *

Group for Metallic Materials and Composite Materials, Technische Universität Ilmenau, Gustav-Kirchhoff-Str. 6, 98693 Ilmenau, Germany
* Correspondence: ferdinandus-sarjanadi.damanik@tu-ilmenau.de (F.S.D.);
  guenther.lange@tu-ilmenau.de (G.L.); Tel.: +49-(0)3677-69-1980 (F.S.D.); +49-(0)3677-69-2881 (G.L.)

Received: 30 May 2020; Accepted: 10 July 2020; Published: 15 July 2020

Abstract: This research studies the effect of multi-wall carbon nanotube (MWCNT) coated nickel to foaming time on the foam expansion and the distribution of pore sizes MWCNT reinforced AlMg4Si8 foam composite by powder metallurgy process. To control interface reactivity and wettability between MWCNT and the metal matrix, nickel coating is carried out on the MWCNT surface. Significantly, different foaming behavior of the MWCNT coated nickel reinforced AlMg4Si8 was studied with a foaming time variation of 8 and 9 min. Digital images generated by the imaging system are used with the MATLAB R2017a algorithm to determine the porosity of the surface and the pore area of aluminum foam efficiently. The results can have important implications for processing MWCNT coated nickel reinforced aluminum alloy composites.

Keywords: aluminum matrix foam composite (AMFC); MWCNT; chemical oxidation; electroless deposition nickel; powder metallurgy; expansion

1. Introduction

The first production of metallic foam was carried out by Benjamin Sosnick in 1948 [1], extensive research and application of metallic foam in many manufacturing sectors such as the automotive and aerospace industries have been carried out. Metal foams have properties such as high strength to weight ratio, high energy absorption capacity, large specific surface, high gas and liquid permeability, and low thermal conductivity [2]. Basic characteristics such as the relative density, cell structure, homogenity of cell morphology and optimum mean average equivalent diameter of pores are things that affect the use of metal foams [3]. Some parameters that affect the quality of the final product foam are: particles (composition, shape, size and volume fraction), gas (composition and purity), particle-surface interaction, matrix alloy composition, foaming temperature and thermal processing conditions (holding time and cooling medium) [4].

In aluminum foam manufacturing processes, commonly known are direct gas foaming, powder metallurgy and casting [5]. Some researchers have investigated adding ceramic particles to aluminum foam, but the main problem for ceramic reinforced aluminum foam is agglomeration and poor bonding between the reinforcement and the matrix, which sometimes results in brittleness deformation and decreased energy absorption efficiency of foam composites [6]. Therefore, it is necessary to develop the type of the new favorable reinforcement and surface treatment of the reinforcement, specifically for the nano-sized reinforcement and to exploit the properties of composite foam. Research of nano and microparticles reinforced closed-cell aluminum foam composites by the powder metallurgy method has been investigated. Compressive strength and the absorption of plastic deformation energy increase about 28 times with the addition of nano-SiC particles when compared to micro-SiC reinforcing aluminum foam [7].

The strengthening effects of Cr_2O_3 and Al_3Ni thermally grown on an aluminum alloy were investigated, the strengths and stiffness of aluminum were enhanced. Al_3Ni layers are also formed on graphite coated nickel reinforced aluminum. The primary Al_3Ni intermetallic formed in the aluminum and nickel reactions is the angular Al_3Ni phase. Layer morphology is an important role in the interface and mechanical properties of composites. The Al_3Ni phase plays a role in binding carbon in the aluminum matrix. These results indicate that the dispersion of graphite in the aluminum matrix is facilitated by a layer of nickel in graphite [8,9]. MWCNT has the potential to be used as a reinforcement in composite materials because MWCNT has superior mechanical properties with tensile strengths of up to 150 GPa and elastic modulus up to 1 TPa, as well as good thermal stability and electrical conductivity [10]. The effect of MWCNT volume fraction (0.1–1.0 wt.%) has been investigated on the compressive behavior and energy absorption of nanocomposite foams. Compressive yield strength increased with the addition of up to 0.5 wt.% MWCNT, but the opposite trend occurred with the addition of 1 wt.% MWCNT [11]. To improve dispersion and wettability and overcome agglomeration and poor bonding problems between MWCNT and aluminum matrix, it can be done with MWCNT surface treatment (chemical oxidation and electroless deposition nickel coating). Chemical oxidation of MWCNT involves impregnation of functional groups (functional groups -COOH) on the surface of nanotubes because the reactivity of oxygen-containing groups is greater than pure carbon [12]. MWCNT oxidation (MWCNT-COOH) can trap much more Sn and Pd following nickel nucleation sites and will help to incorporate MWCNT into a metal matrix with strong bonding. Ni ions easily accept electrons and electrodeposit selectively on the defect sites rather than on other normal sites on the outer surface of MWCNT [13]. The effect of surface treatment on MWCNT (chemical oxidation followed by nickel coating on MWCNT) reinforced AlMg4Si8 by powder metallurgy has been studied, showing uniformly distribution and good bonding on MWCNT reinforced aluminum foam [14].

The goal of this the present research is to investigate the effect of MWCNT coated Nickel to foaming time on the foam expansion and the distribution of pore sizes of the MWCNT reinforced AlMg4Si8 foam composite (AMFC-aluminum matrix foam composite) by the powder metallurgy method. Determination of surface porosity, pore area and foam shape of the aluminum foam was obtained through an imaging system with the MATLAB R2017a algorithm.

2. Materials and Methods

2.1. Materials

Aluminum alloy AlMg4Si8 was used for this study and the average particle size and purity of the metal, MWCNT and foaming agent powders can be seen in Table 1. Figure 1 show field emission scanning electron microscopes (FE-SEM) image morphology of aluminum, magnesium, silicon, TiH_2 and MWCNT in this study.

Table 1. Characteristics of powders used as starting materials.

Powder	Supplier	Purity	Size
Al	TLS Technik (Bitterfeld, Germany)	99.7	<150 μm
Mg	Laborladen.de (Hüfingen, Germany)	99.8	<40 μm
Si	-	>99.95	<40 μm
TiH_2	Alfa Aesar	99	<45 μm
MWCNT	Sigma Aldrich	>90	5–9 μm

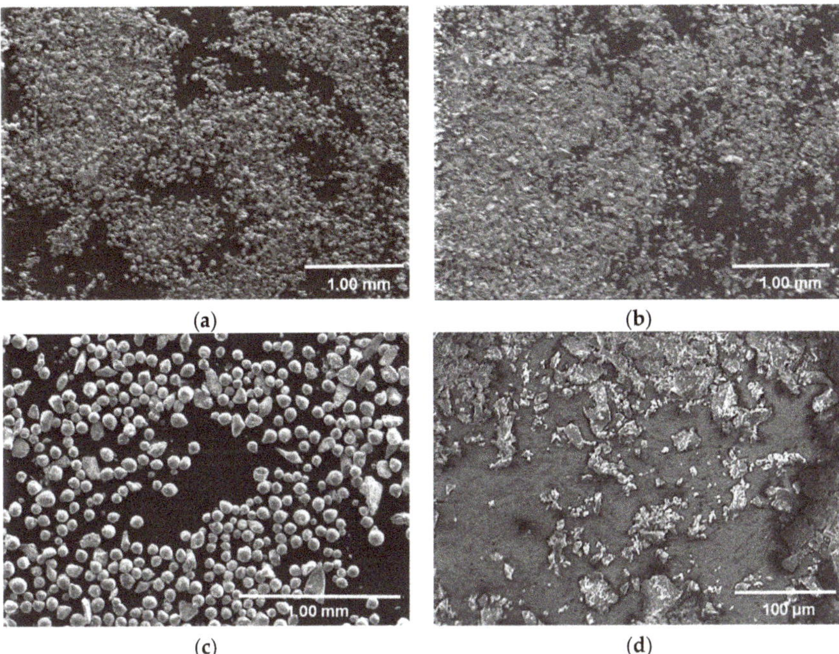

(a)   (b)   (c)   (d)

Figure 1. Field emission scanning electron microscopes micrographs of (**a**) Spheroidal powder particles Al; (**b**) Angular powder particles Mg; (**c**) Spheroidal powder particles Si; (**d**) Angular powder particles TiH$_2$ (TU Ilmenau; FG MWV).

2.2. Method

MWCNT were oxidized in piranha reagent (3 H$_2$SO$_4$: 1 hydrogen peroxide) for 6 h. After the treatments, all the samples were thoroughly washed with demineralized water [15]. After this treatment, the procedure of electroless plating nickel can be divided into three steps: sensitization, activation and plating. MWCNT-oxidized were immersed in a sensitization solution and sonicated for 30 min. The Sn^{2+}-sensitized MWCNT were further stirred in an activating solution for 30 min. The activated MWCNT were washed with deionized water and introduced into an electroless plating bath. After 10 min of plating, the plated MWCNT were washed with distilled water then with methanol [16].

Aluminum closed-cell foams were prepared using by powder metallurgy process. Before mixing, MWCNT coated nickel were sonicated with ethanol in a bath sonicator for 30 min. The process started with the mixing of appropriate amounts of basic ingredients, AlMg4Si8 powder, MWCNT (0.5 wt.%) and TiH$_2$ (0.5 wt.%) as blowing agents, inside a v-shaped cylinder for 20 min. AlMg4Si8, 0.5 wt.% TiH$_2$ and 0.5 wt.% MWCNT were prepared inside a stainless steel mold, then axially compacted (cold pressing) with the pressure of 450 MPa for 3 min. It was heated in a preheated furnace at 300 °C for 2 h. The precursor (volume: 19.2325 cm^3) was foamed at 750 °C for 8 min. FE-SEM techniques were employed to characterize the samples to examine the dispersion of the MWCNT within the AlMg4Si8 foam matrix [17]. The sample was investigated using inexpensive digital systems/devices camera. In this study, the following steps show how the images were acquired, processed and calculated the area of each pore on the camera digital and measured the pores size, shape and size distribution using MATLAB R2017a (with appropriate MATLAB commands added). Flowchart (Figure 2) outlines the algorithm of the image processing and analyses executed in the MATLAB program for images obtained with each imaging system [18].

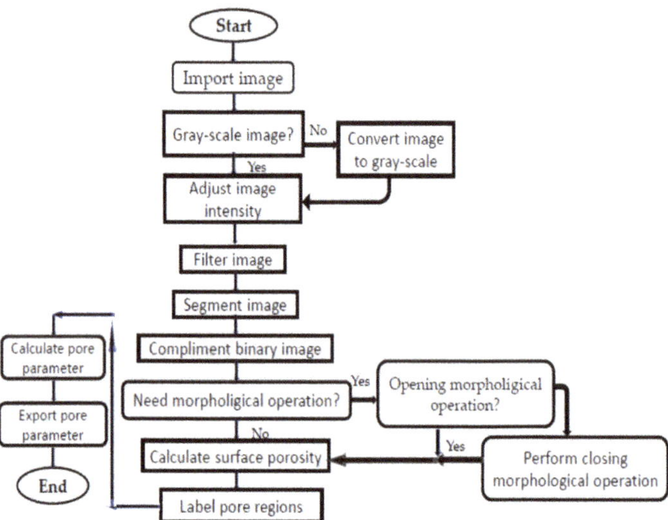

Figure 2. Algorithm of image processing and analyses executed in MATLAB R2017a Reproduced from [18], with permission from Polymers, 2019.

3. Result

3.1. The Effect of Multi-Wall Carbon Nanotube Coated Nickel to Foaming Time on Morphology of AlMg4Si8 Foam

Figure 3 shows the images of produced samples using different reinforcement and foaming time, macroscopic images of aluminum foam were taken using a camera. The results demonstrate that the effect of surface treatment of the MWCNT influence the expansion and microstructure of AlMg4Si8 foam nanocomposite and the cell number density decrease with the addition of reinforcement MWCNT coated nickel in AlMg4Si8. The foam expansion of AlMg4Si8 foam without MWCNT has the highest expansion and shows relatively uniform cells compared to MWCNT coated nickel reinforced AlMg4Si8. The results show the effect of MWCNT coated nickel to foaming time on the foamability of MWCNT reinforced aluminum foam. MWCNT reinforced aluminum foam for 8 min show a smaller size of pore compared to MWCNT reinforced aluminum foam for 9 min and the volumetric expansion coefficient of MWCNT reinforced aluminum foam for 8 min (c) is 0.00415 1/°C, lower than the MWCNT reinforced aluminum foam for 9 min (d), which is 0.00623 1/°C.

Figure 3. Macrostructure of aluminum foam expansion affected by reinforcement and foaming time of (**a**) Precursor material Multi-wall Carbon Nanotube reinforced AlMg4Si8; (**b**) AlMg4Si8 foam; (**c**) MWCNT coated nickel reinforced AlMg4Si8 foam for 8 min; (**d**) MWCNT coated nickel reinforced AlMg4Si8 foam for 9 min. (TU Ilmenau; FG MWV).

Figure 4 shows that the distribution of MWCNT coated nickel in aluminum foam are uniform on the cell wall. FE-SEM analysis results show that all reinforcement (MWCNT) are evenly distributed in the aluminum foam matrix (the top, middle and bottom layers). The presence of MWCNT coated nickel is in the cell wall which means good the wettability at the MWCNT coated nickel reinforced aluminum foam interface.

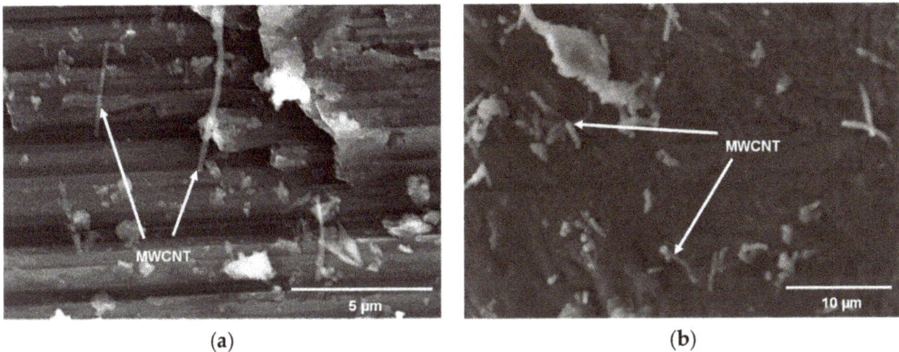

(a)   (b)

Figure 4. (**a**) Field emission scanning electron microscopes image of dispersion MWCNT on the plateau border; (**b**) Show microstructural observed by field emission scanning electron microscopes about the uniform dispersion of MWCNT in the pore cell-wall. (TU Ilmenau; FG MWV).

Table 2 shows foam density, % porosity, expansion and pore number of foam. The porosity of foam material were calculated using the following equation:

$$\% \text{ Porosity } = 1 - \frac{\rho \text{ foam}}{\rho \text{ precursor}} \times 100\% \tag{1}$$

The density of precursors and foam were measured by weighing specimens and geometry, that all the calculations are in comparison with the bulk aluminum, where $\rho1$ and $\rho2$ are densities of foam and precursor. The volumetric expansion coefficient was measured by volume foam compare with the precursor, where α = volumetric thermal expansion coefficient, V0 = the initial volume, ΔV = volume change and ΔT = temperature change. Furthermore, the relative density of precursors and foams and the volumetric expansion coefficient were calculated using the following equations:

$$\rho R = \frac{\rho1}{\rho2} \tag{2}$$

$$\alpha = \frac{\Delta V}{V0.\Delta T} \tag{3}$$

Table 2. Structure of the produced aluminum foam.

Foam Sample	Time (min)	Density (g/cm^3)	Porosity (%)	The Volumetric Expansion Coefficient (1/$^\circ$C)	Pore Number
AlMg4Si8 (a)	8	0.5573	72.32	0.00693	2605
MWCNT(Ni)-AlMg4Si8 (b)	8	0.8368	67.81	0.00415	726
MWCNT(Ni)-AlMg4Si8 (c)	9	0.5581	78.55	0.00623	484

3.2. The Pore Size Distribution of Multi-Wall Carbon Nanotube Coated Nickel Reinforced AlMg4Si8

To compare the porosity of the surface and the pore area of aluminum foam are used the MATLAB algorithm from digital images produced by the imaging system shown in Figure 5. The imaging system shows a plot diagram of the frequency distribution bar of the pore area, the number of large pores relative to the small pores in the image by each imaging system. A summary of the number of pores detected in all analyzed samples affected by the reinforcement and foaming time is shown in Table 2. The structure of AlMg4Si8 without reinforcement show relatively uniform cells than the MWCNT coated nickel reinforced AlMg4Si8 and imaging systems show an inhomogeneous distribution of large and small pores throughout the aluminum foam, and this is caused by the influence of the type of MWCNT as a reinforcement and foaming time used in the foaming process/manufacturing.

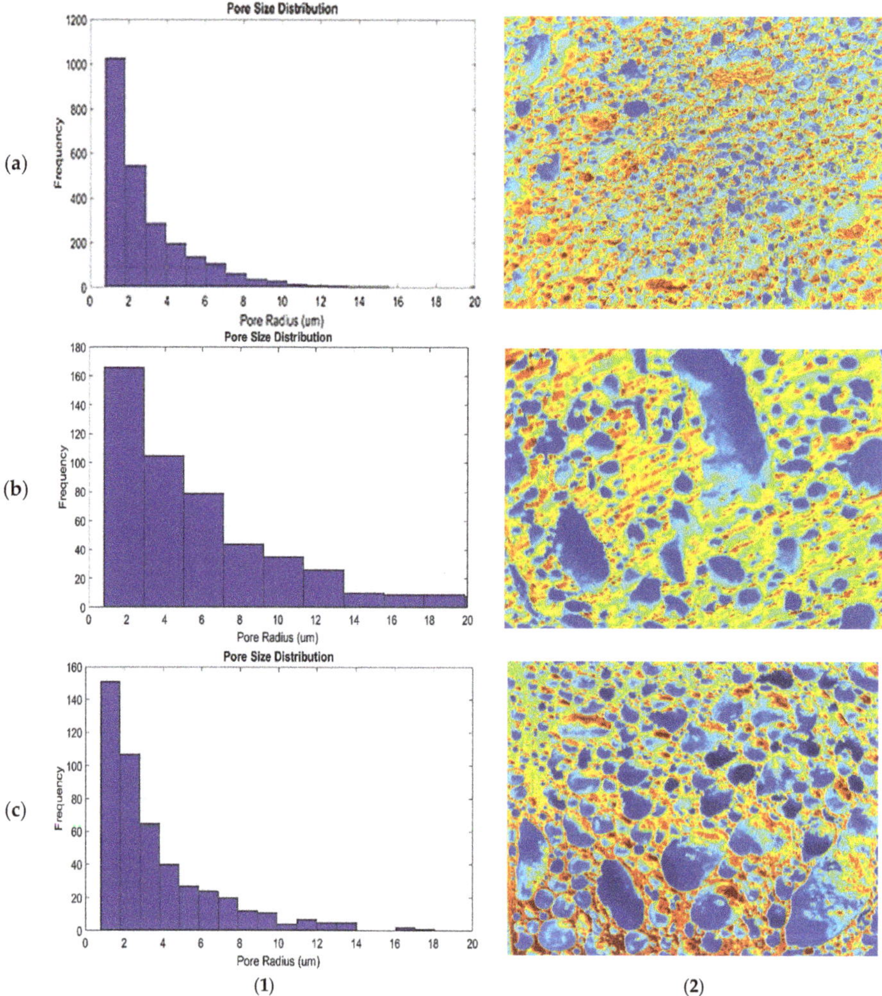

Figure 5. Frequency distribution of pore area (**1**), and Scaled labeled images (**2**) for (**a**) AlMg4Si8 foam for 8 min (**b**) MWCNT coated nickel reinforced AlMg4Si8 foam for 8 min (**c**) MWCNT coated nickel reinforced AlMg4Si8 foam for 9 min.

4. Discussion

Several studies on the effect of pore size on the mechanical properties of the aluminum foam and its alloys have found that porosity and pore size distribution have a completely linear relationship on mechanical properties. Furthermore, the fabrication method and composition used during foam making will affect the foam structure and pore morphology, which will also determine the mechanical properties [19,20].

To discuss the effect of reinforcement MWCNT coated nickel and foaming time on expansion and pore size has to take into account the following aspect. Chemical oxidation and electroless deposition nickel in MWCNT can improve the distribution and wettability of MWCNT as reinforcement in an aluminum foam nanocomposite. Nickel coating on MWCNT can increase the bond between MWCNT coated nickel and aluminum matrix. While MWCNT coated nickel were much more strongly bonded with the aluminum matrix. Figure 4 shows that the MWCNT coated nickel is in a foam cell and has better wettability with the aluminum foam matrix. This investigation shows that the MWCNT surface structure affects the interface between MWCNT and aluminum matrix in the aluminum foam matrix nanocomposite.

Digital images produced by the imaging system are used with the MATLAB algorithm to determine the surface porosity and pore area of aluminum foam. The average diameter of the pores MWCNT coated nickel reinforced AlMg4Si8 increase than AlMg4Si8 without reinforcement. The porosity, the volumetric expansion coefficient and density of all samples are listed in Table 2. According to Table 2, samples a, b and c showed the porosity of each sample (72.32, 67.81, and 78.55%, respectively). The volumetric expansion coefficient (0.00693, 0.00415 and 0.00623 1/°C, respectively) and density (0.5573, 0.8368 and 0.5581 g/cm^3, respectively). Increasing the foaming time MWCNT coated nickel reinforced AlMg4Si8 from 8 to 9 min resulted in a rapid increase of the volumetric expansion coefficient (0.00415 to 0.00623 1/°C) and consequently a rapid decrease in density (0.8368 to 0.5581 g/cm^3). Table 2 shows at the same time for 8 min, the volumetric expansion coefficient of AlMg4Si8 (without MWCNT) is 0.00693 1/°C, while the volumetric expansion coefficient of the MWCNT coated nickel reinforced AlMg4Si8 is 0.00415 1/°C. Increasing the foaming time from 8 to 9 min resulted in an increase of the volumetric expansion coefficient (0.00415 to 0.00623 1/°C), and consequently a rapid decrease of the relative density (0.8368 to 0.5581 g/cm^3).

According to data from Table 2, the comparison of the structure of the aluminum foam produced are:
Pores Number: AlMg4Si8 (8 min) > MWCNT(Ni)–AlMg4Si8(8 min) > MWCNT(Ni)–AlMg4Si8 (9 min)
% Porosity: MWCNT(Ni)–AlMg4Si8 (8 min) < AlMg4Si88 (8 min) < MWCNT(Ni)–AlMg4Si8 (9 min)
Pore size: AlMg4Si8 (8 min) < MWCNT(Ni)–AlMg4Si8(8 min) < MWCNT(Ni)–AlMg4Si8(9 min)
Density: AlMg4Si8 (8 min) < MWCNT(Ni)–AlMg4Si8(9 min) < MWCNT(Ni)–AlMg4Si8(8 min)
The volumetric expansion coefficient: AlMg4Si8 (8 min) > MWCNT(Ni)–AlMg4Si8 (9 min) > MWCNT(Ni)–AlMg4Si8 (8 min)

The results show that the effect of MWCNT surface treatments as reinforcement of the foamability of AlMg4Si8. During the processing of aluminum foam composites, chemical reactions occur at the interface between the matrix and the reinforcement in several systems. Chemical reactions and the types of reaction products formed depend on the processing temperature, pressure and atmosphere, matrix composition and surface chemistry of the reinforcement. The physical and mechanical properties of aluminum foam composites are greatly influenced by the degree of the chemical reaction. The inhomogeneous distribution of large and small pores in all-aluminum foams is shown by the aluminum foam composite imaging system (Figure 5), and this is influenced by MWCNT coated nickel as the reinforcement used and the foaming time in the manufacturing process. The structure of AlMg4Si8 without reinforcement show cells that are relatively uniform compared to MWCNT coated nickel reinforced AlMg4Si8. This shows that MWCNT coated nickel is collected inside the cell wall during foaming and there has been the wettability between MWCNT and aluminum matrix. When aluminum wets the MWCNT coated nickel, the presence of MWCNT can increase the system viscosity and will increase the cell wall thinning rate and as consequence, larger cells will form in aluminum foam and reduce the number of pores.

5. Conclusion

AlMg4Si8 foam is produced by a powder metallurgical process in which titanium hydride is a blowing agent. By varying foaming time and compare with AlMg4Si8 foam (without MWCNT), different stages of early foam formation could be prepared. Influence of MWCNT coated nickel to the foaming time and distribution of pore on the foaming behavior of MWCNT coated nickel reinforced aluminum alloys and the influence of time has been investigated, that:

Uniformly dispersion of MWCNT coated nickel in the aluminum matrix and increase the interfacial bonding strength between MWCNT coated nickel and metal matrix.

Surface treatment on MWCNT will increase the cell wall thinning rate and as a consequence, larger cells will form in aluminum foam. The average diameter of the pores MWCNT coated nickel reinforced AlMg4Si8 is greater than AlMg4Si8 foam.

MWCNT coated nickel reinforced AlMg4Si8 with time for 9 min has a higher volumetric expansion coefficient (0.00623 1/°C) than the foaming time for 8 min (0.00415 1/°C). However, the volumetric expansion coefficient of AlMg4Si8 without reinforcement with a foaming time of 8 minutes still has the highest volumetric expansion coefficient value (0.00693 1/°C).

Author Contributions: Conceptualization, F.S.D. and G.L.; methodology, F.S.D. and G.L.; validation, F.S.D. and G.L.; software, F.S.D.; formal analysis, F.S.D.; investigation, F.S.D; data curation, F.S.D.; writing—original draft preparation, F.S.D.; writing—review and editing, F.S.D. and G.L.; visualization, F.S.D. and G.L; supervision, G.L. All authors have read and agreed to the published version of the manuscript.

Funding: This work is supported and funded by Indonesia Endowment Fund for Education (LPDP-Lembaga Pengelola Dana Pendidikan).

Acknowledgments: SEM analysis facilities were made accessible by The Institute of Micro- and Nanotechnologies Technology center TU Ilmenau. We thank our colleagues from The Department of Metal and Composite Materials Technical University Ilmenau, who provided powders and contributed to the discussions that greatly assisted the research.

Conflicts of Interest: The authors declare no conflict of interest. The funders had no role in the design of the study; in the collection, analyses, or interpretation of data; in the writing of the manuscript; or in the decision to publish the results.

References

1. Cambroneroa, L.E.G.; Ruiz-Romana, J.M.; Corpasb, F.A.; Ruiz Prieto, J.M. Manufacturing of Al-Mg-Si alloy foam using calcium carbonate as foaming agent. *J. Mater. Process. Technol.* **2009**, *209*, 1803–1809. [CrossRef]
2. Degischer, H.; Kriszt, B. *Handbook of Cellular Metals: Production, Processing, Applications*; Wiley-VCH Verlag GmbH & Co. KGaA: Vienna, Austria, 2002.
3. Duarte, I.; Ventura, E.; Olhero, S.; Ferreira, J.M. An effective approach to reinforced closed-cell Al-alloy foams with multiwalled carbon nanotubes. *Carbon* **2015**, *95*, 589–600. [CrossRef]
4. Hipke, T.; Lange, G. *Taschenbuch Für Aluminiumschäume*, 1st ed.; Beuth Verlag GmbH: Berlin, Germany, 2014; ISBN 978-3-410-22071-8.
5. Helwig, H.M.; Garcia-Moreno, F.; Banhart, J. A study of Mg and Cu additions on the foaming behavior of Al-Si alloys. *J. Mater. Sci.* **2011**, *46*, 5227–5236. [CrossRef]
6. Kennedy, A.R.; Asavavisitchai, S. Effects of TiB2 particle addition on the expansion, structure and mechanical properties of PM Al foams. *Scr. Mater.* **2004**, *50*, 115–119. [CrossRef]
7. Afshar, M.; Mirbagheri, M.H.; Movahedi, N. Effect of SiC particle size on the mechanical properties of closed aluminum foams. *Mater. Test.* **2017**, *59*, 571–574. [CrossRef]
8. Sharma, S.K.; Li, F.X.; Ko, G.D.; Kang, K.J. Strengthening effect of Cr2O3 thermally grown on alloy 617 foils at high temperature. *J. Nucl. Mater.* **2010**, *405*, 165–170. [CrossRef]
9. Ip, S.W.; Sridhar, R.; Toguri, J.M.; Stephenson, T.F.; Warner, A.E.M. Wettability of nickel coated graphite by aluminum. *Mater. Sci. Eng.* **1998**, *244*, 31–38. [CrossRef]

10. Koo, J.H. *Polymer Nano Composite, Processing, Characterization, and Application*; Mc Graw-Hill: New York, NY, USA, 2016.

11. Zhang, Z.; Ding, J.; Xia, X.; Sun, X.; Song, K.; Zhao, W.; Liao, B. Fabrication and characterization of closed-cell aluminum foams with different contents of multi-walled carbon nanotubes. *Mater. Des.* **2015**, *88*, 359–365. [CrossRef]

12. Cui, L.J.; Geng, H.Z.; Wang, W.Y.; Chen, L.T.; Gao, J. Functionalization of multi-wall carbon nanotubes to reduce the coefficient of the friction and improve the wear resistance of multi-wall carbon nanotube/epoxy composites. *Carbon* **2013**, *54*, 277–282. [CrossRef]

13. Shin, A.; Han, J.H. Effects of acid treatment of carbon on electroless copper plating. *Korean Inst. Surf. Eng.* **2016**, *49*, 265–273. [CrossRef]

14. Damanik, F.S.; Lange, G. Uniformly dispersion of multi-walled carbon nanotubes in $AlMg_4Si_8$ foam by powder metallurgy. In Proceedings of the 3rd Asia Pacific Conference on Research in Industrial and Systems Engineering, Depok, Indonesia, 16–17 June 2020.

15. Noharaa, L.B.; Filhob, G.P.; Noharac, E.L.; Kleinked, M.U.; Rezendee, M.C. Evaluation of MWCNT coated nickel surface treated by chemical and cold plasma processes. *Mater. Res.* **2005**, *8*, 281–286. [CrossRef]

16. Kamath, R.; Chittappa, H.C.; Kumar, N. Electroless Nickel Coating of Short MWCNT coated nickel. In Proceedings of the International Conference on Emerging Trends in Engineering (ICETE -2014) @NITTE, Warangal, Telangana, India, 15–17 May 2014; pp. 169–173.

17. Duarte, I.; Oliveira, M. Aluminum alloy foams: Production and properties. In *Powder Metallurgy*; IntechOpen: London, UK, 2012.

18. Yunus, S.; Sefa-Ntiri, B.; Anderson, B.; Kumi, F.; Mensah-Amoah, P.; Sackey, S.S. Quantitative pore characterization of polyurethane foam with cost-effective imaging tools and image analysis: A proof-of-principle study. *Polymers* **2019**, *11*, 1879. [CrossRef] [PubMed]

19. Wang, X.H.; Li, J.S.; Hu, R.U.; Kou, H.C.; Lian, Z. Mechanical properties of porous titanium with different distributions of pore size. *Trans. Nonferrous Met. Soc. Chin.* **2013**, *23*, 2317–2322. [CrossRef]

20. Bekoz, N.; Oktay, E. Mechanical properties of low alloy steel foams: Dependency on porosity and pore size. *Mater. Sci. Eng. A* **2013**, *576*, 82–90. [CrossRef]

Article

Mechanical, Thermal, and Acoustic Properties of Aluminum Foams Impregnated with Epoxy/Graphene Oxide Nanocomposites

Susana C. Pinto [1], Paula A.A.P. Marques [1], Matej Vesenjak [2], Romeu Vicente [3], Luís Godinho [4], Lovre Krstulović-Opara [5] and Isabel Duarte [1,*]

[1] Department of Mechanical Engineering, TEMA, University of Aveiro, 3810-193 Aveiro, Portugal; scpinto@ua.pt (S.C.P.); paulam@ua.pt (P.A.A.P.M.)
[2] Faculty of Mechanical Engineering, University of Maribor, 2000 Maribor, Slovenia; matej.vesenjak@um.si
[3] Department of Civil Engineering, University of Aveiro, 3810-193 Aveiro, Portugal; romvic@ua.pt
[4] Department of Civil Engineering, ISISE, University of Coimbra, 3030-788 Coimbra, Portugal; lgodinho@dec.uc.pt
[5] Faculty of Electrical Eng., Mech. Eng. and Naval Architecture, University of Split, 21000 Split, Croatia; Lovre.Krstulovic-Opara@fesb.hr
* Correspondence: isabel.duarte@ua.pt; Tel.: +350-234-370-830

Received: 9 October 2019; Accepted: 9 November 2019; Published: 12 November 2019

Abstract: Hybrid structures with epoxy embedded in open-cell aluminum foam were developed by combining open-cell aluminum foam specimens with unreinforced and reinforced epoxy resin using graphene oxide. These new hybrid structures were fabricated by infiltrating an open-cell aluminum foam specimen with pure epoxy or mixtures of epoxy and graphene oxide, completely filling the pores. The effects of graphene oxide on the mechanical, thermal, and acoustic performance of epoxy/graphene oxide-based nanocomposites are reported. Mechanical compression analysis was conducted through quasi-static uniaxial compression tests at two loading rates (0.1 mm/s and 1 mm/s). Results show that the thermal stability and the sound absorption coefficient of the hybrid structures were improved by the incorporation of the graphene oxide within the epoxy matrix. However, the incorporation of the graphene oxide into the epoxy matrix can create voids inside the epoxy resin, leading to a decrease of the compressive strength of the hybrid structures, thus no significant increase in the energy absorption capability was observed.

Keywords: open-cell aluminum foam; epoxy resin; graphene oxide; hybrid structures; mechanical; thermal and acoustic properties

---

1. Introduction

Metal foams have gained special attention in the last years due to their outstanding properties [1]. The first reference to metallic foams is a French patent published in 1925 by Meller; however, its commercialization only started three decades later in the United States of America (USA). After the stagnation period, and with the development of new technologies that allowed the decrease of cost and difficulties in the manufacturing process, there was a boost in their development that is still ongoing [2]. With these recent developments, metal foams have become commercially available in a wide variety of structures [3]. There are many interesting combinations of different properties, such as high stiffness associated with low specific weight or high compression strength combined with good energy absorption efficiency, making them suitable for automotive industry, ship building, aerospace industry, civil engineering, and medicine [1–5]. Thus, it is important to predict their mechanical behavior prior their application. While the behavior of closed-cell foams is difficult to predict, since often, non-regular structures are obtained by the existent manufacturing methods [6], open-cell foams, used mainly for functional applications, frequently present regular structures—therefore, their behavior is easier to estimate [7]. Cellular metals with regular structures have been developed, e.g., open-cell metal foams [8], advanced pore morphology (APM) foam elements [9–11], and metal hybrid hollow structures [12], to be used as fillers of thin-walled structures [13]. The low compressive strength of open-cell structures limits their application. In this context, the uprising research of the hybrid structure concept, resulting from filling open-cell structures with stronger materials that provide suitable resistance, has brought new insights in this field [8,14,15] Epoxy-based materials possess competitive advantages, such as high compressive resistance, low cost, and simple processing. However, epoxy materials are characterized by low fracture toughness, poor crack resistance, and brittle fracture under loading [16]. Besides this, epoxy—as many organic materials—presents a risk of high flammability [17]. It is reported that the addition of fillers (or a combination of different fillers) to the epoxy matrix can improve or confer important features to the final product [17]. Among the most commonly used, the following can be pointed out: Phosphorous, nitrogen-based fillers, nanoclays, carbon nanotubes, and graphene-based materials [18]. The inclusion of graphene oxide, a two-dimensional (2D) structure formed by a flat monolayer of carbon atoms arranged in a hexagonal lattice and decorated by a variety of functional groups (such as epoxides, carbonyls, hydroxyls), as well as its reduced form (with no or less oxygen functionalities and recovering the sp2 hybridization configuration), has been the subject of considerable interest and analysis [19]. It is documented that graphene and its derivatives greatly improve the fatigue characteristics and toughness of epoxy resins, as well as induce fire-resistance [20]. Several methods (e.g., ball milling, solvent-assisted, ultrasonication, or surface modification) [20] have been explored for preparation of these type of nanocomposite materials in order to ensure the good dispersion of nanofillers into the polymer matrix. The work herein follows the previous pioneering studies performed by Duarte et al. [8,21], where they evaluated the quasi-static and dynamic crush compressive performance of simple multifunctional hybrid structures based on an open-cell aluminum foam impregnated with polymers (e.g., epoxy resin) and their performance as fillers of square thin-walled tubes. Results demonstrate that the polymer–aluminum hybrid structure stabilizes the tube and prevents the unstable global bending or mixed buckling mode. Studies show that the aluminum foam–epoxy hybrid structure-filled structures are more efficient in terms of crashworthiness [21]. Herein, the graphene oxide at a low content load of 0.25 wt.% was studied as reinforcement of the epoxy matrix, used as a void filler of the open-cell foam. Additionally, its effects on the mechanical, thermal, and acoustic properties were evaluated.

2. Materials and Methods

2.1. Materials

An AlSi7Mg0.3 open-cell foam (OCF) block made of AlSi7Mg0.3 with pore sizes of 10 ppi (pores per inch) fabricated by the replication casting method was supplied by Mayser GmbH & Co. KG (Lindenberg, Germany) (formely M-pore). The OCF was fully filled by the epoxy resin (EP).

A commercial EP resin solvent-free mid-viscosity glue was supplied by company KGK Ltd. (Barilović, Croatia), and when cured at 20 °C for 25 min had a density of 1100 kg/m^3, compression strength of 90.4 MPa, bending strength of 38.9 MPa, and tensile strength of 14.8 MP. It was composed of two reagents, component A (polymer) and component B (hardener), and the mixing ratio was A:B = 2:1 (mass ratio) or 9:5 (volume ratio). The component A had a density of 1.12 kg/dm^3, while the component B had a density of 1 kg/dm^3. Graphene oxide (GO) was acquired from Graphenea Inc. (San Sebastián, Spain) in an aqueous dispersion with 4 mg/mL. The GO sheets had a particle size <10 μm and >95% of GO content was a monolayer (nanometric in thickness). The GO had a carbon content of 49–56% and oxygen content of 41–50%, which was determined by X-ray photoelectron spectroscopy (XPS). Before incorporation in the EP, diluted GO suspension was freeze-dried to be used in powder form.

2.2. Fabrication of the Specimens

The dense pure EP and EP reinforced with the graphene oxide (EP/GO) specimens were prepared by pouring the pure EP or EP/GO mixtures into an empty thin-walled tube. The aluminum foam–epoxy hybrid structure (OCF–EP) and aluminum foam–EP/GO nanocomposite hybrid structure specimens (OCF–EP/GO) were fabricated by pouring the pure EP or EP/GO mixtures into an empty thin-walled tube containing an open-cell aluminum foam specimen ($22 \times 22 \times 25$ mm^3). The inner surfaces of the thin-walled tubes were previously cleaned, polished, and greased to facilitate the specimen removal. The components A and B were mixed together well at the mass ratio of 2:1. In the case of the EP/GO specimens, the EP component A was first mixed together well with the GO with a mechanical stirrer at 1000 rpm for 1 h. After that, the mixture was submitted for 30 min to an ultrasonic bath and another 1 h to mechanical stirring. Then, component B was added and mixed with a mechanical stirrer at 250 rpm for 10 min to avoid bubble formation, followed by 20 min of degassing under vacuum. The specimens were extracted from the thin-walled tubes after the filler polymer was completely cured (at room temperature for 7 days).

Preliminary uniaxial quasi-static tests at 0.1 mm/s and Vickers hardness were carried out to assess the dispersion of the GO into the EP. For this purpose, three compositions of GO were tested: 0.1, 0.25, and 0.5 wt.%. These concentrations of GO were in the final epoxy (A + B) mixture. Figure 1 shows the fabricated specimens: OCF (Figure 1a), pure EP (Figure 1b), EP/GO (Figure 1c), OCF-EP (Figure 1d), and EP/GO-OCF (Figure 1e).

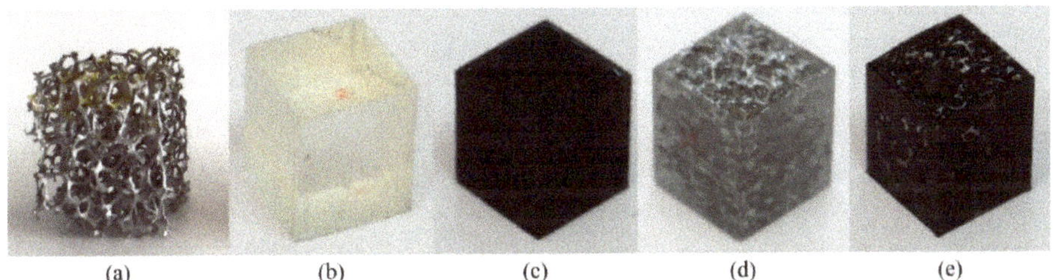

(a)   (b)   (c)   (d)   (e)

Figure 1. Fabricated specimens ($22 \times 22 \times 25$ mm^3): (**a**) Open-cell foam (OCF), (**b**) epoxy resin (EP), (**c**) EP/graphene oxide (GO), (**d**) OCF–EP and (**e**) OCF–EP/GO.

The densities of the materials were: 105.1 kg/m^3 (standard deviation: 2.1 kg/m^3), 1105 kg/m^3 (standard deviation: 29.7 kg/m^3), 1130.3 kg/m^3 (standard deviation: 6.1 kg/m^3), 1162.4 kg/m^3 (standard deviation: 14.3 kg/m^3), and 1169.3 kg/m^3 (standard deviation: 7.7 kg/m^3) for the OCF, EP, EP/GO, OCF–EP, and OCF–EP/GO, respectively.

2.3. Scanning Electron Microscopy and X-ray Micro Computed Tomography Characterisation

To investigate the dispersion of the GO into the EP matrix, scanning electron microscopy (SEM) analysis was applied. The TM 400 Plus (Hitachi, Tokyo, Japan) was used at an accelerating voltage of 15 kV and in backscattering electron mode.

The specimens ($22 \times 22 \times 25$ mm^3) were analyzed in a micro-computed tomography (μCT) equipment from SkyScan 1275 (Bruker, Belgium) with penetrative X-rays of 80 kV and 125 μA, 1 mm Al filter, in a high resolution mode with a pixel size of 18 μm and 45 ms of exposure time. NRecon (Bruker, Belgium), CTVox (Bruker, Belgium), and CTAn (Bruker, Belgium) softwares were used for slicing, three-dimensional (3D) reconstruction, and morphometric analysis.

2.4. Mechanical Characterisation

Uniaxial compression tests were used to study the quasi-static compressive response of the hybrid structures and the individual components (OCF, EP, and EP/GO). The servo-hydraulic dynamic

INSTRON 8801 testing machine (maximum load 50 kN, Instron, Norwood, Massachusetts, USA) and the electro-mechanical RAAGEN testing machine (with maximum load 250 kN, RAAGEN, Ankara, Turkey) were used for quasi-static tests at 0.1 mm/s and 1 mm/s loading rates, respectively. The quasi-static tests were recorded by using a high-resolution video camera. A high cooled middle-wave infrared (IR) thermal camera FLIR SC 5000 [22] was also used to study the deformation and failure modes of the specimens subjected to the loading rate of 1 mm/s. The IR camera detected IR energy emitted from objects, converted it to temperature, and displayed an image of temperature distribution from the specimen surface. Temperature was measured in real time from a fixed distance without contacting the specimen. The load–displacement data were converted to the engineering stress–strain data. The energy absorption density (EAD) and specific energy absorption (SEA) were determined according to the ISO 13314: 2011 [23]. Vickers hardness measurements were carried out on samples using a diamond Vickers indenter Tester HMV-2000 (Shimadzu, Kyoto, Japan) where a 0.025 kgf load was applied for 10 s.

2.5. Thermal and Flammibility/Fire Retarcdancy Characterization

The thermal stability of the EP and EP/GO was assessed by a thermogravimetric (TG) analyzer (TGA-50, Netzsch, Selb, Germany) at a scanning rate of 10 °C/min, in the temperature range of 30–800 °C, under 50 mL/min of synthetic air (80% of N_2 and 20% O_2). The dynamic scanning calorimetry (DSC) method was used in order to measure cure heat (ΔH). The DSC analysis was carried out using the Perkin Elmer 4000 device (Perkin Elmer, Buckinghamshire, UK). Immediately after adding the hardener, the EP and EP/GO resins were submitted to a heating cycle from 30 to 220 °C with a heat rate of 10 °C/min under nitrogen atmosphere. The thermal conductivity properties were evaluated with the Hot Disk TPS 2500 S instrument (Hot Disk, Gothenburg, Sweden) in the transient mode, at room temperature (20 °C), in accordance with the standards ISO 22007-2.2 [24] and ASTM D7984 [25]. The sensor, a thin nickel foil in a double spiral pattern, which was embedded in between two thin layers of Kapton polyimide protective films, was sandwiched between two identical samples ($22 \times 22 \times 25$ mm^3). Three measurements were performed for each specimen type. During the measurement, electrical heat current passed through the nickel spiral and created an increase in temperature. The heat generated dissipated through the sample on either side at a rate which depended on the thermal transport characteristics of the material. By recording the temperature versus time with the sensor, these characteristics could accurately be calculated by a computational algorithm based on Hot Disk Thermal analyzer software.

The fire retardancy tests were based on the direct observation of the response of the specimens when subjected to a flame. Each test consisted of applying an ethanol flame to the specimen's bottom using the set-up in the vertical sample position. A test lasted for 3 s, plus a subsequent application of 3 s if the specimen self-extinguished.

2.6. Acoustic Characterization

The value of the sound absorption coefficient was estimated from measurements made with an impedance tube according to the standard ASTM E 1050 [26]. The cylindrical specimens had a diameter of 37 mm and a height of 22 mm. The tests consisted of placing the specimens at the end of an impedance tube with a diameter of 37 mm [27] and a length of 700 mm. The sound source, an RG noise generator (Wandel & Goltermann, Eningen, Germany) that emits a random noise signal, connected to a loudspeaker, was positioned at the opposite end of the impedance tube. Two microphones were placed inside the tube, separated by 30 mm between the sound source and the specimen, allowing to analyze a frequency range between 115 and 4500 Hz. The measurement allowed calculation of the sound absorption coefficient (α), which refers to the ratio between the amount of sound energy that is dissipated or absorbed by a given material and the sound wave that is passed to the material. The higher the sound absorption coefficient value, the greater the effectiveness of the insulation. Another parameter is the noise reduction coefficient (NRC) indicator, which is the arithmetic mean value of the sound absorption coefficients at frequencies of 250, 500, 1000, and 2000 Hz, and is used to compare and evaluate the performance of commercial materials.

3. Results and Discussion

3.1. Weight Ratio of the Graphene Oxide

In order to assure an uniform dispersion of the GO within the EP, a preliminary study was performed by using quasi-static uniaxial compression and Vickers hardness (HV) tests. For that, three compositions of GO were tested: 0.1 wt.%., 0.25 wt.%., and 0.5 wt.%. Visually, it was observed that until 0.25 wt.%, the dispersion of the GO was satisfactory, with no aggregates observed. Above 0.25 wt.% of GO, the formation of several aggregates was observed and persisted even after vigorous mechanical stirring. Figure 2 shows the average compressive stress–strain curves (Figure 2a) and the HV values (Figure 2b) measured for dense unreinforced EP and reinforced EP with different amounts of GO. Results clearly indicate that the diagram shape of the stress–strain curves of the unreinforced and reinforced EP are similar. The reinforced EP shows an inferior stress–strain behavior when compared to the EP without GO. However, stress–strain curves of the reinforced EP with 0.25 wt.% and 0.5 wt.% are more stable when compared to the unreinforced (pure) EP, in which the first peak stress values are similar (~80 MPa). Furthermore, after the second peak stress, the reinforced EP compositions show a lower decrease in stress in comparison to the unreinforced EP specimens. This could be attributed to the development of pores in the region near to the GO region. The specimen with 0.25 wt.% of GO showed a HV value of 17.7, which is 10% higher than the value 16.1 obtained for the pure EP. The values are in agreement with the one reported by Ho and co-workers [28]. It appears that the GO additions significantly increase the variability in the hardness, since also high standard deviation was observed. The high variability of the hardness may be attributed to the GO dispersion in the polymer matrix, which can affect the final properties. Based on these preliminary results, the 0.25 wt.% of GO was selected as the composition for the specimens in subsequent studies, and is denoted from now on by EP/GO and OCF–EP/GO.

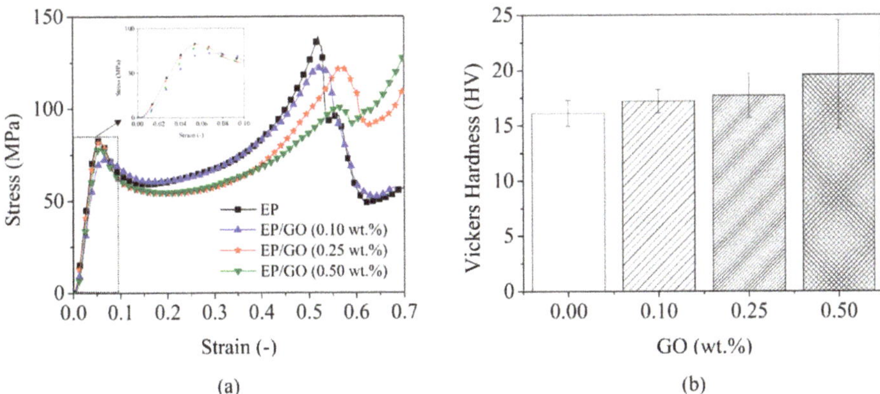

Figure 2. Quasi-static compressive stress–strain curves (**a**) and Vickers hardness (VH) (**b**) of the unreinforced EP and reinforced EP with different compositions of GO.

3.2. Microstructure

The representative scanning electron microscopy (SEM) images in Figure 3 show fractured cross-section of the EP and EP/GO specimens, which were obtained by breaking of the frozen specimens under liquid nitrogen. The results show that the GO sheets are well dispersed and randomly oriented within the polymer. No aggregates are visible in the SEM images. As shown in Figure 3a, the fracture surface of the unreinforced EP exhibits low river patterns and a smooth surface. Additionally, the cracks spread randomly, revealing a weak resistance to the crack initiation and propagation within EP. The failure process of unreinforced EP is a typical brittle fracture pattern, as reported by other authors [29]. Contrary, the EP/GO composites exhibit a rougher fracture surface (Figure 3b) and several river patterns, in which numerous and deep cracks can be observed. It seems that a continuous network of the GO was not formed due to the low content of GO. Furthermore, at higher magnification (Figure 3c), the presence of GO agglomerates within the EP matrix is not observed.

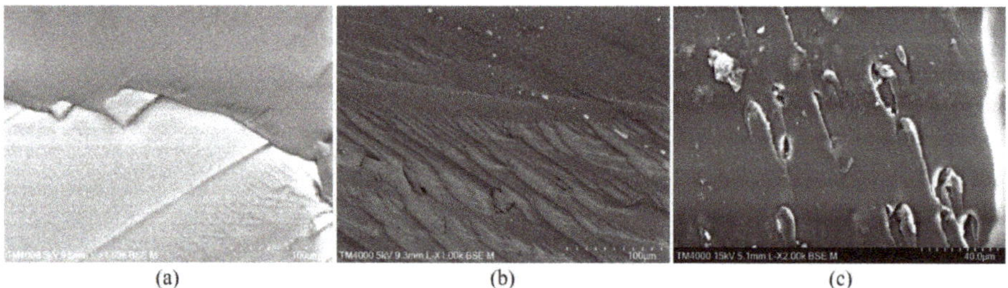

(a) (b) (c)

Figure 3. SEM images of the EP (magnification factor: 1000) (**a**) and EP/GO matrices (magnification factor: 1000 (**b**) and 2000 (**c**)).

The µCT results reveal that the values of the total porosity of OCF–EP/GO and OCF–EP are 0.008% and 0.048%, respectively. Despite that both types of specimens present very low values, the OCF–EP/GO has a higher percentage of total porosity that can be attributed to the formation of the voids in the EP matrix due to the GO. Figure 4 presents a 2D slice and 3D volume rendered image of the OCF–EP/GO hybrid structures, using the NRecon and CTvox, respectively.

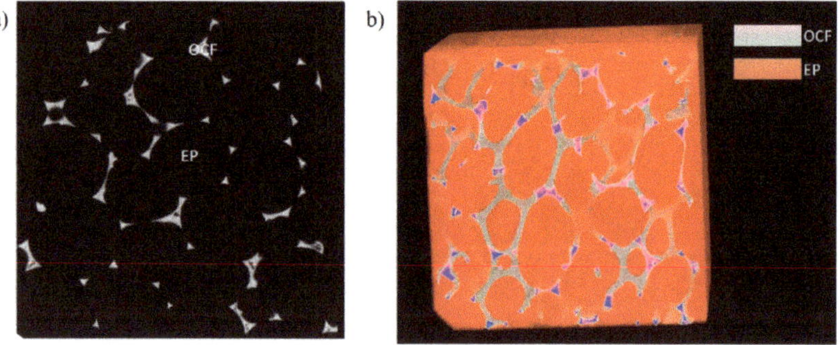

Figure 4. Micro-computed tomography images: (**a**) Two-dimensional (2D) slice and (**b**) three-dimensional (3D) volume rendering of OCF–EP/GO.

3.3. Mechanical Properties

This study is the continuation of the study conducted by Duarte et al. [8], in which the OCF voids were filled with EP. Herein, however, GO was added to the EP matrix to increase the toughness and provide thermal stability. The EP and EP/GO specimens, as well as their respective hybrid structures (OCF–EP and OCF–EP/GO), were subjected to uniaxial compression loading at two different deformation rates. Figures 5 and 6 show the deformation stages during the quasi-static loading at 0.1 mm/s and 1 mm/s, respectively.

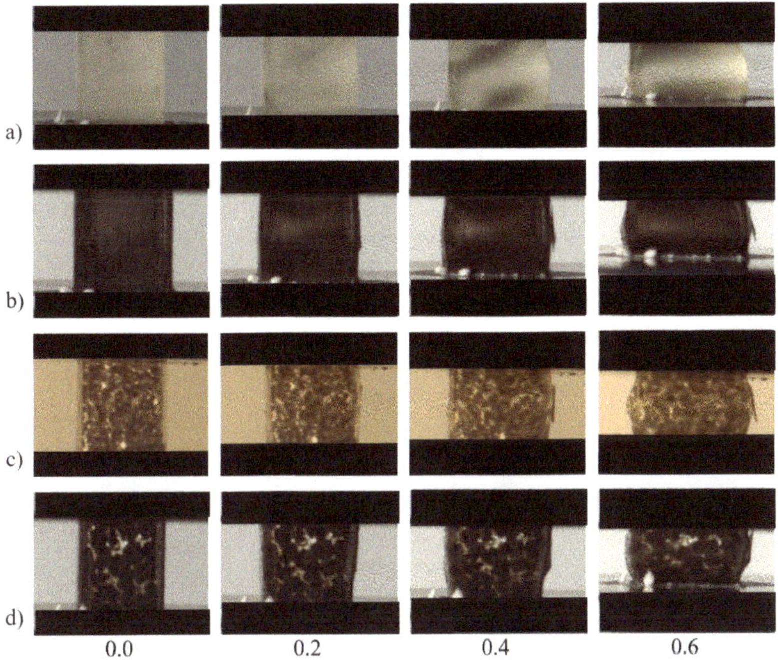

Figure 5. The quasi-static deformation (0.1 mm/s) sequences showing the progressive deformation of the: (**a**) EP, (**b**) EP/GO, (**c**) OCF–EP, and (**d**) OCF–EP/GO.

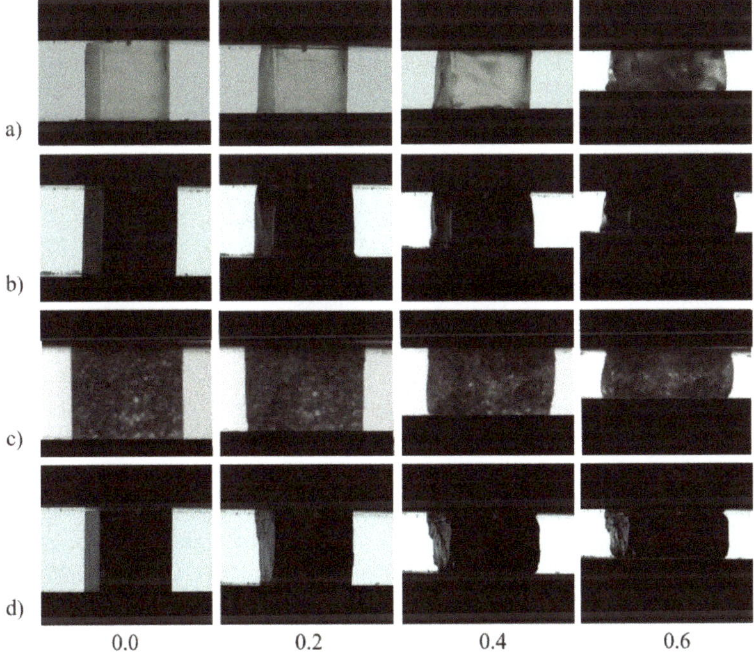

Figure 6. The quasi-static deformation (1 mm/s) deformation sequences showing the progressive deformation of the: (**a**) EP, (**b**) EP/GO, (**c**) OCF–EP, and (**d**) OCF–EP/GO.

The EP, EP/GO, OCF–EP, and OCF–EP/GO exhibited symmetric deformation, i.e., barreling in the center of the specimen during the slower loading rate (0.1 mm/s). Contrary, at loading rate of 1 mm/s, asymmetric deformation was observed. However, the incorporation of 0.25 wt.% of GO had no significant influence on the deformation mode for both tested velocities. Furthermore, the IR thermography (Figure 7) also allowed to study the deformation and failure modes of the hybrid structures (OCF–EP and OCF–EP/GO) at low crosshead rates (1 mm/s), as it was also confirmed in [30]. The regions subjected to higher strain were located near the metal foam skeleton and the specimens started to fail by development of cracks in the EP matrix. These cracks progressed, and aggravated their amplitude and group until the final failure.

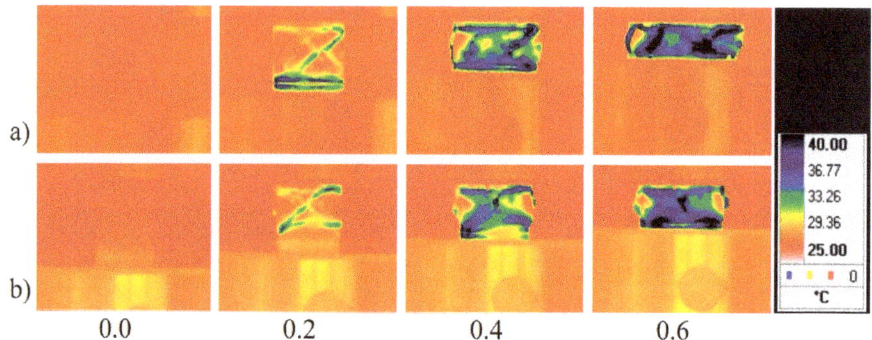

Figure 7. The IR image sequences at different strain increments showing the dynamic progressive deformation of the: (**a**) OCF–EP and (**b**) OCF–EP/GO subjected to a crosshead rate of 1 mm/s.

Figure 8 shows the quasi-static compressive stress–strain curves of the tested specimens (EP, EP/GO, OCF–EP, and OCF–EP/GO) subjected to 0.1 mm/s (Figure 8a) and 1 mm/s (Figure 8b).

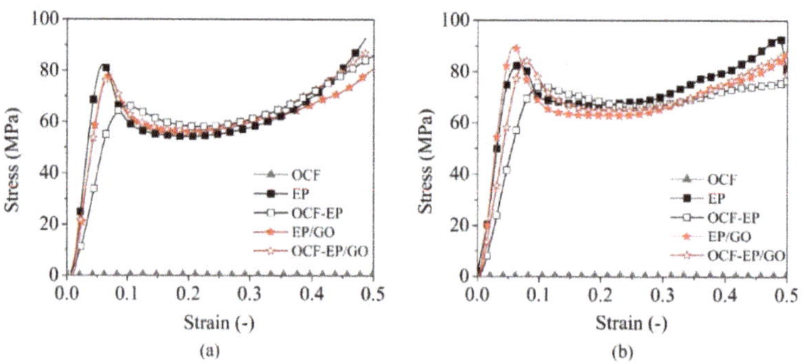

Figure 8. Stress-strain relationship (each curve represents an average of three specimens tested) at different crosshead rates: (**a**) 0.1 mm/s and (**b**) 1 mm/s.

Quasi-static compressive stress–strain curves (Figure 8a,b) show very similar diagram shapes between EP, EP/GO, OCF–EP, and OCF–EP/GO for both loading rates. The EP and EP/GO specimens, as well as the corresponding hybrid structures (OCF–EP and OCF–EP/GO), do not exhibit a typical metal foam stress–strain behavior. The compressive stress–strain curves of the EP, EP/GO, OCF–EP, and OCF–EP/GO specimens are very similar to the compressive response obtained from the empty aluminum tubes, in-situ closed-cell foam-filled tubes and polymer-open-cell aluminum foam-filled tubes [21,31,32]. The initial quasi-linear elastic region up to the peak stress is followed by a rapid stress drop region to a given minimum value of stress, and then by the stress fluctuation region until the abrupt final densification. These results are also in agreement with the previous results [8], in which the OCF–EP specimens reach high values of the initial peak stress (~75 MPa). The experiments conducted at 0.1 mm/s revealed that the first peak stress values of 82.5 MPa and 78.6 MPa were obtained for strain values of 0.060 and 0.067 for the EP and EP/GO, respectively. It is noteworthy to state that the stress values for the EP are in agreement with those in the literature [8]. The EP is a polymer composed of long carbon chains with strong chemical bonds between the carbon atoms and oxygen (epoxy groups). Thus, the EP is a tough and brittle polymer, resulting in high stress values during loading [33]. The decrease in the peak stress at the incorporation of GO, even at low loading velocity, can be attributed to some voids created in the EP and to the fact that GO can decrease the high crosslinking of the EP chains during curing process. Simultaneously, the strain at the peak stress is higher for the EP/GO, suggesting a higher mobility of molecule chains. The stress–strain behavior of the OCF structure ($22 \times 22 \times 25$ mm^3) shows a significant lower stiffness (~0.31 MPa), which was already described in [8].

The compressive behavior of the hybrid structures is mainly governed by the filling polymer, EP and EP/GO, since similar profile curves were obtained with and without the OCF. The peak stress values of OCF–EP were 66.21 MPa (0.1 mm/s) and 73.95 MPa (1 mm/s), while the peak stress values of OCF–EP/GO were 77.39 MPa (0.1 mm/s) and 84.25 MPa (1 mm/s). Interestingly, slightly inferior peak stress values were obtained since the OCF structure acts like a defect (confirmed by IR tomography), since the EP has a much higher stiffness than the OCF. In comparison with the OCF, the hybrid structures present a significant increase in strength and a different stress–strain response. Comparing the OCF with the filled ones, an increase of the peak stress of more than 40 times was registered. In the case of the hybrid structures, high peak stresses occurred for low strain rates, followed by a minor decrease in stress until 0.25 of strain, after which the stress started to increase again up to the densification. The decrease after the first peak is related to the initial formation and propagation of cracks, which is a typical behavior of the brittle cured EP. All specimens are strain rate sensitive. Globally, the specimens subjected to 1 mm/s show a higher stiffness in comparison to the ones subjected to 0.1 mm/s.

The energy absorption density (EAD) to strain relationship for the tested specimens is shown in Figure 9. The EAD curves of the hybrid structures were compared to the EAD curves of the pure EP specimens. The EAD curves were calculated by integrating the compressive stress–strain curve from undeformed state up to the strain corresponding to the second stress peak of each test. Specific energy absorption (SEA) was obtained by dividing energy absorption with the mass of each corresponding specimen.

From Figure 9, it can be observed that the EP and EP/GO significantly contribute to the increase of the energy absorption in comparison to the OCF (Figure 9a) and that the increase is linear for all samples at both crosshead rates. The linear energy absorption curves indicate that these hybrid structures are ideal materials to be used as energy absorbers. The hybrid structure specimens present very high energy absorption capacity (Figure 9b) when compared to the OCF, and even if compared to the closed-cell aluminum foam [30]. It can be observed that the polymer filler, either EP or EP/GO, greatly contributes to the increase of the energy absorption capacity of the OCF specimens. However, there is a negligible difference between EP and EP/GO with or without OC skeleton (OCF–EP and OCF–EP/GO). Comparing the materials with and without the OCF, the values of EAD and SEA are similar, which indicates that the mechanical properties are governed by the polymer. However, the higher energy absorption capacity (Figure 9c,d) compensates for the increase in higher mass of the specimens. The incorporation of GO does not reveal a significant increase in the mechanical properties. From the diagrams, it was also observed that the energy absorption capacity is also strongly strain rate sensitive.

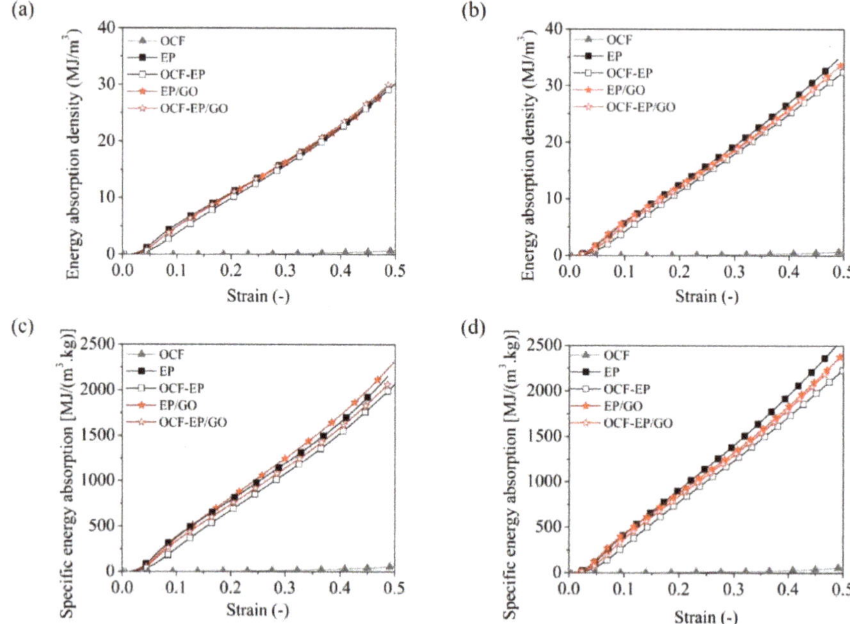

Figure 9. Average energy absorption and specific absorption energy capacity at crosshead rate of 0.1 mm/s (**a,c**) and 1 mm/s (**b,d**).

3.4. Thermogravimetric Properties

Figure 10 contains the thermogravimetric (TG) and derivative thermogravimetric (DTG) curves of the EP and EP/GO specimens under synthetic air atmosphere. The EP and EP/GO specimens show the typical degradation curve of the EP in air atmosphere, with expected three mass loss steps [34]. Both specimens exhibit similar thermal response, since the plotted curves are practically overlapping and suggesting that the presence of GO did not significantly change the degradation mechanism of the EP matrix and that GO is well mixed within the EP matrix.

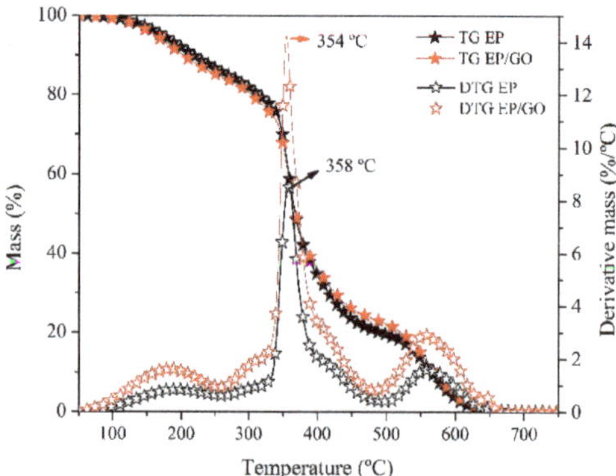

Figure 10. Thermograms of the EP and EP/GO matrices.

The first mass loss was observed between 100–250 °C and represents approximately 20% of its total mass initiated by degradation of the weakest and unstable linkages, oxygen functional groups (hydroxyl, carboxyl, and carbonyl groups), and the unreacted epoxy or any other impurity traces that began to decompose. The unstable oxygen functional groups transformed into CO, CO_2, and water vapor. As temperature further increased, the main major mass loss (300–380 °C) occurred in which the DTG peaks are 358 and 354 °C for EP and EP/GO, respectively. This can be attributed to the degradation of the main epoxy chain. The final weight lost, 450–600 °C, can be assigned to the degradation of the skeleton of the carbon atoms. The inclusion of GO 0.25 wt.% increased T50 by 1 °C and decreased T10 by 14 °C (Figure 10). This reduction can be attributed to the defects generated in the EP network caused by possible agglomerated GO sheets (although not observed in the SEM analysis) or weak interaction between GO and the EP (weaker than the crosslinking of the EP monomer). A similar response was detected by Chhetri et al. [35]. However, in the temperature range between 420–520 °C, the thermal stability of the EP/GO was slightly improved in comparison to the EP. This might be attributed to a good dispersion of the GO into the EP matrix, which helped to develop efficient interfacial interaction within the nanocomposites. Furthermore, the high aspect ratio of the GO structure may also serve as barrier to prevent the diffusion of small gaseous molecules produced by the thermal degradation. The residue after 750 °C was less than 1% (0.25% for the EP and 0.30% for the EP/GO), since both specimens decomposed into gas such as CO_2 and water vapor.

3.5. Calorimetric Properties

Figure 11 shows the DSC heat flow curves obtained for the EP and EP/GO mixtures, immediately after adding the hardener. The heat of reaction (ΔH) was measured from the areas enclosed by the thermograms. During the curing reaction, the epoxy started to polymerize, cross-link, and then harden. The onset curing temperatures for both mixtures were located around 55 °C and the specimens were fully cured at 200 °C. The peak at which the cure reaction happened slightly decreased with the incorporation of GO, from 93.3 °C for EP to 92.4 °C to EP/Go. Although this temperature decrease is less than 1%, it is suggested that GO acts as a catalyst, accelerating the epoxy curing reaction. A similar behavior was observed by Park and Kim [36], where a decrease of 2 °C was found with the incorporation of 0.5 wt.% of graphene nanoplatelets. These authors suggested that the interactions between the amino groups present in EP resin and the remaining hydroxyl groups in graphene nanomaterial are the main reason for the lowering in peak temperature. Regarding the ΔH associated with the cure reactions, the ΔH value decreased from 312.6 for EP to 290.0 J/g for EP/GO, suggesting that GO interferes with the cure reaction of EP. The steric hindrance effect of GO resulting from the high surface area of GO nanosheets could have contributed to the lower cross-linking degree in the EP mixture [37,38]. Galpaya et al. [39] reported declines of 14%, from 137.1 to 117.3 J/g, in ΔH of curing in epoxy matrices with the incorporation of only 0.3 wt.% of GO.

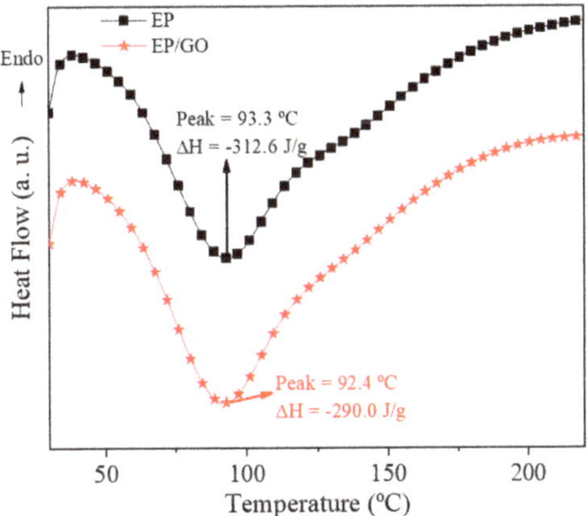

Figure 11. Dynamic scanning calorimetry (DSC) curves of the EP and EP/GO matrices.

3.6. Thermal Properties

Figure 12 shows the thermal conductivity values of the tested specimens. The thermal conductivity values obtained for the EP specimens (0.20 W/m·K) are in accordance with the values reported in literature (0.15–0.20 W/m·K) [40]. With the incorporation of 0.25 wt.% of GO, a slight increase of approximately 10% relative to the pure EP specimens was observed. The study by Aradhana et al. [41] reported an increase of more than 200% in thermal conductivity with 0.5 wt.% of GO within the EP. An increase of 20% and 90% in thermal conductivity was obtained by Kim et al. [42] for 1 wt.% and 3 wt.%, respectively. Both studies stated the importance and the key role of GO dispersion for the increase of thermal conductivity. SEM analysis can be helpful to understand the reason for a smaller increase of the thermal conductivity in this study. The content of GO used in this work is lower than in the described studies and might be too low to allow the formation of a conductive network that would provide an effective pathway for the phonon movement in the insulator matrix, leading to higher thermal conductivity values. On the other hand, even a low level of GO in the EP could result in the formation of aggregates, due to the high content of oxygen functionalities and ability to form intra/intermolecular bonds that restrict heat transport. Aggregates reduce the aspect ratio, decrease the contact area, and trap or scatter phonons [35]. The thermal conductivity value of the hybrid structures raised up to 0.32 and 0.40 W/m·K without and with GO, respectively. The presence of the interconnected OCF structure facilitated the heat transfer and promoted a substantial increase of approximately 60% in thermal conductivity. The presence of the GO seems to reduce the interfacial resistance and enhance heat flow [43,44]. In this case, the enhancement was approximately 100% compared to the EP/GO.

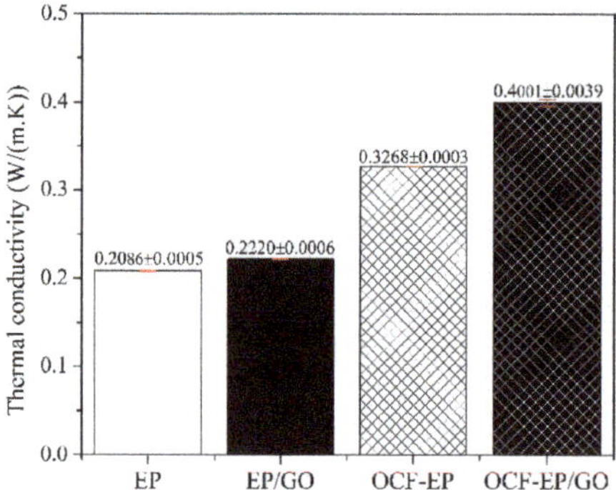

Figure 12. Thermal conductivity of the EP, EP/GO, OCF–EP, and OCF–EP/GO.

3.7. Acoustic Properties

The sound absorption coefficient values are represented in Figure 13, displaying the measured values for samples EP, OCF–EP, EP/GO, and OCF–EP/GO. In general, it should be said that the values of sound absorption coefficient registered for all samples are quite low (lower than 0.3) throughout the frequency range, which is attributed to the non-porous and high stiff structure of the EP matrix.

Indeed, when the sound absorption curve determined for the EP sample was analyzed, very low values were registered, generally below 0.05, and with only two small peaks (at 500 Hz and 3500 Hz) exhibiting a value around 0.1. The isolated addition of both the OCF skeleton and of the GO reinforcement seemed to be beneficial in what concerns sound absorption, although with a concentrated effect around the first peak occurring at a lower frequency. Further improvement was obtained when both the OCF skeleton and the GO reinforcement were used (OCF–EP/GO), for which case, a maximum peak around 0.25 at 750Hz was registered, and with higher absorption at a somewhat broader frequency range (values above 0.1 were registered between approximately 500 Hz and 1500 Hz). Interestingly, high frequency sound absorption was very low (below 0.1 in all cases), a behavior that can, once again, be strongly related to the low porosity and high stiffness of all samples. This behavior contrasts with the typical sound absorption of porous and fibrous materials (broadly used in acoustic applications), which may exhibit peak sound absorption values around 1.0, and allow significant high frequency sound absorption (see, for example, the reference book by Cox and d'Antonio [45]).

To better understand the above-described behavior, it is important to bear in mind that sound waves propagate fast in solid materials, but not as much in air and liquids. When the sound wave hits a solid surface, the material vibrates, the molecules that compose the material collide between each other, and the kinetic energy resulting from those collisions is passed from molecule to molecule. The transfer/transmission of energy is higher when molecules are closer to each other and have stronger bonding [5,6]. Chemically, the EP is composed of long carbon chains with very strong chemical bonds between the carbon atoms [33], and therefore lacks in chain mobility and high transmission of energy. The incorporation of GO introduces voids into the EP matrix, increasing the chain mobility, and when the sound propagates through the voids, oscillations of air molecules occur, leading to a frictional loss. Furthermore, the inclusion of GO can contribute to a rougher surface, which also results in a higher sound absorption coefficient. The hybrid structures present a higher sound absorption coefficient, since the OCF structure allows for a higher chain mobility, while in the filling process, some voids are additionally introduced at the metal–polymer interface. The noise reduction potential of cellular metals was studied by Hinze et al. [46] using the impedance method. They demonstrated that the acoustic absorption mainly depends on pore morphology, porosity, and the thickness of the specimen. They reported that the acoustic absorption is only achieved if sound waves can cross through the absorber. The reflection of sound waves from the absorber surface is reduced if the porosity of absorbers is high (80 to 95%). Also, similar sound absorption levels with material containing bigger pores are achieved if the thickness of the absorber is high.

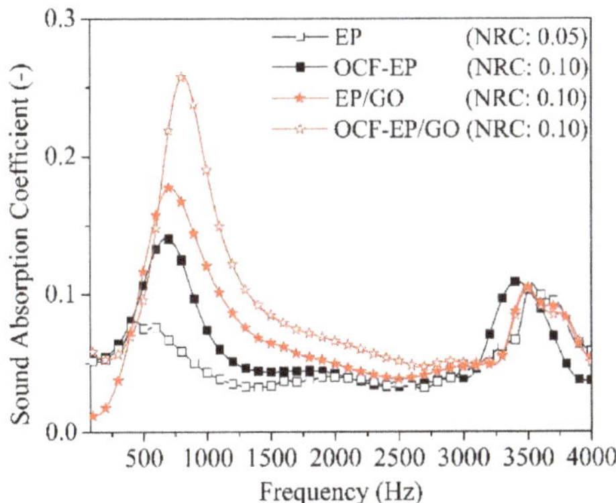

Figure 13. Sound absorption coefficient of the OC, EP, EP/GO, OCF–EP, and OCF–EP/GO.

3.8. Fire-Retardant Properties

The EP presents high flammability, as reported in literature [47]. Several authors have focused on the enhancement of polymers' fire-retardant properties. The incorporation of graphene and its derivatives can improve fire-retardant properties by forming a char layer on the surface of the polymer that avoids or delays the escape of volatiles from the decomposition process by creating a complex path. As such, it reduces the volatiles in gaseous phase available for fire propagation and returns heat back to the surface of the polymer [48,49]. Figure 14 shows the burning behavior of the EP and EP/GO. For the EP specimen, the flame self-extinguishes after 5 s; in the case of EP/GO, the flame self-extinguishes after only 3 s, suggesting a better (flame-retardant) behavior of GO. Due to the fact that GO presents no toxicity or environmental issues, it presents great potential to be one of the most promising fire-retarding fillers for epoxy nanocomposites [36].

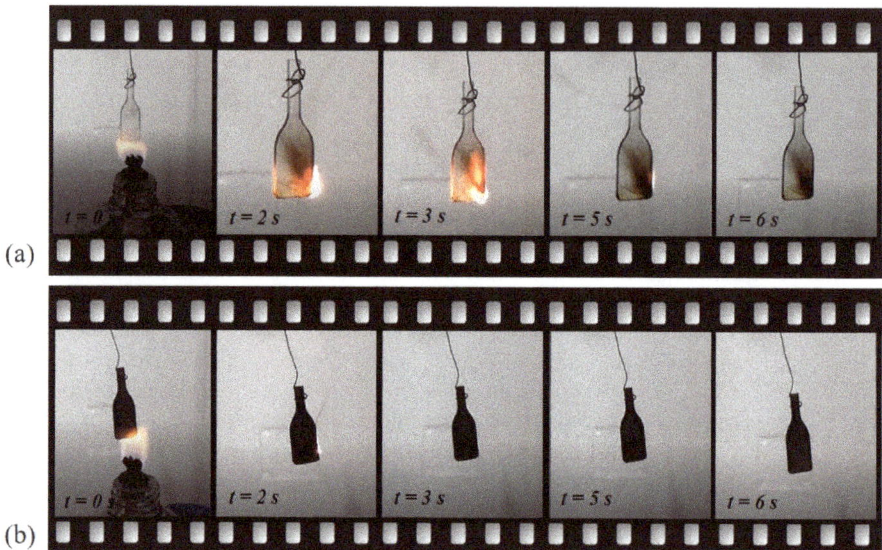

Figure 14. Fire-retardant behavior of the: (**a**) EP and (**b**) EP/GO.

4. Conclusions

This work analyzes the influence of graphene oxide (GO) that was carefully dispersed in the epoxy resin (EP) matrix on the mechanical, thermal, and sound absorbing properties. For that purpose, the EP and GO dispersed within the EP (EP–GO) and the hybrid structures, the OCF–EP and OCF–EP/GO (EP and EP–GO infiltrated in the open-cell aluminum foam (OCF)), were fabricated and tested. The motivation of this work was to study the influence of the incorporation of GO within the EP to obtain superior mechanical, thermal, and acoustic properties for advanced multifunctional structures. The main results of the study allow to point out the following conclusions:

1. The EP, EP–GO, and the epoxy–aluminum hybrid structures (OCF–EP and OCF–EP/GO) are sensitive to strain rate.
2. The presence of the EP and EP/GO decreases the oscillations in stress plateau that are usually observed in the OCF, since the shape of the stress–strain diagram is governed by the characteristics of the polymer filler.
3. The presence of the EP increases the compressive strength and energy absorption of the OCF.
4. The use of GO as a reinforcement of the EP matrix decreases the compressive strength at quasi-static uniaxial mechanical tests, thus no significant increase in the energy absorption capability was observed.
5. GO induces thermal stability to the EP, as observed by TGA and fire-retardant tests.
6. The thermal conductivity increases with the addition of GO, and the hybrid structures present even higher thermal conductivity due to the presence of the OCF skeleton.
7. Although the sound absorption of the specimens was low, it was noted that the nanofillers, as well as the aluminum structure, increase the sound absorption coefficient, especially at low frequencies.

Author Contributions: Conceptualization, S.C.P., I.D., P.A.A.P.M., and R.V.; methodology, S.C.P., I.D., P.A.A.P.M., M.V., L.K.-O., and L.G.; formal analysis, S.C.P., I.D., and P.A.A.P.M.; investigation, S.C.P., I.D., P.A.A.P.M., M.V., L.K.-O., L.G., and R.V.; writing—original draft preparation, S.C.P. and I.D.; writing—review and editing, S.C.P., I.D., M.V., R.V., and P.A.A.P.M.; supervision, I.D., P.A.A.P., and R.V.

Funding: This research was funded by the Portuguese Foundation for Science and Technology (FCT) under the grant number SFRH/BD/111515/2015 and the UID/EMS/00481/2019-FCT and CENTRO-01-0145-FEDER-022083—Centro

Portugal Regional Operational Programme (Centro2020), under the PORTUGAL 2020 Partnership Agreement, through the European Regional Development Fund. Authors also acknowledge the financial support from the Slovenian Research Agency (research core funding No. P2-0063 and bilateral project (Slovenia – Croatia) No. BI-HR/018-19-012).

Conflicts of Interest: The authors declare no conflict of interest.

References

1. Banhart, J. Manufacture, characterisation and application of cellular metals and metal foams. *Prog. Mater. Sci.* **2001**, *46*, 559–632. [CrossRef]
2. Lefebvre, L.-P.; Banhart, J.; Dunand, D.C. Porous metals and metallic foams: Current status and recent developments. *Adv. Eng. Mater.* **2008**, *10*, 775–787. [CrossRef]
3. Duarte, I.; Vesenjak, M.; Vide, M.J. Automated continuous production line of parts made of metallic foams. *Metals* **2019**, *9*, 531. [CrossRef]
4. Banhart, J.; Baumeister, J. Production methods for metallic foams. *MRS Symp. Proc.* **2011**, *521*, 121–132. [CrossRef]
5. Banhart, J. Metal Foams: Production and stability. *Adv. Eng. Mater.* **2006**, *8*, 781–794. [CrossRef]
6. Ulbin, M.; Vesenjak, M.; Borovinšek, M.; Duarte, I.; Higa, Y.; Shimojima, K.; Ren, Z. Detailed analysis of closed-cell aluminum alloy foam internal structure changes during compressive deformation. *Adv. Eng. Mater.* **2018**, *20*, 1800164. [CrossRef]
7. Duarte, I.; Peixinho, N.; Andrade-campos, A.; Valente, R. Special issue on cellular materials. *Sci. Technol. Mater.* **2018**, *30*, 1–3. [CrossRef]
8. Duarte, I.; Vesenjak, M.; Krstulović-Opara, L.; Ren, Z. Crush performance of multifunctional hybrid structures based on an aluminium alloy open-cell foam skeleton. *Polym. Test.* **2018**, *67*, 246–256. [CrossRef]
9. Stöbener, K.; Lehmhus, D.; Avalle, M.; Peroni, L.; Busse, M. Aluminum foam-polymer hybrid structures (APM aluminum foam) in compression testing. *Int. J. Solids Struct.* **2008**, *45*, 5627–5641. [CrossRef]
10. Duarte, I.; Vesenjak, M.; Krstulović-Opara, L.; Ren, Z. Compressive performance evaluation of APM (Advanced Pore Morphology) foam filled tubes. *Compos. Struct.* **2015**, *134*, 409–420. [CrossRef]
11. Duarte, I.; Ferreira, J.M.F. 2D Quantitative analysis of metal foaming kinetics by hot-stage microscopy. *Adv. Eng. Mater.* **2014**, *16*, 33–39. [CrossRef]
12. Friedl, O.; Motz, C.; Peterlik, H.; Puchegger, S.; Reger, N.; Pippan, R. Experimental investigation of mechanical properties of metallic hollow sphere structures. *Metall. Mater. Trans. B* **2008**, *39*, 135–146. [CrossRef]
13. Duarte, I.; Krstulović-Opara, L.; Vesenjak, M. Axial crush behaviour of the aluminium alloy in-situ foam filled tubes with very low wall thickness. *Compos. Struct.* **2018**, *192*, 184–192. [CrossRef]
14. Ashby, M.F.; Bréchet, Y.J.M. Designing hybrid materials. *Acta Mater.* **2003**, *51*, 5801–5821. [CrossRef]
15. Ashby, M.F. *Materials Selection in Mechanical Design*, 4th ed.; Butterworth-Heinemann: Oxford, UK, 2010.
16. Mia, X.; Zhong, L.; Wei, F.; Zeng, L.; Zhang, J.; Zhang, D.; Xu, T. Fabrication of halloysite nanotubes/reduced graphene oxide hybrids for epoxy composites with improved thermal and mechanical properties. *Polym. Test.* **2019**, *76*, 473–480. [CrossRef]
17. Kausar, A.; Rafique, I.; Anwar, Z.; Muhammad, B. Recent developments in different types of flame retardants and effect on fire retardancy of epoxy composite. *Polym. Plast. Technol. Eng.* **2016**, *55*, 1512–1535. [CrossRef]
18. Weil, E.D.; Levchik, S. A review of current flame retardant systems for epoxy resins. *J. Fire Sci.* **2004**, *22*, 25–40. [CrossRef]
19. Bortz, D.R.; Heras, E.G.; Martin-Gullon, I. Impressive Fatigue life and fracture toughness improvements in graphene oxide/epoxy composites. *Macromolecules* **2012**, *45*, 238–245. [CrossRef]
20. Domun, N.; Hadavinia, H.; Zhang, T.; Sainsbury, T.; Liaghat, G.H.; Vahid, S. Improving the fracture toughness and the strength of epoxy using nanomaterials—A review of the current status. *Nanoscale* **2015**, *7*, 10294–11329. [CrossRef] [PubMed]
21. Duarte, I.; Krstulović-Opara, L.; Dias-de-Oliveira, J.; Vesenjak, M. Axial crush performance of polymer-aluminium alloy hybrid structure filled tubes. *Thin-Walled Struct.* **2019**, *138*, 124–136. [CrossRef]
22. Krstulović-Opara, L.; Vesenjak, M.; Duarte, I.; Ren, Z.; Domazet, Z. A Infrared thermography as a method for energy absorption evaluation of metal foams. *Mater. Today Proc.* **2016**, *3*, 1025–1030. [CrossRef]

23. *ISO 13314: 2011: Mechanical Testing of Metals—Ductility Testing—Compression Test for Porous and Cellular Metals*; International Organization for Standardization: Geneva, Switzerland, 2011.

24. *ISO 22007-2.2. Plastics—Determination of Thermal Conductivity and Thermal Diffusivity—Part 2: Transient Plane Heat Source (Hot Disc) Method*; International Organization for Standardization: Geneva, Switzerland, 2008.

25. *ASTM D7984 Standard Test Method for Measurement of Thermal Effusivity of Fabrics Using a Modified Transient Plane Source (MTPS) Instrument Thermal*; ASTM International: West Conshohocken, PA, USA, 2016.

26. *ASTM E 1050 Standard Test Method for Impedance and Absorption of Acoustical Materials Using a Tube, Two Microphones and a Digital Frequency Analysis System*; ASTM International: West Conshohocken, PA, USA, 2012.

27. da Silva, J. Experimental Methods for Determining the Sound Absorption Coefficient of Building Materials. Master's Thesis, University of Coimbra, Coimbra, Portugal, 2008.

28. Ho, M.-W.; Lam, C.-K.; Lau, K.; Ng, D.H.L.; Hui, D. Mechanical properties of epoxy-based composites using nanoclays. *Compos. Struct.* **2006**, *75*, 415–421. [CrossRef]

29. Tang, J.; Zhou, H.; Liang, Y.; Shi, X.; Yang, X.; Zhang, J. Properties of graphene oxide/epoxy resin composites. *J. Nanomater.* **2014**, *2014*, 696859. [CrossRef]

30. Krstulović-Opara, L.; Surjak, M.; Vesenjak, M.; Tonković, Z.; Kodvanj, J.; Domazet, Z. Comparison of infrared and 3D digital image correlation techniques applied for mechanical testing of materials. *Infrared Phys. Technol.* **2015**, *73*, 166–174. [CrossRef]

31. Duarte, I.; Krstulović-Opara, L.; Vesenjak, M. Characterisation of aluminium alloy tubes filled with aluminium alloy integral-skin foam under axial compressive loads. *Compos. Struct.* **2015**, *121*, 154–162. [CrossRef]

32. Duarte, I.; Vesenjak, M.; Krstulović-Opara, L.; Ren, Z. Static and dynamic axial crush performance of in-situ foam-filled tubes. *Compos. Struct.* **2015**, *124*, 128–139. [CrossRef]

33. Park, Y.T.; Qian, Y.; Chan, C.; Suh, T.; Nejhad, M.G.; Macosko, C.W.; Stein, A. Epoxy toughening with low graphene loading. *Adv. Funct. Mater.* **2015**, *25*, 575–585. [CrossRef]

34. Ma, S.; Liu, W.; Hu, C.; Wang, Z.; Tang, C. Toughening of epoxy resin system using a novel dendritic polysiloxane. *Macromol. Res.* **2010**, *18*, 392–398. [CrossRef]

35. Chhetri, S.; Adak, N.C.; Samanta, P.; Murmu, N.C.; Kuila, T. Functionalized reduced graphene oxide/epoxy composites with enhanced mechanical properties and thermal stability. *Polym. Test.* **2017**, *63*, 1–11. [CrossRef]

36. Park, J.K.; Kim, D.S. Effects of an Aminosilane and a Tetra-Functional Epoxy on the Physical Properties of Di-Functional Epoxy/Graphene Nanoplatelets Nanocomposites. *Polym. Eng. Sci.* **2014**, *54*, 969–976. [CrossRef]

37. Teng, C.; Ma, C.M.; Lu, C.; Yang, S.; Lee, S.; Hsiao, M.; Yen, M.; Chiou, K.; Lee, T. Thermal Conductivity and Structure of Non-Covalent Functionalized Graphene/Epoxy Composites. *Carbon* **2011**, *49*, 5107–5116. [CrossRef]

38. Prolongo, M.G.; Salom, C.; Sanchez-Cabezudo, C.A.M.; Masegosa, R.M.; Prolongo, S.G. Influence of Graphene Nanoplatelets on Curing and Mechanical Properties of Graphene/Epoxy Nanocomposites. *J. Therm. Anal. Calorim.* **2016**, *125*, 629–636. [CrossRef]

39. Galpaya, D.; Wang, M.; George, G.; Motta, N.; Waclawik, E.; Yan, C. Preparation of Graphene Oxide/Epoxy Nanocomposites with Significantly Improved Mechanical Properties. *J. Appl. Phys.* **2014**, *116*, 53518. [CrossRef]

40. Dixon, C.; Strong, M.R.; Zhang, S.M. Transient plane source technique for measuring thermal properties of silicone materials used in electronic assemblies. *Int. Microelectron. Packag. Soc.* **2000**, *23*, 494–500.

41. Aradhana, R.; Mohanty, S.; Nayak, S.K. Comparison of mechanical, electrical and thermal properties in graphene oxide and reduced graphene oxide filled epoxy nanocomposite adhesives. *Polymer (Guildf)* **2018**, *141*, 109–123. [CrossRef]

42. Kim, J.; Yim, B.; Kim, J.; Kim, J. The Effects of Functionalized Graphene Nanosheets on the Thermal and Mechanical Properties of Epoxy Composites for Anisotropic Conductive Adhesives (ACAs). *Microelectron. Reliab.* **2012**, *52*, 595–602. [CrossRef]

43. He, X.; Huang, Y.; Liu, Y.; Zheng, X.; Kormakov, S.; Sun, J.; Zhuang, J.; Gao, X.; Wu, D. Improved thermal conductivity of polydimethylsiloxane/short carbon fiber composites prepared by spatial confining forced network assembly. *J. Mater. Sci.* **2018**, *53*, 14299–14310. [CrossRef]

44. Zhang, Y.-F.; Ren, Y.-J.; Guo, H.-C.; Bai, S. Enhanced thermal properties of PDMS composites containing vertically aligned graphene tubes. *Appl. Therm. Eng.* **2019**, *150*, 840–848. [CrossRef]

45. Cox, T.; d'Antonio, P. *Acoustic Absorbers and Diffusers: Theory, Design and Application*; CRC Press: Boca Raton, FL, USA, 2016.

46. Hinze, B.; Rösler, J.; Lippitz, N. Noise reduction potential of cellular metals. *Metals* **2012**, *2*, 195–201. [CrossRef]

47. Yu, B.; Shi, Y.; Yuan, B.; Qiu, S.; Xing, W.; Hu, W.; Song, L.; Lo, S.; Hu, Y. Enhanced thermal and flame retardant properties of flame-retardant-wrapped graphene/epoxy resin nanocomposites. *J. Mater. Chem. A* **2015**, *3*, 8034–8044. [CrossRef]

48. Hamdani, S.; Longuet, C.; Perrin, D.; Lopez-Cuesta, J.-M.; Ganachaud, F. Flame retardancy of silicone-based materials. *Polym. Degrad. Stab.* **2009**, *94*, 465–495. [CrossRef]

49. Sang, B.; Li, Z.; Li, X.; Yu, L.; Zhang, Z. Graphene-based flame retardants: A review. *J. Mater. Sci.* **2016**, *51*, 8271–8395. [CrossRef]

 *metals*

Article

Hybrid Structures Made of Polyurethane/Graphene Nanocomposite Foams Embedded within Aluminum Open-Cell Foam

Susana C. Pinto [1], Paula A. A. P. Marques [1], Romeu Vicente [2], Luís Godinho [3] and Isabel Duarte [1],*

[1] Department of Mechanical Engineering, TEMA, University of Aveiro, 3810-193 Aveiro, Portugal; scpinto@ua.pt (S.C.P.); paulam@ua.pt (P.A.A.P.M.)

[2] Department of Civil Engineering, RISCO, University of Aveiro, 3810-193 Aveiro, Portugal; romvic@ua.pt

[3] Department of Civil Engineering, ISISE, University of Coimbra, 3030-788 Coimbra, Portugal; lgodinho@dec.uc.pt

* Correspondence: isabel.duarte@ua.pt; Tel.: +350-234-370-830

Received: 15 May 2020; Accepted: 5 June 2020; Published: 9 June 2020

Abstract: This paper focuses on the development of hybrid structures containing two different classes of porous materials, nanocomposite foams made of polyurethane combined with graphene-based materials, and aluminum open-cell foams (Al-OC). Prior to the hybrid structures preparation, the nanocomposite foam formulation was optimized. The optimization consisted of studying the effect of the addition of graphene oxide (GO) and graphene nanoplatelets (GNPs) at different loadings (1.0, 2.5 and 5.0 wt%) during the polyurethane foam (PUF) formation, and their effect on the final nanocomposite properties. Globally, the results showed enhanced mechanical, acoustic and fire-retardant properties of the PUF nanocomposites when compared with pristine PUF. In a later step, the hybrid structure was prepared by embedding the Al-OC foam with the optimized nanocomposite formulation (prepared with 2.5 wt% of GNPs (PUF/GNPs2.5)). The process of filling the pores of the Al-OC was successfully achieved, with the resulting hybrid structure retaining low thermal conductivity values, around 0.038 $W \cdot m^{-1} \cdot K^{-1}$, and presenting an improved sound absorption coefficient, especially for mid to high frequencies, with respect to the individual foams. Furthermore, the new hybrid structure also displayed better mechanical properties (the stress corresponding to 10% of deformation was improved in more than 10 and 1.3 times comparatively to PUF/GNPs2.5 and Al-OC, respectively).

Keywords: open-cell foam; polyurethane foam; hybrid structures; graphene-based materials; nanocomposites

---

1. Introduction

In recent years, porous materials have attracted a huge interest from both academia and industry because they may find applications in a variety of fields, such as energy storage [1], catalysis [2], drug release [3], sound and thermal insulation [4], environmental remediation [5] and others. Giving the International Union of Pure and Applied Chemistry (IUPAC) definition, porous materials can be categorized based on their pore sizes: microporous (pore size <2 nm), mesoporous (2–50 nm), and macroporous (>50 nm) [6]. According to these categories, the properties of the porous materials and their subsequent applications will differ. Moreover, they can be found in three-dimensional (3D) and two-dimensional (2D) structures. Common 3D porous materials are sponges, foams, wood, and bone. Two-dimensional porous materials include separation membranes, filter paper, textiles, and so on.

Depending on the previously referred characteristics, and additionally based on their chemical composition, the porous materials can present multifunctionality. Multifunctional materials offer multiple characteristics that can be translated into excellent performance in existing applications and also open up avenues for untouched application fields [7]. These can exist naturally or can be engineered, with the latter being usually obtained by combining two or more materials. New functionalities arise from the synergistic combination of the individual materials properties [8].

One interesting example of porous materials are aluminum open-cell foams (Al-OC). This type of foam is characterized by a low weight, high thermal and electrical conductivities, and high internal surface area. Furthermore, they are recyclable and non-flammable [9]. However, they present a low compressive strength, when compared, for example, with closed pore cell foams. To overcome this drawback, Al-OC foams can be combined with other materials, such as silicone [10,11], epoxy [10,12], or polyurethane [13]. Although the mechanical performance of the composite foams is compensated, their final weight is increased, which is not desirable for certain applications requiring lightweight structures. In this context, filling these metallic skeletons with lightweight porous materials can be an interesting alternative to bulk polymers. Reinfried M. et al. [14] explored the concept of hybrid foams, which consist of two different interpenetrating embedded foam-material classes. The idea behind the hybrid foams is to overcome the individual shortcomings of single-material foams by combining foams of two different material classes and therefore achieving synergistic property combinations that are relevant and beneficial for future applications.

Recently, our research group explored the concept of lightweight multifunctional hybrid structures by combining Al-OC foams with cellulose/graphene foams [15]. We reported the impregnation of a cellulose/graphene foam into an Al-OC foam, creating a hybrid structure with higher mechanical properties (increase in stress of 100 times) with respect to the cellulose foam. This multifunctional hybrid foam presented also high sound absorption coefficient (near 1 between 1000–4000 Hz) and low thermal conductivity.

To further explore these types of structures, in the present work we considered the incorporation of polyurethane foams (PUF) into Al-OC ones. PUF are known for their excellent thermal and acoustic insulation properties, low thermal conductivity, good mechanical and chemical stability and low manufacturing cost [16]. The PUF represents three quarters of the production of polyurethane (PU) materials and the major market sectors include insulation materials in buildings, shock absorbers for vehicles, packaging, footwear, and furniture [17,18]. However, due to their high flammability, the improvement of their fire-retardancy properties became crucial [19]. Taking this into account, PUF precursors like polyols have been synthetized with specific chemical functional groups to confer fire retardancy [16,20]. In addition, nanoclays (montmorillonite) [21], titanium dioxide [22], iron oxide magnetic nanoparticles [23], expandable graphite [19,24], and carbon nanostructures [25–29] have also been employed to confer flame retardancy. Often, the combinations of different fillers are used to access improved fire-retardancy behavior due to a synergetic effect [30,31]. Interestingly, the graphene-based materials' addition to polymeric matrices has been reported to provide, besides fire-retardancy, the ability to improve the mechanical properties [32,33] and sound absorption features [34,35].

Pursuing the goal to contribute to the development of lightweight multifunctional materials, this work presents, in a first step, the effect of the addition of two graphene-based materials, graphene oxide (GO) and graphene nanoplatelets (GNPs) on the PUF properties. The focus was on the fire retardancy, mechanical, acoustic, and thermal properties. After the characterization of the different PUF nanocomposites, a selected composition was incorporated in an Al-OC, creating a lightweight multifunctional hybrid structure that was further characterized.

2. Materials and Methods

2.1. Materials

The raw materials employed in PUF synthesis were: (i) methylene diphenyl diisocyanate (MDI) (VORANATE M229 from Dow Chemicals, Estarreja, Portugal), with average functionality of 2.7 and

NCO content of 31.1%; and (ii) polyol, with a hydroxyl value of 239 mg KOH/g (VORACOR CR1112 from Dow Chemicals, Estarreja, Portugal).

Graphene oxide (GO) (4 mg/mL aqueous dispersion) was purchased from Graphenea (San Sebastián, Spain) and graphene nanoplatelets (GNPs) in powder were acquired from Cheaptubes (Cambridgeport, MA, USA). Silicone oil was acquired from Sigma-Aldrich (Darmstadt, Germany). The AlSi7Mg0.3 open-cell foams with pore sizes of 10 ppi (pores per inch) were supplied by Mayser GmbH & Co. KG (Lindenberg, Germany).

2.2. Sample Preparation

Pristine PUF and PUF nanocomposites were prepared by a two-step procedure, as schematized in Figure 1, route A. First, the pre-polymer (polyol), silicone oil (5 wt%), water as blowing agent (5 wt%) and GO or GNPs (1.0, 2.5 and 5.0 wt%) were placed in glass beaker and homogenized for 30 s using a mechanical stirrer at high speed. Next, the proper amount of MDI to obtain a $R_{NCO/OH}$ = 0.80 (ratio between NCO groups of isocyanates and OH groups) was added and the mixture was homogenized again for 10 s. The PUF and PUF nanocomposites were obtained by free expansion in the cup mold at room temperature. The foams are hereafter referred as PUF, PUF/xGNPs and PUF/xGO, where x refers to the carbon nanostructure content.

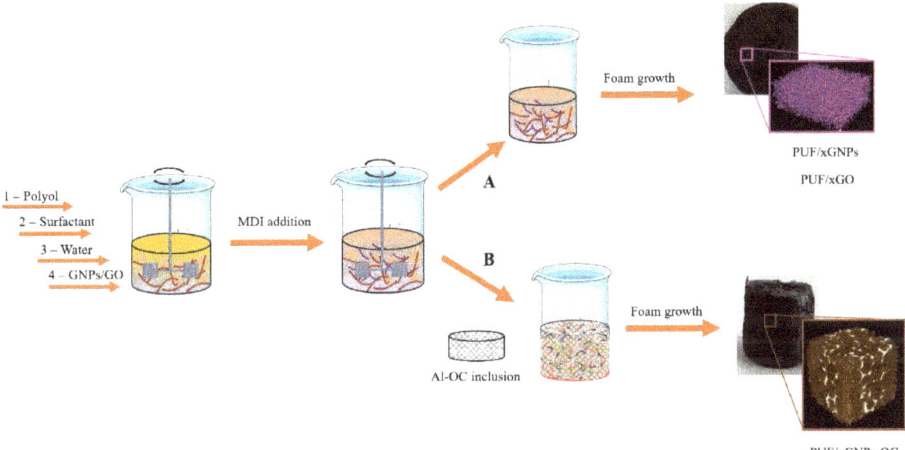

Figure 1. Scheme describing the preparation of: (**A**) polyurethane foam (PUF) nanocomposites; and (**B**) hybrid structures.

For the preparation of the Al-OC hybrid (the interconnected porous metallic foam with the optimized PUF nanocomposite) the metallic foam was placed into the cup containing the nanocomposite mixture right after the addition of MDI. The pores of the Al-OC were filled during the nanocomposite foam expansion (Figure 1, route B), hereafter referred as PUF/GNPs2.5-OC. Before characterization, the samples were settled to rest for 24 h at room temperature to ensure complete reaction.

2.3. Sample Characterization

Attenuated total reflection–Fourier transformed infrared spectroscopy (ATR-FTIR) spectra were collected using a Perkin Elmer FTIR System Spectrum BX Spectrometer (Buckinghamshire, UK) equipped with a single horizontal Golden Gate ATR cell, in the range 4000 to 500 cm^{-1} and running 64 scans with a resolution of 4 cm^{-1}. Scanning electron microscopy (SEM) analysis was performed in a TM 4000 Plus (Hitachi, Tokyo, Japan) scanning electron microscope at accelerating voltage of 15.0 kV. Samples ($10 \times 10 \times 10$ mm^3) were analyzed in a X-ray microcomputed tomography (µCT) equipment from SkyScan 1275 (Bruker μCT, Kontich, Belgium) with penetrative X-rays of 30 kV and 125 μA, in high-resolution mode with a pixel size of 8 μm and 450 ms of exposure time. NRecon and CTVox software (Bruker, Kontich, Belgium) were used for 3D reconstruction and CTan software (Bruker, Kontich, Belgium) was used in morphometric analysis (total porosity, pore size distribution and cell-wall thickness). The apparent density and porosity were determined geometrically for three specimens of each PUF nanocomposite.

The thermal conductivity properties were evaluated with a Hot Disk TPS 2500 S instrument (Gothenburg, Sweden), at 20 °C; in accordance with the standard ISO 22007-2.2 and ASTM D7984, the specimens were cube shaped with ($25 \times 25 \times 25$ mm^3). The value of the sound absorption coefficient was estimated from measurements made with an impedance tube according standard ASTM E 1050 [36] for cylindrical shaped specimens with 37 mm of diameter and 22 mm of thickness. The thermal stability of the nanocomposites foams was assessed by thermogravimetric analysis (TGA) using a thermogravimetric analyzer (Netzsch Jupiter, Selb, Germany) at a scanning rate of 10 °C/min, in the temperature range of 30–800 °C, under a synthetic air atmosphere (80% N_2 and 20% O_2). The fire retardancy test was based in the direct observation of the response of the specimens when submitted to a flame. The test consisted in applying an ethanol flame, at the specimen's bottom using the set-up in vertical sample position, for 3 s, plus a subsequent application (3 s) if the specimen self-extinguished. The tests were conducted in half cylindrical shaped specimens, with 30 mm of diameter and 10 mm of thickness.

The mechanical testing machine (Shimadzu MMT, Kyoto, Japan; maximum load 101 N) was used to study the quasi-static compressive response of PUF nanocomposites ($10 \times 10 \times 10$ mm^3) under a strain rate of 1 mm/min up. The uniaxial compression test of Al-OCF and hybrid PUF structures were performed in a Shimadzu-AGS-X-10kN (Kyoto, Japan) testing machine at a speed of 6 mm/min.

3. Results

3.1. PUFs Nanocomposites

The success of PUF and PUF nanocomposites preparation relies on the appropriate reaction between the precursors, namely the extinction of isocyanate groups through the reaction with hydroxyl groups of the polyol and urethane formation; this was confirmed by FTIR analysis (Figure S1). The addition of GO or GNP, at the used concentrations, did not prevent the progression of foam formation. However, the GO and GNPs did not disperse in the same way in the polymer matrix, as can be easily observed in the left column of Figure 2. The photographs show a homogeneous black color when GNPs were added, suggesting a good interaction between these nanofillers and the polymer matrix. On the contrary, black spots were observed in the PUF when the GO was used. The GO nanosheets were directly obtained from the chemical exfoliation of graphite and contain several oxygen chemical functionalities, which is what makes the GO highly hydrophilic. The GNPs were also obtained from graphite exfoliation, but without the use of chemical oxidants, thus resulting in a non-oxidized surface [37], GNPs being hydrophobic. This difference in the chemical surface structure of the carbon nanostructures determines their dispersion in the polymeric matrix, which have hydrophobic domains enabling GNPs dispersion. This section will provide a concise and precise description of the experimental results, their interpretation, as well as the experimental conclusions that can be drawn.

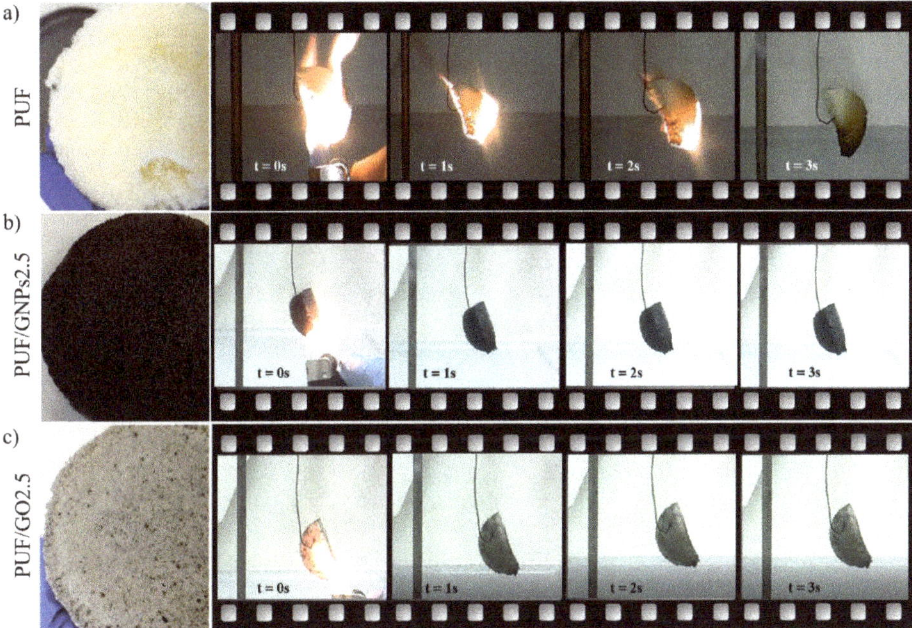

Figure 2. Flame behavior of: (**a**) PUF, (**b**) PUF/GNPs2.5, and (**c**) PUF/GO2.5.

3.1.1. Thermal Stability

Although the dispersion of both nanofillers (GNPs and GO) was not the same, their flame-retardant action was remarkable. A control test with pristine PUF showed flame propagation at some extension with smoke release, even if no dripping was observed, (Figure 2a). On the contrary, PUF nanocomposites prepared with 2.5 wt%, showed fire retardant properties, with the flame extinguished after 1 s. Importantly, the specimens maintained their shape after burning without dripping or smoke release (Figure 2b,c). Similar behavior was observed for the compositions with 5.0 wt% of the GO or GNPs. For the lower nanofillers amount tested (1 wt%), the results were not as good, since at 3 s there was still flame propagation, mainly for PUF/GNPs1.0, with smoke release in both cases (Figure S2). Carbon nanostructures are known to play a key role either in slowing down the flame propagation or even in providing self-extinction to thermoplastic or thermosetting polymeric matrices [38]. Their positive effect in the thermal stability of PUF nanocomposites may be attributed to the high specific area and layered structure of the nanofillers, which tend to form a dense and continuous char layer acting like a physical barrier at the PUF surface. This barrier becomes an obstacle to the release of the volatile degradation products, preventing or causing the delay of the degradation of the whole composite [39]. It is remarkable that, although the GO nanosheets are not as good dispersed as the GNPs in the polymer foam (photographs in the left side of Figure 2b,c), their effect on the flame retardancy was as efficient or even more efficient than the GNPs. GO's benefits in the fire retardancy efficacy over graphene have been referred to and are attributed to the GO oxygen functionalities that decompose and dehydrate at quite low temperatures. This causes a cool down of the polymer substrate during the combustion process and simultaneously release gaseous species that dilute the oxygen atmosphere near the ignition zone [30,39,40]. Graphene and GO are described mainly as co-flame retardants, at loadings from 1.0 to 10 wt% [41]. Also, functionalized graphene designed and prepared from expandable graphite and phosphorus-containing compounds has been described as flame retardants and smoke suppressor of PUF at 6.1 wt% [30]. It is worth mentioning that in our study, the pristine PUF and PUF nanocomposites with 1 wt% of nanofillers also kept their shape after burning. However, due to the better results obtained for 2.5 and 5.0 wt%, the PUF with 1.0 wt% of nanofillers were excluded in the further characterization results.

The thermogravimetric (TG) and derivative thermogravimetric (DTG) curves are shown in Figure S3a,b, respectively. The pristine PUF and PUF nanocomposites showed similar TGA curves, suggesting that GNPs and GO do not significantly influence the decomposition of PUF. This finding is most likely related to the small amount of nanofillers used. The initial decomposition temperature corresponding to 5% ($T_{5\%}$), and 50% ($T_{50\%}$) of mass degradation, the maximum-rate degradation temperature (T_{max}), and mass residue at 750 °C are listed in Table 1. The results show that GNPs had a positive effect on $T_{5\%}$ on PUF, while GO was detrimental to the early thermal stability. These results are caused by the earlier degradation of the GO oxygen functionalities at low temperature, thus accelerating the degradation of the PUF matrix. The oxygen functionalities of GO may interact with the PUF precursors during the foam formation thus interfering with the crosslinking reaction, as reported by Gama et al. [19]. In fact, at high temperatures there is the complete burning of GO, proved by the lower percentage of residue at 750 °C [42]. The sample PUF/GNPs2.5 presents the higher thermal stability with $T_{50\%}$ of 404 °C. It was reported by Liu et al. [43] that graphene can act as a heat source and accelerate the decomposition of PUF. As so, the 2.5 wt% seems to have a more positive effect on the thermal properties of PUF than 5.0 wt%.

Table 1. Experimental data of TGA analysis.

Samples	$T_{5\%}$ (°C)	$T_{50\%}$ (°C)	T_{max} (°C)	Residue 750 °C (%)
PUF	173.6	349.8	312.3	5.68
PUF/GNPs2.5	197.7	404.6	314.7	5.18
PUF/GNPs5.0	188.5	392.9	316.5	5.12
PUF/GO2.5	165.4	360.7	311.4	2.42
PUF/GO5.0	149.2	355.6	312.9	1.88

3.1.2. Morphology

SEM images of pristine PUF and PUF nanocomposites show an inhomogeneous open-cell structure composed by quasi-spherical interconnected pores for all specimens. However, depending on the presence or absence of nanofillers, some small differences were noticed. As shown in Figure 3a insets, the GNPs and GO sheets are located, and sometimes wrapped, in the PUF cell walls. It is reported that carbon nanostructures have a nucleating effect during foam formation, thus altering the PUF morphology [44]. Usually, the presence of such fillers decreases the average cell size and increase foam density improving damping properties, flame-retardancy, and mechanical properties [45]. By comparing PUF/xGNPs with PUF/xGO SEM images, GO seems to promote thicker cell walls and joints which can be related with the agglomeration of GO or from the affinity of polyol and MDI with GO. The GO nanosheets are located between adjacent cavities of the foam, and the cell wall sticks together around them, producing thicker cell walls. This effect is accentuated with the increase in the quantity of GO: PUF/GO5.0 presented bigger pore size and thicker cell walls than PUF/GO2.5. On the contrary, the increase of GNP content in the PUF promotes a pore size decrease.

The 3D reconstruction performed by μCT analysis (Figure 3b) confirms the high porosity of the specimens, with porosity values of 94.4% for PUF, 93.8% for PUF/GNPs2.5, 93.5% for PUF/GO2.5, 91.9% for PU/GNPs5.0, and 91.8% for PUF/GO5.0. The μCT images also demonstrated that the morphology of the samples is heterogeneous with wide range of pore size and wall thickness. The mean cell sizes for pristine PUF, PUF/GNPS2.5, PUF/GO2.5, and PUF/GNPs5.0 are in the range 168–328 μm, while for PUF/GO5.0 these are located between 328–468 μm (Figure 3c). PUF/GO5.0 nanocomposites have also higher percentage of thicker wall cells (Figure 3d). It is worth mentioning that, due to the limitation of resolution of μCT, only pores and cell walls thickness above 8 μm were detected.

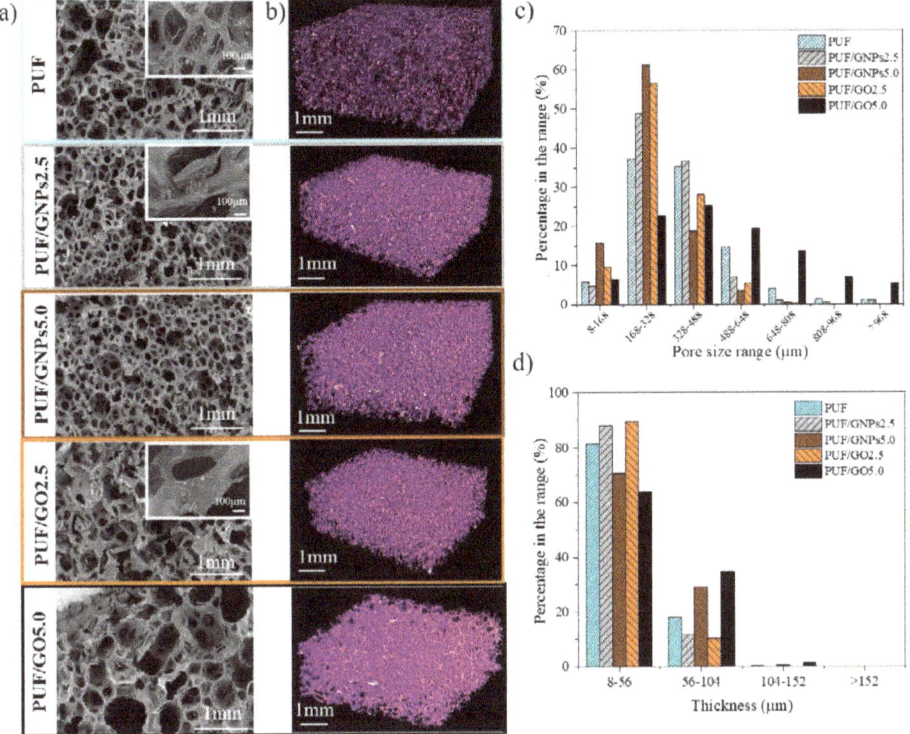

Figure 3. (**a**) SEM images, (**b**) 3D μCT rendering, (**c**) pore size distribution, and (**d**) cell wall thickness distribution for PUF, PUF/GNPs2.5, PUF/GNPs2.5, PUF/GO2.5, and PUF/GO5.0.

3.1.3. Mechanical Properties

The mechanical properties of the foams, the apparent density, the compressive strength at 10% of compression, and the compressive modulus are summarized in Table 2. The results indicate that the foams' densities increase with either GO and GNPs loading. In addition, the compressive strengths and compressive modulus steadily grow with the increase of both nanofillers content. It is reported that the addition of nanofillers to cellular materials can have different effects depending on several aspects, the loading content, the size and shape, the compatibility between the nanofillers and the matrix [33,46]. For the same fillers loading, the GO provides higher compressive modulus and higher resistivity to undergo load without total collapse of pores (higher stress plateau) than the GNPs. From the SEM images it was observed that GO induced thicker cell walls, while PUF/GNPs were similar to PUF. The improvement of compressive modulus and strength by the addition of carbon nanostructures to PUF was also reported by other experimental works, where phosphorus-functionalized GO [30] and expandable graphite [19] were used. Furthermore, MDI has isocyanate terminal groups (NCO) that can react with carboxylic and hydroxyl groups of GO leading to the chemical cross-linking between GO and PU matrix via the formation of amides or carbamate esters, thus improving the mechanical response [47].

From the average stress–strain compressive curves (Figure 4a), the foams exhibit the typical behavior of cellular materials. An elastic region at low strain values (5% of strain), where the stress–strain curve is linear, followed by a near constant stress until 55–60% of strain, is designated the stress plateau. Finally, densification takes place, with the complete collapse of the cells and the formation of a compact material, like a thin film, which is characterized by an abrupt increase of stress.

Table 2. Apparent density, compressive modulus, and compressive strength for 10% of deformation (n = 5).

Samples	Apparent Density (kg/m³)		Compressive Modulus (kPa)		Compressive Strength (kPa) (Strain: 0.1)	
	Mean Value	Standard Deviation	Mean Value	Standard Deviation	Mean Value	Standard Deviation
PUF	44	1.1	247	3.6	21.7	0.7
PUF/GNPs2.5	47	0.8	570	12.8	28.3	5.8
PUF/GNPs5.0	56	2.0	966	81.9	48.3	5.4
PUF/GO2.5	50	1.2	598	46.4	35.3	0.3
PUF/GO5.0	60	3.6	1510	165.4	56.3	5.0

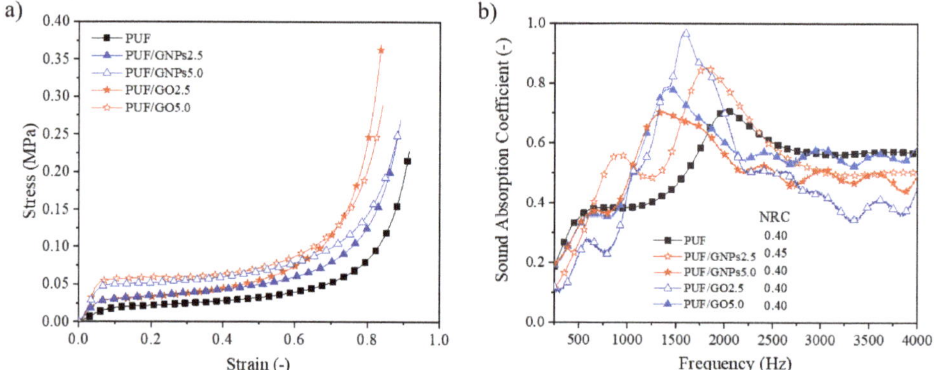

Figure 4. (a) Average stress–strain compressive curves; (b) the sound absorption coefficient of the different PU based composite foams.

3.1.4. Sound Absorption

Figure 4b shows the sound absorption coefficient between 100 and 4000 Hz for the PUF and PUF with 2.5 and 5.0 wt% of GNPs and GO. Globally, the sound absorption coefficient increases with frequency until 1500 Hz, followed by a small decrease and finally a subsequent stabilization between 2000 and 4000 Hz. The sound absorption coefficient is strongly influenced by the morphology of the foams, namely the density (associated to cell-wall thickness) and pore features (size, quantity, interconnectivity, tortuosity) [48]. It was reported that a smaller interconnected pore structure gives a better sound absorption coefficient due to the high airflow resistivity provided by cell walls [49]. Furthermore, higher porosity, associated with low density offers less resistance to sound-wave dissipation which results in a low sound absorption coefficient. Many factors contribute or influence the sound absorption, and therefore the overall values are a balance of all factors [17,48,50,51]. The incorporation of GNPs and GO increases the compressive modulus (with the later presenting higher values), and thus the cells have greater propension to undergo cell stretching, bending, and buckling without deformation. Also, by SEM and μCT analysis it was observed that the incorporation of GNPs and GO fillers decreases the porosity and pore size. The PUF/GO2.5 has the higher sound absorption curve between 1250 and 1750 Hz, reaching the value of 1 and absorbing more than approximately 50% than PUF. In fact, between 1000 and 1750 Hz, all the PUF nanocomposites have superior sound absorption coefficient comparatively to pristine PUF, which covers the sensitive frequency region of the human ear [34]. One parameter often used to describe sound absorption is the noise reduction coefficient (NRC). This parameter corresponds to the average of the sound absorption coefficients at the octave bands of 250, 500, 1000, and 2000 Hz, and rounding the result to the nearest multiple of

0.05. However, the use of NRC has its limitations since equal NRC values do not necessarily translate the same curve profile. Thus, it is worthwhile to use all the available data, especially for sound absorption at frequencies below 250 Hz or above 2000 Hz. The NRC values are gathered in Figure 4b and results showed that, although PU/GO2.5 and PU/GO5.0 have the same NRC that PU/GNPs5.0, they have distinct profiles of sound absorption vs. frequency, in this case, with PU/GO2.5 having better performance between 1200–2000 Hz. Gama et al. [17] obtained similar NRC values around 0.40–0.45 for PUF with a similar density (around 40 kg/m^3) and a higher thickness (approximately 40 mm).

3.1.5. Thermal Conductivity

To evaluate the possible application of these types of foams as thermal insulation materials, their thermal conductivity was evaluated (Table 3). A good thermal insulation material should present low thermal conductivity. The results obtained for all specimens (0.035–0.037 W·m^{-1}·K^{-1}) are in the range of most widely used commercial insulation materials, 0.030–0.040 W·m^{-1}·K^{-1} as reported in the literature [18,52,53]. The thermal diffusivity decreases with the addition of carbon nanostructures, except for the PUF/GO5.0, as listed on Table 3. This could be due to the barrier effect created by the nanofillers that difficult the heat flow.

Table 3. Thermal conductivity, thermal diffusivity, and specific heat (n = 5).

Samples	Thermal Conductivity (W·m^{-1}·K^{-1})		Thermal Diffusivity (mm^2·s^{-1})		Specific Heat (MJ·m^{-3}·K^{-1})	
	Mean Value	Standard Deviation	Mean Value	Standard Deviation	Mean Value	Standard Deviation
PUF	0.0352	0.0004	0.8577	0.0065	0.0410	0.0003
PUF/GNPs2.5	0.0361	0.0001	0.8037	0.0004	0.0448	0.0001
PUF/GNPs5.0	0.0393	0.0002	0.7672	0.0119	0.0486	0.0010
PUF/GO2.5	0.0355	0.0001	0.8276	0.0060	0.0429	0.0003
PUF/GO5.0	0.0366	0.0001	1.001	0.0060	0.0360	0.0002

From the combination of different materials, it was expected to create different multifunctional structures with high strength and reduced weight as a result of the synergetic effect of the individual materials which could be applied in different fields. In this sense, following the characterization of PUF nanocomposites regarding their fire retardancy, mechanic, acoustic, and thermal insulator properties, the formulation PUF/GNPs2.5 was selected to be incorporated in the Al-OC skeleton as a filling material and denoted as PU/GNPs2.5-OC. However, it is worth mentioning that all the developed PUF nanocomposites were suitable to be incorporated inside Al-OC.

3.2. Hybrid Structures

The effect of filling the voids of Al-OC foams with bulky polymers has been reported [10–12,54], showing enhanced mechanical properties due to the pore filling which improves the compressive strength and energy absorption capacity of the hybrid foams. However, the use of bulky polymers to fill the voids of the Al-OC foams can be disadvantageous when lightweight structures are required. Here, we wanted to explore the properties of the hybrid structure resulting from filling the Al-OC foam with a porous one, thus not compromising the lightness of the final structure. An easy and simple process to prepare this hybrid was followed (Figure 1, route B).

3.2.1. Structure

Figure 5a,b show the Al-OC and the PU/GNPs2.5-OC photographs, respectively, and in Figure 5c a 3D reconstruction of the hybrid structure obtained by μCT analysis is presented. It is worth mentioning that the weight ratio of Al-OC and PUF/GNPs 2.5 in the hybrid structure was 48.2% and

51.8%, respectively. It was observed that, during the PUF nanocomposite expansion, when in contact with Al-OC structs, some of the foam cells started to collapse. Because of that, the PUF/GNPs2.5 structure inside the Al-OC foam presented a different porosity and morphology from the optimized one. Although the pores were not uniform and the control of the porosity during the foaming was not possible, it ensured that the procedure were performed under the same operating conditions. In addition, the hybrid structures were reproducible as a whole, taking into account not only the filling material but the also the Al-OC. Though no chemical bonding between the metal surface and polymer was observed or even expected, an excellent form-fitting connection of the PUF/GNPs2.5 to the rough metal surface was visible by direct observation. The PUF/GNPs2.5 density was increased in the vicinity of the metal surface, visible in the color scheme (green color, Figure 5c).

Figure 5. Specimens (**a**) Al-OC; (**b**) PUF/GNPs2.5-OC; and (**c**) μCT PUF/GNPs2.5-OC.

3.2.2. Mechanical Properties

The stress–strain curve is shown in Figure 6a and the energy absorption (EAD) and specific energy absorption (SEA) are illustrated in Figure S4. As described earlier for pristine PUF and PUF nanocomposites, these hybrid structures also present the typical behavior of foams, with three distinct regions: elastic, stress plateau, and densification [10]. Comparing the PU/GNPs2.5-OC compressive response with Al-OC, a similar behavior can be observed; however, with higher stress peak and stress plateau values and the densification occurring earlier. The EAD is higher for hybrid structures, with improvements of 27% comparative to the Al-OC foam and 13 times lower than the PUF/GNPs2.5. However, the SEA is lower, suggesting that the increase in EAD does not compensate the increase in weight, as found by Reinfried [14], that combined a steel open-cell foam with an expanded polystyrene foam. The values of apparent density, stress peak, EAD and SEA, and thermal conductivity are gathered in Table 4.

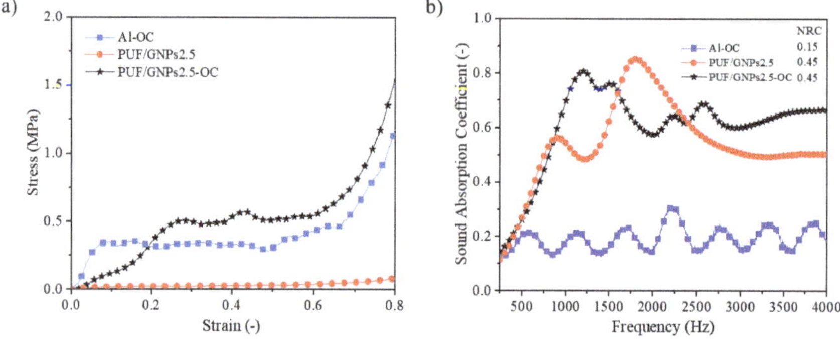

Figure 6. (**a**) Average stress–strain compressive curves; (**b**) sound absorption of Al-OC, PUF/GNPs2.5 and PUF/GNPs2.5-OC.

Table 4. Apparent density, stress peak, energy absorption, specific energy absorption, and thermal conductivity (n = 3).

Samples	Apparent Density (kg·m^{-3})		Stress Peak (MPa)		EAD (MJ·m^{-3}) Strain: 0.8		SEA (MJ·m^{-3}·kg^{-1}) Strain: 0.8		Thermal Conductivity (W·m^{-1}·K^{-1})	
	Mean Value	Std. Dev.	Mean Value	Std. Dev.	Mean Value	Std. Dev.	Mean Value	Std. Dev.	Mean Value	Std. Dev.
Al-OC	110	1.6	0.312	0.015	0.317	0.004	249	1.5	1.1781	0.0002
PUF/GNPs2.5	50	1.2	0.028	0.004	0.028	0.004	445	31	0.0355	0.0001
PUF/GNPs2.5-OC	116	6.4	0.396	0.005	0.401	0.002	127	8.8	0.0377	0.0001

Contrary to the hybrid structures composed by the Al-OC impregnated with dense polymers, epoxy [12] and polydimethylsiloxane [11], whose compressive mechanical behavior is governed by the dense materials, in this case it is the metal skeleton (Al-OC) that controls the compressive response under loading, as reported in a similar study describing the Al-OC filled with bacterial cellulose nanocomposite foam [15]. However, it should be noted that the weight of these structures is much lighter, and a good compromise between weight and strength can be achieved.

3.2.3. Sound Absorption

The sound absorption ability of the hybrid structure (evaluated by the sound absorption coefficient) present higher sound absorption values (peak at 1200 Hz with a value of 0.8 and followed by a nearly constant value of 0.7 for higher frequencies) (Figure 6b) when compared with the filling material. The increase of stiffness provided by the Al-OC structure can improve the absorption at low frequencies. Increased sound absorption can be obtained for thick specimens [55,56]. The NRC value for PUF/GNPs2.5-OC was 0.45, three times higher than for pristine Al-OC. These values are comparable with other cellular materials reported in the literature with similar thickness (around 22 mm) and density (120 kg/m^3) [57].

3.2.4. Thermal Conductivity

The values of thermal conductivity obtained for the hybrid PUF/GNPs2.5-OC are similar to those obtained for PUF and its nanocomposites, around 0.038 (W·m^{-1}·K^{-1}), suggesting that Al-OC contribution is negligible in the overall thermal conductivity value. This may be due to the small (about 1 mm) layer of filler foam (PUF and PUF/GNPs2.5) surrounding the metal foam that blocks the heat transfer. Another aspect is related to the highly porous structure of Al-OC, so the amount of solid material (Al-SiO3) that would effectively increase the conductivity value is very small. The same trend was observed in our previous works [11,12,15], in which the contribution of Al-OC was small regarding the filling materials. The polymer filler (e.g., polydimethylsiloxane [11], epoxy [12] and cellulose nanocomposites [15]) acts as an insulator, preventing effective heat transfer. On the other hand, some authors [58,59] have demonstrated that the open-cell metal foams can be used to enhance the low thermal conductivity of the pure phase change materials (PCM), composite PCMs and paraffin for application to many situations (e.g., latent heat thermal energy storage system, heat sink and heat exchanger). Furthermore, Fiedler et al. reported [60] that the polymeric adhesives exhibit a low thermal conductivity, and thus form thermal barriers between advanced pore morphology (APM) foam elements (sphere-like closed-cell aluminum foam), resulting a distinct decrease of the thermal conductivity of adhesively bonded APMs. Globally, hybrid PUF/GNPs2.5-OC can be used as a thermal insulation material [18,56].

4. Conclusions

In the present study, hybrid structures were prepared by impregnating an Al-OC foam with other cellular material, a PUF nanocomposite. Prior to the incorporation into the open structure of Al-OC foams, the standalone PUF nanocomposites were prepared with different amounts of GNPs and

GO and fully characterized. The presence of the carbon nanostructures in the PUF nanocomposites provided PUF with excellent fire retardancy, better mechanical strength, and thermal and acoustic insulating properties. The formulation with 2.5 wt% of GNPs was considered the more promising one. The process of filling the Al-OC open structure with the optimized PUF nanocomposite was successfully achieved, and the resulting hybrid structure maintain a low thermal conductivity value (0.038 W·m^{-1}·K^{-1}), high sound absorption coefficient specially at mid to high frequencies (NRC 0.45) and better mechanical behavior (the stress corresponding to 10% of deformation was improved in more than ten times). Therefore, these hybrid structures are thoroughly multifunctional materials with potential applications in the construction, automotive and, aeronautic sectors.

Supplementary Materials: The following are available online at http://www.mdpi.com/2075-4701/10/6/768/s1: Figure S1: Normalized FT-IR spectra of MDI, Polyol, PUF, PUF/GNPs2.5 and PUF/GO2.5; Figure S2: Flame response of PUF/GNPs1.0 and PUF/GO1.0; Figure S3: (a) Thermogravimetric (TG) and (b) derivative thermogravimetric (DTG) curves of PUF with different graphene based materials (GNPs and GO) additives under oxidative atmosphere; Figure S4: (a) EAD and (b) SEA curves of Al-OC; PUF/GNPs2.5 and PUF/GNPs2.5-OC.

Author Contributions: Conceptualization, S.C.P., P.A.A.P.M., I.D. and R.V.; methodology, S.C.P., I.D., P.A.A.P.M. and L.G.; formal analysis, S.C.P., I.D., P.A.A.P.M.; investigation, S.C.P., I.D., P.A.A.P.M., L.G. and R.V.; writing—original draft preparation, S.C.P. and P.A.A.P.M.; writing—review and editing, S.C.P., I.D., R.V. and P.A.A.P.M.; supervision, I.D.; P.A.A.P. and R.V. All authors have read and agreed to the published version of the manuscript.

Funding: This research was funded by Foundation for Science and Technology: UIDP/00481/2020-FCT and SFRH/BD/111515/2015, and Centro Portugal Regional Operational Program (Centro2020), under the PORTUGAL 2020 Partnership Agreement, through the European Regional Development Fund: CENTRO-01-0145-FEDER-022083.

Acknowledgments: This work was supported by the projects UIDB/00481/2020 and UIDP/00481/2020-FCT-Portuguese Foundation for Science and Technology and CENTRO-01-0145-FEDER-022083—Centro Portugal Regional Operational Programme (Centro2020), under the PORTUGAL 2020 Partnership Agreement, through the European Regional Development Fund and FCT scholarship grant SFRH/BD/111515/2015.

Conflicts of Interest: The authors declare no conflict of interest.

References

1. Zhao, C.Y.; Lu, W.; Tian, Y. Heat transfer enhancement for thermal energy storage using metal foams embedded within phase change materials (PCMs). *Sol. Energy* **2010**, *84*, 1402–1412. [CrossRef]
2. Chen, X.; Liang, Y.; Wan, L.; Xie, Z.; Easton, C.D.; Bourgeois, L.; Wang, Z.; Bao, Q.; Zhu, Y.; Tao, S.; et al. Construction of porous N-doped graphene layer for efficient oxygen reduction reaction. *Chem. Eng. Sci.* **2019**, *194*, 36–44. [CrossRef]
3. Ulker, Z.; Erkey, C. An emerging platform for drug delivery: Aerogel based systems. *J. Control. Release* **2014**, *177*, 51–63. [CrossRef] [PubMed]
4. Zhu, G.; Xu, H.; Dufresne, A.; Lin, N. High-adsorption, self-extinguishing, thermal, and acoustic-resistance aerogels based on organic and inorganic waste valorization from cellulose nanocrystals and red mud. *ACS Sustain. Chem. Eng.* **2018**, *6*, 7168–7180. [CrossRef]
5. Cao, J.; Wang, Z.; Yang, X.; Tu, J.; Wu, R.; Wang, W. Green synthesis of amphipathic graphene aerogel constructed by using the framework of polymer-surfactant complex for water remediation. *Appl. Surf. Sci.* **2018**, *444*, 399–406. [CrossRef]
6. Sing, K.S. Characterization of porous solids: An introductory survey. *Stud. Surf. Sci. Catal.* **1991**, *62*, 1–9. [CrossRef]
7. Wu, J.; Xu, F.; Li, S.; Ma, P.; Zhang, X.; Liu, Q.; Fu, R.; Wu, D. Porous Polymers as Multifunctional Material Platforms toward Task-Specific Applications. *Adv. Mater.* **2018**, *31*, 1802922. [CrossRef]
8. Ferreira, A.D.B.; Nóvoa, P.R.; Marques, A.T. Multifunctional material systems: A state-of-the-art review. *Compos. Struct.* **2016**, *151*, 3–35. [CrossRef]
9. Duarte, I.; Ferreira, J.M.F. Composite and Nanocomposite Metal Foams. *Materials* **2016**, *9*, 79. [CrossRef]
10. Duarte, I.; Vesenjak, M.; Krstulović-Opara, L.; Ren, Z. Crush performance of multifunctional hybrid foams based on an aluminium alloy open-cell foam skeleton. *Polym. Test.* **2018**, *67*, 246–256. [CrossRef]
11. Pinto, S.C.; Marques, P.A.; Vesenjak, M.; Vicente, R.; Godinho, L.; Krstulović-Opara, L.; Duarte, I. Characterization and physical properties of aluminium foam-polydimethylsiloxane nanocomposite hybrid structures. *Compos. Struct.* **2019**, *230*, 111521. [CrossRef]

12. Pinto, S.; Marques, P.A.; Vesenjak, M.; Vicente, R.; Godinho, L.; Krstulović-Opara, L.; Duarte, I. Mechanical, thermal, and acoustic properties of aluminum foams impregnated with epoxy/graphene oxide nanocomposites. *Metals* **2019**, *9*, 1214. [CrossRef]

13. Li, A.; Li, A.; He, S.; Xuan, P. Cyclic compression behavior and energy dissipation of aluminum foam–polyurethane interpenetrating phase composites. *Compos. Part A Appl. Sci. Manuf.* **2015**, *78*, 35–41. [CrossRef]

14. Reinfried, M.; Stephani, G.; Luthardt, F.; Adler, J.; John, M.; Krombholz, A. Hybrid foams—A new approach for multifunctional applications. *Adv. Eng. Mater.* **2011**, *13*, 1031–1036. [CrossRef]

15. Pinto, S.C.; Silva, N.H.; Pinto, R.J.; Freire, C.S.; Duarte, I.; Vicente, R.; Vesenjak, M.; Marques, P.A. Multifunctional hybrid structures made of open-cell aluminum foam impregnated with cellulose/graphene nanocomposites. *Carbohydr. Polym.* **2020**, *238*, 116197. [CrossRef]

16. Kausar, A. Polyurethane composite foams in high-performance applications: A review. *Polym. Technol. Eng.* **2017**, *57*, 346–369. [CrossRef]

17. Gama, N.V.; Silva, R.; Carvalho, A.; Ferreira, A.; Barros-Timmons, A. Sound absorption properties of polyurethane foams derived from crude glycerol and liquefied coffee grounds polyol. *Polym. Test.* **2017**, *62*, 13–22. [CrossRef]

18. Gama, N.V.; Soares, B.; Freire, C.S.; Silva, R.; Neto, C.P.; Barros-Timmons, A.; Ferreira, A. Bio-based polyurethane foams toward applications beyond thermal insulation. *Mater. Des.* **2015**, *76*, 77–85. [CrossRef]

19. Gama, N.V.; Silva, R.; Mohseni, F.; Davarpanah, A.; Amaral, V.S.; Ferreira, A.; Barros-Timmons, A. Enhancement of physical and reaction to fire properties of crude glycerol polyurethane foams filled with expanded graphite. *Polym. Test.* **2018**, *69*, 199–207. [CrossRef]

20. Xu, W.; Wang, G.; Zheng, X. Research on highly flame-retardant rigid PU foams by combination of nanostructured additives and phosphorus flame retardants. *Polym. Degrad. Stab.* **2015**, *111*, 142–150. [CrossRef]

21. Piszczyk, Ł.; Danowska, M.; Mietlarek-Kropidłowska, A.; Szyszka, M.; Strankowski, M. Synthesis and thermal studies of flexible polyurethane nanocomposite foams obtained using nanoclay modified with flame retardant compound. *J. Therm. Anal. Calorim.* **2014**, *118*, 901–909. [CrossRef]

22. Dong, Q.; Chen, K.; Jin, X.; Sun, S.; Tian, Y.; Wang, F.; Liu, P.; Yang, M. Investigation of flame retardant flexible polyurethane foams containing DOPO immobilized titanium dioxide nanoparticles. *Polymers* **2019**, *11*, 75. [CrossRef] [PubMed]

23. Nikje, M.M.A.; Moghaddam, S.T.; Noruzian, M. Preparation of novel magnetic polyurethane foam nanocomposites by using core-shell nanoparticles. *Polímeros* **2016**, *26*, 297–303. [CrossRef]

24. Huang, J.; Tang, Q.; Liao, W.; Wang, G.; Wei, W.; Li, C. Green preparation of expandable graphite and its application in flame-resistance polymer elastomer. *Ind. Eng. Chem. Res.* **2017**, *56*, 5253–5261. [CrossRef]

25. Pandey, G.; Thostenson, E.T. Carbon nanotube-based multifunctional polymer nanocomposites. *Polym. Rev.* **2012**, *52*, 355–416. [CrossRef]

26. Caglayan, C.; Gurkan, I.; Gungor, S.; Cebeci, H. The effect of CNT-reinforced polyurethane foam cores to flexural properties of sandwich composites. *Compos. Part A Appl. Sci. Manuf.* **2018**, *115*, 187–195. [CrossRef]

27. Colloca, M.; Gupta, N.; Porfiri, M. Tensile properties of carbon nanofiber reinforced multiscale syntactic foams. *Compos. Part B Eng.* **2013**, *44*, 584–591. [CrossRef]

28. Njuguna, J.; Pielichowski, K.; Desai, S. Nanofiller-reinforced polymer nanocomposites. *Polym. Adv. Technol.* **2008**, *19*, 947–959. [CrossRef]

29. Pinto, D.; Bernardo, L.; Amaro, A.; Lopes, S. Mechanical properties of epoxy nanocomposites using titanium dioxide as reinforcement—A review. *Constr. Build. Mater.* **2015**, *95*, 506–524. [CrossRef]

30. Cao, Z.-J.; Liao, W.; Wang, S.-X.; Zhao, H.-B.; Wang, Y.-Z. Polyurethane foams with functionalized graphene towards high fire-resistance, low smoke release, superior thermal insulation. *Chem. Eng. J.* **2019**, *361*, 1245–1254. [CrossRef]

31. Gharehbaghi, A.; Bashirzadeh, R.; Ahmadi, Z. Polyurethane flexible foam fire resisting by melamine and expandable graphite: Industrial approach. *J. Cell. Plast.* **2011**, *47*, 549–565. [CrossRef]

32. Yan, D.-X.; Xu, L.; Chen, C.; Tang, J.; Ji, X.; Li, Z. Enhanced mechanical and thermal properties of rigid polyurethane foam composites containing graphene nanosheets and carbon nanotubes. *Polym. Int.* **2012**, *61*, 1107–1114. [CrossRef]

33. Piszczyk, Ł.; Kosmela, P.; Strankowski, M. Elastic polyurethane foams containing graphene nanoplatelets. *Adv. Polym. Technol.* **2017**, *37*, 1625–1634. [CrossRef]

34. Kim, J.M.; Kim, H.; Kim, J.; Lee, J.W.; Kim, W.N. Effect of graphene on the sound damping properties of flexible polyurethane foams. *Macromol. Res.* **2017**, *25*, 190–196. [CrossRef]

35. Lee, J.; Jung, I. Tuning sound absorbing properties of open cell polyurethane foam by impregnating graphene oxide. *Appl. Acoust.* **2019**, *151*, 10–21. [CrossRef]

36. Silva, J. da Metodologias Experimentais Para a Determinação do Coeficiente de Absorção de Sonora em Materiais de Construção. Master's Thesis, University of Coimbra, Coimbra, Portugal, July 2008.

37. Colomer, J.-F.; Marega, R.; Traboulsi, H.; Meneghetti, M.; Van Tendeloo, G.; Bonifazi, D. Microwave-assisted bromination of double-walled carbon nanotubes. *Chem. Mater.* **2009**, *21*, 4747–4749. [CrossRef]

38. Cheng, Y.; Zhou, S.; Hu, P.; Zhao, G.; Li, Y.; Zhang, X.-H.; Han, W. Enhanced mechanical, thermal, and electric properties of graphene aerogels via supercritical ethanol drying and high-temperature thermal reduction. *Sci. Rep.* **2017**, *7*, 1439. [CrossRef]

39. Malucelli, G. The role of graphene in flame retardancy of polymeric materials: Recent advances. *Curr. Graphene Sci.* **2018**, *2*, 27–34. [CrossRef]

40. Idumah, C.I.; Hassan, A.; Affam, A.C. A review of recent developments in flammability of polymer nanocomposites. *Rev. Chem. Eng.* **2015**, *31*, 149–177. [CrossRef]

41. Sang, B.; Li, Z.; Li, X.-H.; Yu, L.-G.; Zhang, Z.-J. Graphene-based flame retardants: A review. *J. Mater. Sci.* **2016**, *51*, 8271–8295. [CrossRef]

42. Wang, Y.; He, Q.; Qu, H.; Zhang, X.; Zhu, J.; Zhao, G.; Lopera, H.C.; Yu, J.; Sun, L.; Bhana, S.; et al. Magnetic graphene oxide nanocomposites: Nanoparticles growth mechanism and property analysis. *J. Mater. Chem. C* **2014**, *2*, 9478–9488. [CrossRef]

43. Liu, H.; Dong, M.; Huang, W.; Gao, J.; Liu, Y.; Zheng, G.; Liu, C.; Shen, C.; Guo, Z. Lightweight conductive graphene/thermoplastic polyurethane foams with ultrahigh compressibility for piezoresistive sensing. *J. Mater. Chem. C* **2017**, *5*, 73–83. [CrossRef]

44. Gama, N.V.; Amaral, C.; Silva, T.; Vicente, R.; Coutinho, J.A.P.; Barros-Timmons, A.; Ferreira, A. Thermal energy storage and mechanical performance of crude glycerol polyurethane composite foams containing phase change materials and expandable graphite. *Materials* **2018**, *11*, 1896. [CrossRef] [PubMed]

45. Jawaid, M.; Bouhfid, R.; Qaiss, A.K. Functionalized graphene nanocomposites and their derivatives: Synthesis, processing and applications. In *Micro and Nano Technologies*; Elsevier Science: Amsterdam, The Nertherlands, 2018; ISBN 9780128145531.

46. Li, Y.; Zou, J.; Zhou, S.; Chen, Y.; Zou, H.; Liang, M.; Luo, W. Effect of expandable graphite particle size on the flame retardant, mechanical, and thermal properties of water-blown semi-rigid polyurethane foam. *J. Appl. Polym. Sci.* **2013**, *131*. [CrossRef]

47. Wang, G.; Fu, Y.; Guo, A.; Mei, T.; Wang, J.; Li, J.; Wang, X. Reduced graphene oxide—Polyurethane nanocomposite foam as a reusable photoreceiver for efficient solar steam generation. *Chem. Mater.* **2017**, *29*, 5629–5635. [CrossRef]

48. Najib, N.; Ariff, Z.M.; Bakar, A.; Sipaut, C. Correlation between the acoustic and dynamic mechanical properties of natural rubber foam: Effect of foaming temperature. *Mater. Des.* **2011**, *32*, 505–511. [CrossRef]

49. Arjunan, A.; Baroutaji, A.; Praveen, A.S.; Olabi, A.G.; Wang, C.J. Acoustic performance of metallic foams. In *Reference Module in Materials Science and Materials Engineering*; Elsevier BV: Amsterdam, The Netherlands, 2019.

50. Gwon, J.G.; Kim, S.K.; Kim, J.H. Sound absorption behavior of flexible polyurethane foams with distinct cellular structures. *Mater. Des.* **2016**, *89*, 448–454. [CrossRef]

51. Sung, G.; Kim, J.W.; Kim, J.H. Fabrication of polyurethane composite foams with magnesium hydroxide filler for improved sound absorption. *J. Ind. Eng. Chem.* **2016**, *44*, 99–104. [CrossRef]

52. Al-Ajlan, S.A. Measurements of thermal properties of insulation materials by using transient plane source technique. *Appl. Therm. Eng.* **2006**, *26*, 2184–2191. [CrossRef]

53. Jelle, B.P. Traditional, state-of-the-art and future thermal building insulation materials and solutions—Properties, requirements and possibilities. *Energy Build.* **2011**, *43*, 2549–2563. [CrossRef]

54. Vesenjak, M.; Krstulović-Opara, L.; Ren, Z. Characterization of irregular open-cell cellular structure with silicone pore filler. *Polym. Test.* **2013**, *32*, 1538–1544. [CrossRef]

55. Lu, T.J.; Hess, A.; Ashby, M.F. Sound absorption in metallic foams. *J. Appl. Phys.* **1999**, *85*, 7528–7539. [CrossRef]

56. *Handbook of Cellular Metals*; Degischer, H., Kriszt, B., Eds.; Wiley: Weinheim, Germany, 2002; ISBN 9783527303397.
57. Pedroso, M.; De Brito, J.; Silvestre, J. Characterization of eco-efficient acoustic insulation materials (traditional and innovative). *Constr. Build. Mater.* **2017**, *140*, 221–228. [CrossRef]
58. Xiao, X.; Zhang, P.; Li, M. Effective thermal conductivity of open-cell metal foams impregnated with pure paraffin for latent heat storage. *Int. J. Therm. Sci.* **2014**, *81*, 94–105. [CrossRef]
59. Li, W.; Qu, Z.; He, Y.; Tao, W. Experimental and numerical studies on melting phase change heat transfer in open-cell metallic foams filled with paraffin. *Appl. Therm. Eng.* **2012**, *37*. [CrossRef]
60. Fiedler, T.; Sulong, M.A.; Vesenjak, M.; Higa, Y.; Belova, I.V.; Öchsner, A.; Murch, G.E. Determination of the thermal conductivity of periodic APM foam models. *Int. J. Heat Mass Transf.* **2014**, *73*, 826–833. [CrossRef]

Article

Fabrication and Mechanical Properties of Rolled Aluminium Unidirectional Cellular Structure

Matej Vesenjak [1,*], Masatoshi Nishi [2], Toshiya Nishi [3], Yasuo Marumo [3], Lovre Krstulović-Opara [4], Zoran Ren [1,5] and Kazuyuki Hokamoto [6]

[1] Faculty of Mechanical Engineering, University of Maribor, 2000 Maribor, Slovenia; zoran.ren@um.si
[2] National Institute of Technology, Kumamoto College, Yatsushiro 866-8501, Japan; nishima@kumamoto-nct.ac.jp
[3] Graduate School of Science and Technology, Kumamoto University, Kumamoto 860-8555, Japan; 24.t0sy@gmail.com (T.N.); marumo@mech.kumamoto-u.ac.jp (Y.M.)
[4] Faculty of Electrical Engineering, Mechanical Engineering and Naval Architecture, University of Split, 21000 Split, Croatia; lovre.krstulovic-opara@fesb.hr
[5] International Research Organization for Advanced Science and Technology, Kumamoto University, Kumamoto 860-8555, Japan
[6] Institute of Pulsed Power Science, Kumamoto University, Kumamoto 860-8555, Japan; hokamoto@mech.kumamoto-u.ac.jp
* Correspondence: matej.vesenjak@um.si; Tel.: +386-2220-7717

Received: 12 May 2020; Accepted: 5 June 2020; Published: 9 June 2020

Abstract: The paper focuses on the fabrication of novel aluminium cellular structures and their metallographic and mechanical characterisation. The aluminium UniPore specimens have been manufactured by rolling a thin aluminium foil with acrylic spacers for the first time. The novel approach allows for the cheaper and faster fabrication of the UniPore specimens and improved welding conditions since a lack of a continuous wavy interface was observed in the previous fabrication process. The rolled assembly was subjected to explosive compaction, which resulted in a unidirectional aluminium cellular structure with longitudinal pores as the result of the explosive welding mechanism. The metallographic analysis confirmed a strong bonding between the foil surfaces. The results of the quasi-static and dynamic compressive tests showed stress–strain behaviour, which is typical for cellular metals. No strain-rate sensitivity could be observed in dynamic testing at moderate loading velocities. The fabrication process and the influencing parameters have been further studied by using the computational simulations, revealing that the foil thickness has a dominant influence on the final specimen geometry.

Keywords: unidirectional cellular structure; porosity; fabrication; explosive compaction; metallography; computational simulation; experimental tests; mechanical properties

---

1. Introduction

There is an ever-increasing demand for new multifunctional lightweight materials in advanced applications in engineering, transportation, and medicine, which can often be met by cellular metals. Their behaviour can be tailored [1] by combining the base material, porosity, morphology (size and shape of the cells, connectivity between cells) and topology (distribution of the cells within the material) according to [2]. Additionally, the behaviour can be tuned either by the partial [3] or full [4] infiltration of polymer filler into the cellular structure. These parameters and the manufacturing procedure [5] have to be carefully chosen to achieve required physical properties (e.g., stiffness, strength, energy absorption, conductivity) of cellular metals [6]. The main advantages of cellular materials and structures are a lightweight design, fire retardancy, efficient energy absorption, isolation and damping [5]. They can be used in various industrial applications (e.g., as sandwich structures, filters, heat exchangers, isolators, dampers, bearings, and energy absorbers [7]) due to their advantageous physical [8], biocompatible [9], and ergonomic characteristics [10].

Nevertheless, the production costs of cellular metals are high in general due to some technological problems that have yet to be solved. The current research and development trends are presently centred on the development of foam formation and stabilisation mechanisms, the investigation of advanced blowing constituents (agents), the optimisation of the production and the decrease of their market value [11]. The technological problems are mainly related to the control of the material structure since most existing technologies do not allow for precise control of the shape, size, and distribution of pores. This results in a scatter of physical and other characteristics of these materials and components. Some of the already existing manufacturing methods for different types of cellular structures—e.g., Metallic Hollow Sphere Structures [12], Kagome structures [13], auxetic structures [14], additively manufactured open-cell structures [15], Lotus-type [16] or Gasar [17] and UniPore structures [18], and syntactic foams [19,20]—allow for a higher level of regularity and reproducibility.

The recent development of unidirectional UniPore structures [21] enabled the production of unidirectional cellular metals with a nearly constant size of cells and the intercellular wall thickness through the length of the specimens. Furthermore, the cells are completely isolated, without gaps between each other [18] and advanced mechanical properties [22]. The fabrication method [23] is based on explosive welding phenomena of metal (e.g., copper [18], aluminium [24]) cylindrical pipes with circular cross-section assembly. The transversely isotropic UniPore cellular structure exhibits a promising combination of mechanical [25] and thermal behaviour [26]. The original fabrication of UniPore structures has shown only moderate welding conditions not forming a continuous wavy interface at some interface sections because of the changing collision angle [18]. Additionally, the fabrication consists of a tedious filling of expensive thin inner pipes (with a pipe diameter smaller than 3 mm and its wall thickness of approx. 0.2 mm) with a polymer to avoid complete compaction and its removal after fabrication. Due to these shortcomings, new fabrication methods have been considered. A new procedure to manufacture UniPore structures was proposed [27]. It consists of rolling a non-expensive copper foil with equally spaced spacer bars (made of acryl) placed on the foil and subsequent compaction by explosive detonation.

Herein, the fabrication and properties of novel rolled aluminium UniPore structures were analysed. The rolling of the non-expensive aluminium foil improved welding conditions and decreased handling time of the specimens before and after fabrication. Various rolled UniPore geometries have been fabricated and characterised by metallographic analysis and (quasi-static and dynamic) mechanical compressive testing for the first time. Additionally, the fabrication process was analysed in detail with computational simulations based on the finite element analysis.

2. Fabrication Method and Specimens

The fabrication method of the rolled aluminium UniPore structures is a convenient and cheaper method for manufacturing the UniPore cellular structures with unidirectional pores by using the aluminium foil (A1100-O). It consists of the following steps: (i) preparation of acrylic spacer bars by cutting the acrylic resin plate into rectangular shaped bars, (ii) positioning of the acrylic spacer bars on the aluminium foil in a uniform pattern with an offset of approximately 3 mm, (iii) tight rolling of the aluminium foil with acrylic resin spacer bars around the aluminium bar as the centre (core), (iv) insertion of the rolled foil into the outer aluminium pipe, (v) central insertion of the aluminium pipe with rolled foil into the PVC round container (height: 270 mm and diameter: 83 mm), (vi) filling the void space between the central aluminium pipe and container wall with the primary explosive (750 g), (vii) explosive ignition by an electric detonator (booster) to achieve explosive compaction of aluminium pipe and foil, (viii) removal of the acrylic bars by heating the recovered specimens.

The schematic illustration of the fabrication method and the explosive compaction (cylindrical) assembly is presented in Figure 1, while the physical properties of the components and the assembled specimens are listed in Table 1. The porosity of the specimens could be altered by changing the thickness of the outer pipe and by reducing the diameter of the inner aluminium bar.

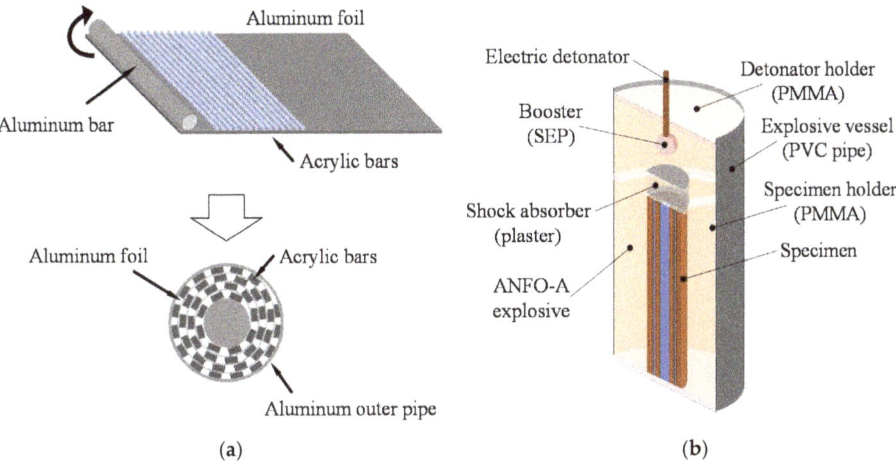

Figure 1. (a) Schematic preparation of the specimens and (b) experimental production assembly.

Table 1. Physical properties of the prepared aluminium UniPore specimens.

Specimen Type	Outer Pipe Diameter (mm)	Length (mm)	Al Core Diameter (mm)	Al Foil Thickness (mm)	Acrylic Bar Thickness (mm)	Acrylic Bar Quantity (-)	Estimated Porosity (%)
No. 1				0.2	0.5	67	16.5
No. 2	30/24	210	10	0.4	0.5	40	9.4
No. 3				0.2	1.0	35	18.1
No. 4				0.4	1.0	31	15.0

The ammonium-nitrate based ANFO-A with the detonation velocity of 2.3 km/s and the bulk density of 530 kg/m^3 was used as the primary explosive for manufacturing the rolled aluminium UniPore structure. The primary explosive was electrically detonated using a booster (10 g SEP explosive). Ignition of the primary explosive caused propagation of the detonation wave through the primary explosive. The detonation gas uniformly radially accelerated the outer aluminium pipe towards the centre of the specimen. The achieved velocity was high enough to allow for welding between surfaces of the outer pipe and aluminium foil. Stable welding conditions and inclination angle were similar to the already known explosive welding mechanism [28]. Furthermore, the explosive welding of clads is being characterised as cold pressure welding. The annealing effect, which would decrease the hardness, does not usually appear, and the increase in hardness is considered to be the result of the work hardening [22]. Figure 2 presents the cross-sections of the recovered specimens, while their geometrical properties (dimensions and porosity) are given in Table 2.

Figure 2. Cross-sections of the fabricated specimens.

Table 2. Properties of the recovered specimens.

Specimen Type	Number of Recovered Specimens (-)	Average Diameter (mm)	Average Height (mm)	Average Mass (g)	Achieved Average Porosity (%)
No. 1	4	26.2	11.0	12.8	15.5
No. 2	4	26.2	11.9	14.9	9.3
No. 3	1	26.3	10.4	12.7	17.5
No. 4	4	26.0	10.4	12.3	15.0

The shape and pore topology can be easily varied and adjusted for specific and individual applications by using the above-described fabrication method porosity (e.g., via wall thickness), dimensions (e.g., diameter).

3. Computational Analysis of the Fabrication Process

3.1. Computational Model

The computational simulations of the high-strain-rate deformation mechanism during fabrication of the aluminium UniPore structures were carried out to analyse the outer pipe's acceleration during the fabrication and the deformation of the recovered specimens in more detail. The computational simulations of all four specimen types (Figure 3) were performed based on the following assumptions and simplifications: (i) two-dimensional computational models were used, (ii) the aluminium pipe, bar and foil were compressed without joining, and (iii) the geometry of the rolled foil was modelled with multiple concentric circles.

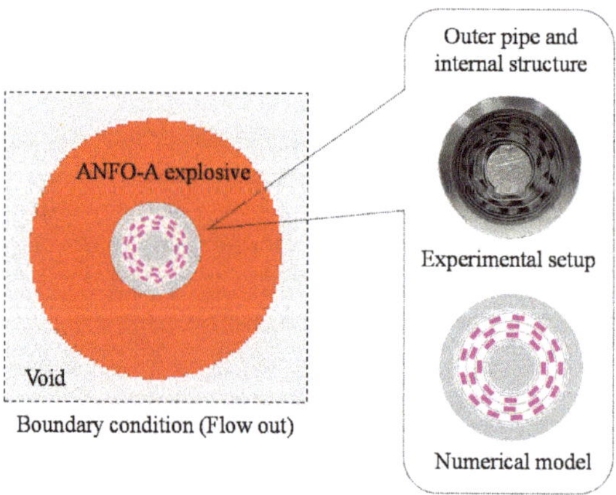

Figure 3. A two-dimensional computational model of the aluminium UniPore specimens.

The aluminium and acrylic resin were discretised by the Lagrangian mesh, while the ANFO-A explosive has been modelled with the Eulerian finite elements. The computational simulation was based on the Euler–Lagrange interaction within the engineering code AUTODYN. To assure reliable results and reasonable computational times a finite element mesh convergence study was performed. The results showed the appropriate element size of 0.2 mm and 0.5 mm for the foil and aluminium bar, respectively. The Johnson–Cook constitutive equation [29] was applied for the aluminium (A1100-O) because it takes into account the strain-rate sensitivity and hardening effects. The relation between the pressure and volume of the aluminium and acrylic resin at a specific temperature was defined using the Mie–Gruneisen equation of state [30]. It is based on the shock Hugoniot equation and can be expressed as [31]:

$$P = p_H + \Gamma \, \rho (e - e_H) \tag{1}$$

$$p_H = \frac{\rho_0 \, c_0^2 \, \mu \, (1 + \mu)}{\left[1 - (s-1) \, \mu\right]^2} \tag{2}$$

$$e_H = \frac{1}{2} \frac{p_H}{\rho_0} \left(\frac{\mu}{1 + \mu} \right) \tag{3}$$

where P represents the pressure, Γ the Gruneisen coefficient, ρ the density, e the internal energy and $\mu = \rho / \rho_0 - 1$. The shock velocity (U_s)

$$U_s = c_0 + s \cdot u_s \tag{4}$$

changes linearly (represented by the Hugoniot relation) with the particle velocity (u_p). The parameters s and c_0 are experimentally determined material constants [31]. Parameters of the Mie–Gruneisen equation of state applied in the computational analysis are given in Table 3.

Table 3. Values for the Mie–Gruneisen equation of state.

Material	Reference Density ρ_0 (kg/m^3)	Gruneisen Coefficient Γ (-)	Speed of Sound c_0 (m/s)	Material Constant s (-)
A1100-O	2707	1.9	5386	1.339
Acrylic resin	1186	0.97	2598	1.516

In the computational simulations, the ANFO-A was described as highly pressurised gas with an initial pressure of 0.939 GPa, a detonation velocity of 2.3 km/s, and a density of 530 kg/m^3 [32].

3.2. Computational Results

The diagrams in Figure 4 shows the of the outer pipe velocity V changes with time. The results based on the acrylic resin bar of 0.5 mm in thickness are represented with dotted lines, while the results based on the acrylic resin bar of 1 mm in thickness are represented with solid ones. It can be observed from the diagrams that the collision velocity is higher than 300 m/s, which is sufficient to obtain explosive welding [33]. The outer pipe velocity strongly depends on the aluminium foil thickness, while the influence of acrylic resin bar thickness is minimal.

The computationally estimated deformation process during the explosive compaction of all four specimen types is shown in Figure 5. The highly localised deformation during the explosive compaction can be observed. The primary effective plastic deformation occurs in areas around the acrylic bars, where the thin aluminium foil is locally bent.

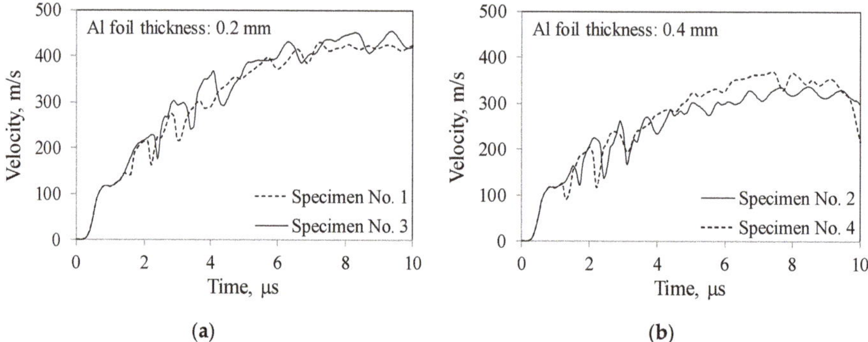

Figure 4. Velocity variation of the outer aluminium pipe during explosive compaction: (**a**) foil thickness 0.2 mm and (**b**) foil thickness 0.4 mm.

Figure 5. Simulation of the deformation process during explosive compaction of the four aluminium UniPore samples (Table 1) with annotated effective plastic strain.

The deformed shapes of the UniPore structures obtained by the computational simulations are presented in Figure 6. The computational results are in an excellent agreement with the actual specimens in terms of deformed shapes, which are shown in Figure 2. The deformation mechanism up to the impact between the foil surfaces and acrylic resin bars was thoroughly investigated by conducted computational analyses. However, it should be noted that the metal jet formation was not directly considered in the computational simulations.

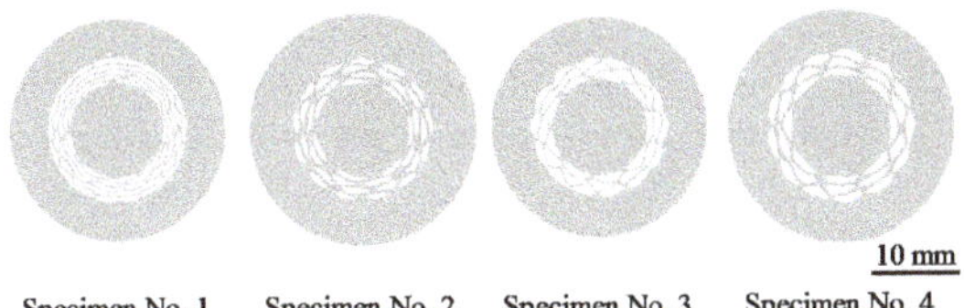

Figure 6. Computationally predicted final deformed shapes of the four analysed specimen configurations (Table 1).

4. Metallographic Analysis

Metallographic analysis of the recovered specimens has been performed to determine the quality and suitability of the fabrication method. The specimens were prepared according to the standard metallographic methods (embedding in the epoxy resin, grinding, polishing, chemically etching). The microstructure of two perpendicular (longitudinal and transversal) cross-sections was analysed by the light microscopy using the optical microscope Nikon LV150N (Nikon, Tokyo, Japan). Figures 7 and 8 show the metallographic images of the longitudinal and transversal cross-section, respectively.

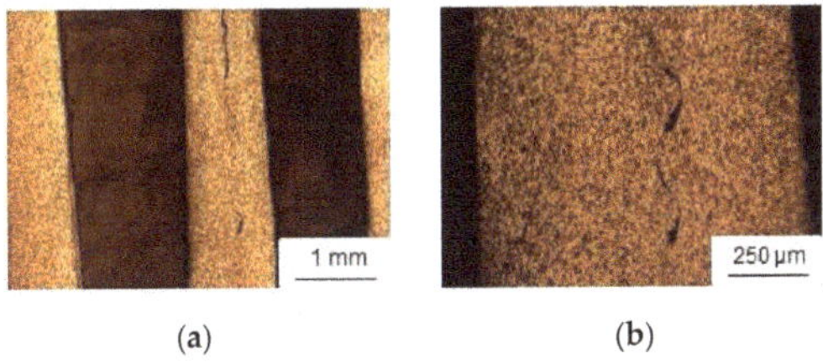

Figure 7. Metallographic images of the longitudinal cross-section of UniPore specimen. (**a**) lower magnification; (**b**) higher magnification.

A wavy interface between two colliding surfaces of the aluminium foil is represented in Figure 7, which assures a strong connection and good bonding between surfaces [33]. From the metallographic analysis, it can be concluded that the foil surfaces were welded at a sufficiently high velocity, despite a few places, where the surfaces might not be bonded completely.

The metallographic images in Figure 8 also show good bonding between the foil surfaces in the transversal cross-section and that the pores are separated and isolated between each other.

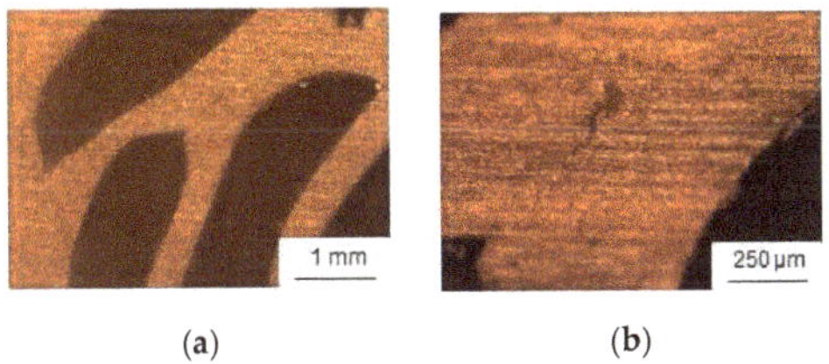

Figure 8. Metallographic images of the transversal cross-section of UniPore specimen. (**a**) lower magnification; (**b**) higher magnification.

5. Compressive Experiments

5.1. Experimental Set-Up

The mechanical behaviour of manufactured rolled aluminium UniPore structures was evaluated in transversal direction by the quasi-static and dynamic (one specimen of the type No. 1, 2 and 3) compressive experimental tests using the universal testing machine (Instron 8801, Instron, Norwood, MA, USA). The velocity of the cross-head during the quasi-static and dynamic loading cases was set to 0.1 mm/s and 284 mm/s, respectively. Displacements and loading forces were measured during the compressive tests. At the same time, the deformation mechanism was captured by an HD video camera Sony HDR-SR8E (Sony, Tokyo, Japan) during the quasi-static tests, and the middle-wave infrared (IR) thermal camera FLIR SC 5000 (FLIR Systems, Wilsonville, OR, USA) during the dynamic tests. The IR thermography allows following the yielding, cracking and failure during the dynamic loading [34]. It has been already successfully implemented for studying the response of various cellular structures.

5.2. Experimental Results

The compressive deformation behaviour of the four specimen types subjected to quasi-static loading conditions is illustrated in Figure 9. A similar deformation behaviour can be noted for all cases. Initially, the porous part of the specimen is compressed, followed by the deformation of the specimen's core. A strong interface bonding can be observed for the specimen No. 1 (aluminium foil thickness: 0.2 mm and acryl bar thickness: 0.5 mm). The bonds between the foil surfaces of the other three specimen types failed at larger strains, especially in case of the specimen No. 4 (aluminium foil thickness: 0.4 mm and acryl bar thickness: 1 mm).

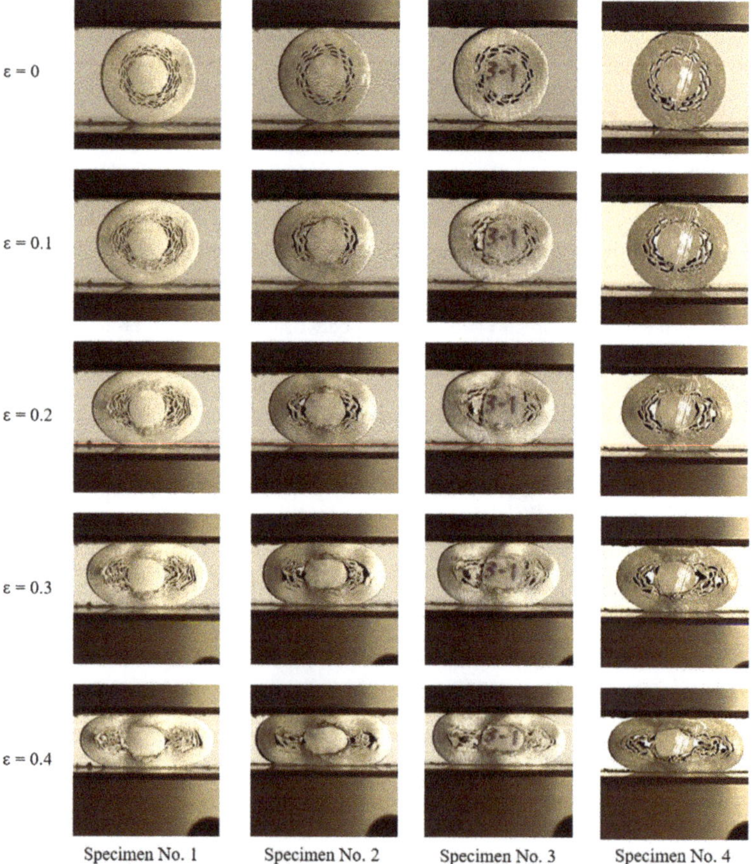

Figure 9. Transversal compressive testing of rolled aluminium UniPore structure.

Figure 10 shows the IR images of the deformation mechanism of specimens (No. 1, No. 2 and No. 4) during the dynamic tests. The interface between the foil surfaces tends to fail again in the specimen No. 4. However, the deformation mechanism seems to be similar for all specimen types. The porous structure starts to yield below and above the aluminium core with the plastification zone spreading through the specimen up to full densification. Furthermore, the explosive welding of clads is characterised as cold pressure welding.

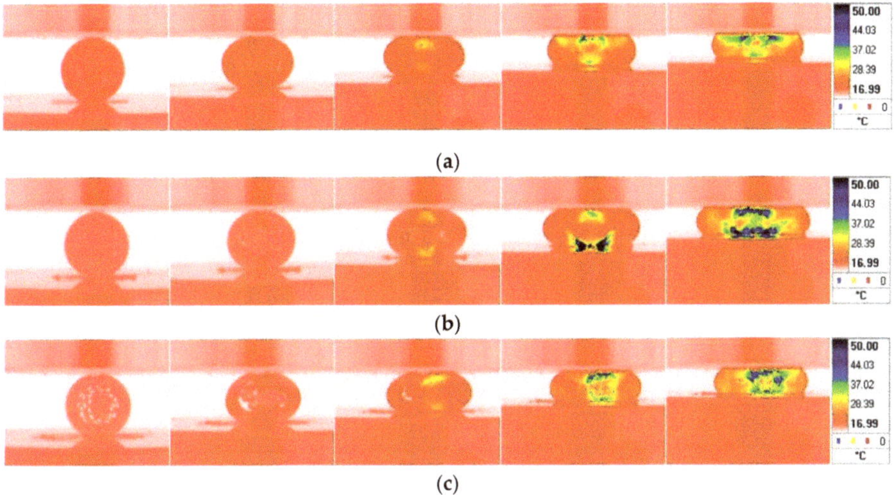

Figure 10. IR thermography images of the dynamic loading sequence (strain increment: ~0.15): (a) Specimen No. 1; (b) Specimen No. 2; (c) Specimen No. 4.

The mechanical response in terms of force-displacement diagrams is shown in Figure 11. A compressive relationship typical for the cellular metals [2] can be observed in all cases. After the initial quasi-elastic region, the yield stress is reached, followed by a short stress plateau region. Then the force starts to build up gradually and reaches the densification displacement at approximately 14 mm (equal to the strain of 0.54), after which the force drastically increases (the porous structure above and below the core is wholly densified).

The measurements show a very consistent response (almost no deviation between specimens of the same type) up to the end of the plateau region. The deviation becomes prominent during the gradual force increase, which can be attributed to bending, buckling, and collapsing of the foil (intercellular walls) and the separation of the bonds between the foil surfaces.

Finally, the quasi-static and dynamic tests provided similar results. Thus, no strain-rate sensitivity was noted for the tested moderate strain-rates.

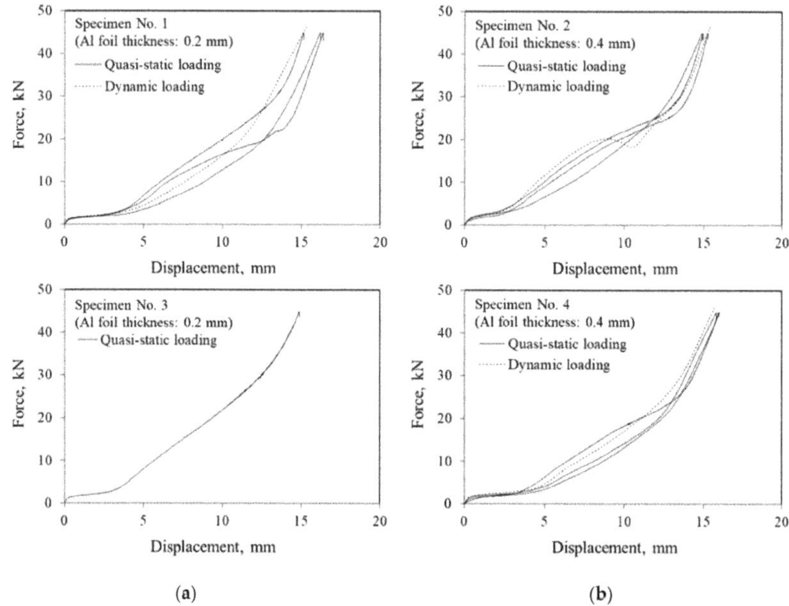

Figure 11. Compressive force-displacement curves in the transversal direction: (**a**) foil thickness 0.2 mm and (**b**) foil thickness 0.4 mm.

6. Conclusions

The existing UniPore structures with unidirectional pores have shown two main disadvantages: (i) moderate welding conditions, which result in a lack of a continuous wavy interface at some bonded sections, and (ii) the fabrication consists of filling the expensive thin inner pipes with a polymer to avoid complete compaction and its removal after fabrication. Explosive compaction has been applied for the first time to fabricate the rolled aluminium UniPore specimens. After rolling an aluminium foil with acrylic spacers (no additional filling and removal step of the polymer was required), the assembly was subjected to explosive compaction, which—due to the mechanism of explosive welding—resulted in a unidirectional cellular structure with longitudinal pores. The fabrication process and the influencing parameters have been studied in detail by use of computational simulations, revealing that the foil thickness has a dominant influence on the finial specimen geometry (shape). The computational results compare well to the actual specimens in terms of the specimens' final shape. The metallographic analysis confirmed a strong bonding between the foil surfaces, which could be observed through the wavy interface, typical in the successful explosive welding. The compressive experiments showed a typical cellular metal response with excellent repeatability and a low deviation up to the end of the plateau region. Due to the moderate loading velocity, the dynamic compressive tests revealed negligible strain-rate sensitivity.

The future work should be focused on analysing a higher porosity by changing the thickness of the outer pipe and the rolled foil layer, and the diameter of the inner aluminium bar. Furthermore, it would be meaningful to perform a full strain-rate sensitivity study of the UniPore structures.

Author Contributions: Conceptualisation, K.H., Y.M. and Z.R.; methodology, K.H., Y.M. and Z.R.; software, M.N.; validation, M.N.; investigation, K.H., T.N., L.K.-O. and M.V.; data curation, M.N., T.N. and M.V.; writing—original draft preparation, K.H., Y.M., M.N., L.K.-O., Z.R. and M.V.; writing—review and editing, K.H., Y.M., M.N., L.K.-O., Z.R. and M.V.; supervision, K.H., Y.M. and Z.R.; project administration, K.H., Y.M. and Z.R. All authors have read and agreed to the published version of the manuscript.

Funding: This research was partially funded by the Slovenian Research Agency–ARRS, research core funding No. P2-0063 and the bilateral research project No. BI-HR/18-19-012.

Acknowledgments: The authors acknowledge the financial support from the Slovenian Research Agency–ARRS.

Conflicts of Interest: The authors declare no conflict of interest.

References

1. Lefebvre, L.P.; Banhart, J.; Dunand, D.C. Porous Metals and Metallic Foams: Current Status and Recent Developments. *Adv. Eng. Mater.* **2008**, *10*, 775–787. [CrossRef]
2. Ashby, M.F.; Evans, A.; Fleck, N.A.; Gibson, L.J.; Hutchinson, J.W.; Wadley, H.N.G. *Metal Foams: A Design Guide*; Elsevier Science: Burlington, MA, USA, 2000.
3. Matsumoto, R.; Kanatani, S.; Utsunomiya, H. Filling of surface pores of aluminum foam with polyamide by selective laser melting for improvement in mechanical properties. *J. Mater. Process. Technol.* **2016**, *237*, 402–408. [CrossRef]
4. Duarte, I.; Vesenjak, M.; Krstulović-Opara, L.; Ren, Z. Crush performance of multifunctional hybrid foams based on an aluminium alloy open-cell foam skeleton. *Polym. Test.* **2018**, *67*, 246–256. [CrossRef]
5. Banhart, J. Manufacture, characterisation and application of cellular metals and metal foams. *Prog. Mater. Sci.* **2001**, *46*, 559–632. [CrossRef]
6. Lehmhus, D.; Vesenjak, M.; de Schampheleire, S.; Fiedler, T.; Schampheleire, S.; Fiedler, T. From Stochastic Foam to Designed Structure: Balancing Cost and Performance of Cellular Metals. *Materials* **2017**, *10*, 922. [CrossRef]
7. Hohe, J.; Hardenacke, V.; Fascio, V.; Girard, Y.; Baumeister, J.; Stöbener, K.; Weise, J.; Lehmhus, D.; Pattofatto, S.; Zeng, H.; et al. Numerical and experimental design of graded cellular sandwich cores for multifunctional aerospace applications. *Mater. Des.* **2012**, *39*, 20–32. [CrossRef]
8. García-Moreno, F. Commercial Applications of Metal Foams: Their Properties and Production. *Materials* **2016**, *9*, 85. [CrossRef] [PubMed]
9. Zhang, L.; Tan, J.; Meng, Z.D.; He, Z.Y.; Zhang, Y.Q.; Jiang, Y.H.; Zhou, R. Low elastic modulus Ti-Ag/Ti radial gradient porous composite with high strength and large plasticity prepared by spark plasma sintering. *Mater. Sci. Eng. A* **2017**, *688*, 330–337. [CrossRef]
10. Harih, G.; Borovinsek, M.; Ren, Z.; Dolsak, B. Optimal Products' Hand-Handle Interface Parameter Identification. *Int. J. Simul. Model.* **2015**, *14*, 404–415. [CrossRef]
11. Duarte, I.; Banhart, J. A study of aluminium foam formation—kinetics and microstructure. *Acta Mater.* **2000**, *48*, 2349–2362. [CrossRef]
12. Öchsner, A.; Augustin, C. (Eds.) *Multifunctional Metallic Hollow Sphere Structures*; Springer: Berlin, Germany, 2009; ISBN 978-3-642-00490-2.
13. Lee, Y.-H.; Lee, B.-K.; Jeon, I.; Kang, K.-J. Wire-woven bulk Kagome truss cores. *Acta Mater.* **2007**, *55*, 6084–6094. [CrossRef]
14. Novak, N.; Vesenjak, M.; Ren, Z. Auxetic Cellular Materials-a Review. *J. Mech. Eng.* **2016**, *62*, 485–493. [CrossRef]
15. Soro, N.; Attar, H.; Wu, X.; Dargusch, M.S. Investigation of the structure and mechanical properties of additively manufactured Ti-6Al-4V biomedical scaffolds designed with a Schwartz primitive unit-cell. *Mater. Sci. Eng. A* **2019**, *745*, 195–202. [CrossRef]
16. Nakajima, H. Fabrication, properties and application of porous metals with directional pores. *Prog. Mater. Sci.* **2007**, *52*, 1091–1173. [CrossRef]
17. Shapovalov, V.I. Porous metals. *MRS Bull.* **1994**, *19*, 24–29. [CrossRef]
18. Hokamoto, K.; Vesenjak, M.; Ren, Z. Fabrication of cylindrical unidirectional porous metal with explosive compaction. *Mater. Lett.* **2014**, *137*, 323–327. [CrossRef]
19. Movahedi, N.; Conway, S.; Belova, I.V.; Murch, G.E.; Fiedler, T. Influence of particle arrangement on the compression of functionally graded metal syntactic foams. *Mater. Sci. Eng. A* **2019**, *764*, 138242. [CrossRef]
20. Katona, B.; Szlancsik, A.; Tábi, T.; Orbulov, I.N. Compressive characteristics and low frequency damping of aluminium matrix syntactic foams. *Mater. Sci. Eng. A* **2019**, *739*, 140–148. [CrossRef]
21. Hokamoto, K.; Torii, S.; Kimura, A.; Kai, S. Metal Pipe Joint Body and Manufacturing Method Thereof. Japanese Patent 5848016, 4 December 2015.

22. Vesenjak, M.; Hokamoto, K.; Anžel, I.; Sato, A.; Tsunoda, R.; Krstulović-Opara, L.; Ren, Z. Influence of the explosive treatment on the mechanical properties and microstructure of copper. *Mater. Des.* **2015**, *75*, 85–94. [CrossRef]

23. Nishi, M.; Oshita, M.; Ulbin, M.; Vesenjak, M.; Ren, Z.; Hokamoto, K. Computational Analysis of the Explosive Compaction Fabrication Process of Cylindrical Uni-directional Porous Copper. *Met. Mater. Int.* **2018**, *24*, 1143–1148. [CrossRef]

24. Hokamoto, K.; Shimomiya, K.; Nishi, M.; Krstulović-opara, L.; Vesenjak, M.; Ren, Z. Fabrication of unidirectional porous-structured aluminum through explosive compaction using cylindrical geometry. *J. Mater. Process. Technol.* **2018**, *251*, 262–266. [CrossRef]

25. Vesenjak, M.; Hokamoto, K.; Sakamoto, M.; Nishi, T.; Krstulović-Opara, L.; Ren, Z. Mechanical and microstructural analysis of unidirectional porous (UniPore) copper. *Mater. Des.* **2016**, *90*. [CrossRef]

26. Fiedler, T.; Borovinšek, M.; Hokamoto, K.; Vesenjak, M. High-performance thermal capacitors made by explosion forming. *Int. J. Heat Mass Transf.* **2015**, *83*, 366–371. [CrossRef]

27. Vesenjak, M.; Hokamoto, K.; Matsumoto, S.; Marumo, Y.; Ren, Z. Unidirectional porous metal fabricated by rolling of copper sheet and explosive compaction. *Mater. Lett.* **2016**, *170*, 39–43. [CrossRef]

28. Meyers, M.A.; Murr, L.E. *Shock Waves and High-Strain-Rate Phenomena Effects in Metals*; Springer: New York, NY, USA, 1981.

29. Mayers, M.A. *Dynamic Behavior of Materials*; Wiley Interscience: New York, NY, USA, 1994.

30. Marsh, S.P. *Los Alamos Scientific Laboratory Report*; LA-4167-MS; University of California Press: Oakland, CA, USA, 1969.

31. Marsh, M.A. *LASL Shock Hugoniot Data*; University of California Press: Berkeley, CA, USA, 1980.

32. Fujita, M.; Chiba, A.; Fukuda, I.; Osaka, H.; Baba, F.; Manabe, T. An analysis of multilayered explosive bonding process. *Kyogyo-kayaku* **1987**, *48*, 176–182. (In Japanese)

33. Crossland, B. *Explosive Welding of Metals and Its Application*; Oxford University Press: Oxford, UK, 1982; Volume 8, p. 233.

34. Krstulović-Opara, L.; Surjak, M.; Vesenjak, M.; Tonković, Z.; Kodvanj, J.; Domazet, Ž. Comparison of infrared and 3D digital image correlation techniques applied for mechanical testing of materials. *Infrared Phys. Technol.* **2015**, *73*, 166–174. [CrossRef]

Article

Comparative Study on the Uniaxial Behaviour of Topology-Optimised and Crystal-Inspired Lattice Materials

Chengxing Yang, Kai Xu and Suchao Xie *

Key Laboratory of Traffic Safety on Track, Ministry of Education, School of Traffic & Transportation Engineering, Central South University, Changsha 410075, China; Chengxing_Yang_Hn@163.com (C.Y.); csu_xukai@csu.edu.cn (K.X.)
* Correspondence: xsc0407@csu.edu.cn; Tel.: +86-1370-7481-946

Received: 10 March 2020; Accepted: 7 April 2020; Published: 8 April 2020

Abstract: This work comparatively studies the uniaxial compressive performances of three types of lattice materials, namely face-centre cube (FCC), edge-centre cube (ECC), and vertex cube (VC), which are separately generated by topology optimisation and crystal inspiration. High similarities are observed between the materials designed by these two methods. The effects of design method, cell topology, and relative density on deformation mode, mechanical properties, and energy absorption are numerically investigated and also fitted by the power law. The results illustrate that both topology-optimised and crystal-inspired lattices are mainly dominated by bending deformation mode. In terms of collapse strength and elastic modulus, VC lattice is stronger than FCC and ECC lattices because its struts are arranged along the loading direction. In addition, the collapse strength and elastic modulus of the topology-optimised FCC and ECC are close to those generated by crystal inspiration at lower relative density, but the topology-optimised FCC and ECC are obviously superior at a higher relative density. Overall, all topology-generated lattices outperform the corresponding crystal-guided lattice materials with regard to the toughness and energy absorption per unit volume.

Keywords: lattice material; topology optimisation; crystal inspiration; mechanical properties; energy absorption

---

1. Introduction

As a mimic of nature cellular material, lattice material is designed with its unit cells arranged periodically along tessellation directions. Metallic lattice materials can provide excellent mechanical properties, e.g., ultra light weight, better durability, high specific strength and stiffness, and a superior energy absorption capability [1–4]. Methods for lattice material design can be generally classified into two categories, i.e., manual generation and mathematical generation [5].

Manual generation means designing a lattice material by using beams and trusses with joints modified to create seamless transitions between unit cell elements [5]. There are numerous manually designed lattice materials and some of them are inspired by crystal structures. Typical examples are simple cubic [6], diamond [7], face-centred cubic (FCC) [8], FCCZ (i.e., FCC with enhanced vertical struts) [9], body-centred cubic (BCC) [10], Kagome [11], F2BCC [12]; Gurtner-Durand [13], Octet-truss [14], Octahedron [13], Octahedron-cross [15], Octahedral [16], tetrakaidecahedron [13], rhombic dodecahedron [17], etc. Additive manufacturing (AM) technology can be employed to fabricate lattice materials [18]. Lozanovski et al. [19,20] numerically investigated the strut defects of lattice materials fabricated by the AM method, and a Monte Carlo simulation-based approach was proposed to predict the stiffness of a lattice material with defects.

The mathematical generation method utilises algorithms and constraints to create a lattice structure [5]. Examples of mathematically generated lattice materials are triply periodic minimal surfaces (TPMS) (e.g., skeletal-TPMS and sheet-TPMS lattices) [17,21,22], all face-centred cubic (AFCC) [23], FCC [1], edge-centre cube (ECC) [1], vertex cube (VC) [1], cuttlebone-like lattice (CLL) [24], etc.) The topology optimisation algorithm can also be used to determine the optimal material distribution for lightweight structure design.

Two common solutions for topology optimisation are the optimality criteria (OC) [25] and the Method of Moving Asymptotes (MMA) [26], and details are referred to [27]. Its applications on engineering structures are reviewed by Sigmund and Maute [28] and Kentli [29]. Topology optimisation methods based on discrete elements are the ground structure approach (GSA) [30], solid isotropic material with penalization (SIMP) method [31–34], homogenization method (HM) [35], evolutionary structural optimization (ESO) [36], level-set method (LSM) [37,38], and the hybrid cellular automata (HCA) algorithm [39]. In terms of lattice material design by topology optimisation, Hu et al. [24] proposed a CLL material that exhibited high compression-resistant capability. However, it can only suffer uniaxial loading rather than multiple loadings from triaxial directions. Xiao et al. [1] proposed three topology-optimised lattice materials, i.e., FCC, VC, and ECC, under three different loading modes, which were similar to manually designed lattices. Yang et al. [40] addressed four different lattice cells under various extreme loading modes, which expanded the library of structural metamaterials. However, the differences between topology-optimised and manually generated lattices have not been assessed.

In the present work, our objective is to systematically explore the similarities and differences of three types of lattice materials guided by two methods, i.e., topology optimisation and crystal inspiration. Section 2 presents the detailed design process and topological geometries, as well as the relationship between relative density and aspect ratio for crystal-inspired lattices. Finite element models are built in Section 3, and the results are presented in Section 4, including deformation modes, mechanical properties, and energy absorption. Finally, some meaningful conclusions are reported in Section 5.

2. Cell Architecture Design

2.1. Topology Optimisation Problem

In this study, two kinds of design methods, i.e., topology optimisation and crystal inspiration, are adopted to generate lattice cells. Topology optimisation is carried out in LS-TaSC^{TM} by integrating with Ls-dyna$^®$ (LSTC, Livermore, CA, USA). The implicit algorithm is employed. The optimisation problem can be mathematically stated as follows:

$$\begin{cases} \min & F(x) = c(x) \\ s.t. & \mathbf{K}(x)\mathbf{u} = \mathbf{f}, \\ & M(x)/M_0 \le M_f, \\ & 0 < x_{min} \le x_i \le 1, \ i = 1 \dots n \end{cases} \quad (1)$$

The goal of the topology optimisation in Equation (1) is to minimise the compliance $c(x)$. The design variable x, with $0 < x_{min} \le x \le 1$ is known as relative density. n denotes the total number of design variables. $\mathbf{f}$ and $\mathbf{u}$ are, respectively, the vectors of nodal force and displacement, while $\mathbf{K}(x)$ is the structure stiffness matrix. M_0 and $M(x)$ represent the initial mass and the current mass of the design domain. A constraint is imposed on the mass fraction M_f. The base material is a high-strength steel referred to Yang et al. [41], with the key parameters presented in Table 1. Three different loading conditions, i.e., face-centred loadings, edge-centred loadings and vertexes loadings, represented by arrows, are illustrated in Table 2. The magnitude of loadings (1 N) is well selected to make sure that the design domain only suffers elastic deformation and the loading points are fixed during the optimisation process. Increasing the value of loading may lead to different optimised cells and it may also cause a problem in designing structures with a high level of porosity [1]. The topology optimisation method is a hybrid cellular automata (HCA) algorithm [42], and the solid isotropic material with penalization (SIMP) model is adopted. Two termination conditions are used to stop the optimisation process: (i) the number of iterations exceeds the maximum number of iterations, or (ii) the change in the topology is smaller than the tolerance [43]. Corresponding optimal cells are also presented in Table 2.

Table 1. Material properties of the base material, data from [41].

Density, ρ_0/(kg·m^{-3})	Young's Modulus, E_0/GPa	Poisson's Ratio, μ	Tangent Modulus, E_{tan}/MPa	Yield Strength, σ_y/MPa	Ultimate Strength σ_u/MPa
7850	206	0.26	517	382	482

Table 2. Lattice materials with various relative densities generated from topology optimisation and crystal inspiration.

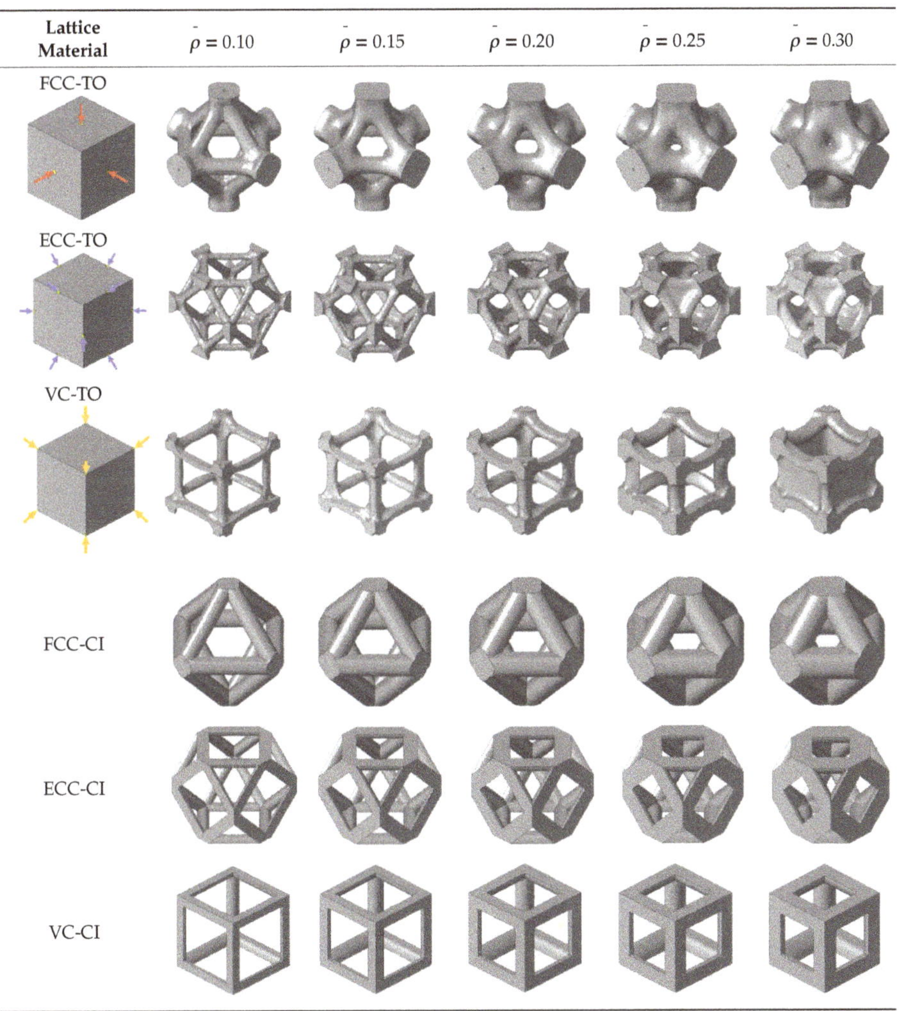

Lattice Material	$\bar\rho = 0.10$	$\bar\rho = 0.15$	$\bar\rho = 0.20$	$\bar\rho = 0.25$	$\bar\rho = 0.30$
FCC-TO					
ECC-TO					
VC-TO					
FCC-CI					
ECC-CI					
VC-CI					

2.2. Analogy with Crystal Structures

Crystal structures are favourable sources of inspiration in manual designs with struts and joints [44]. As shown in Figure 1, a simple cube ($7.5 \times 7.5 \times 7.5$ mm^3) is composed of six faces, twelve edges, and eight vertexes. Therefore, three lattice materials can be generated by, respectively, connecting the face-centre points, edge-centre points, and vertexes. The obtained lattice materials are referred to as the face-centred cube (FCC), edge-centred cube (ECC), and vertex cube (VC), which is consistent with Xiao et al. [1].

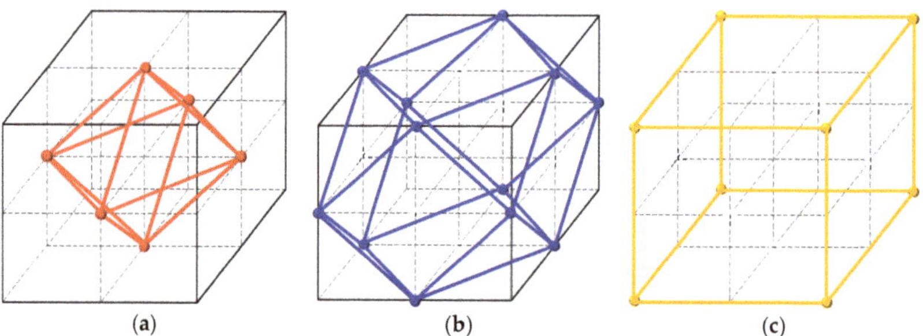

(a)   (b)   (c)

Figure 1. Schematic illustration of manually designed lattice materials: (**a**) FCC (face-centre cube); (**b**) edge-centre cube (ECC); (**c**) vertex cube (VC).

To mathematically analyse a lattice material, the individual struts shown in Figure 2 are considered. The materials overlapping in joints are removed to calculate the actual volume occupied by the lattice material. It is assumed that the lattice is composed of several struts and each strut contains one cylinder and two cones. Thus, the relative density ($\overline{\rho}$) of three cells can be obtained via Equations (3) and (4), respectively.

$$\overline{\rho} = \rho_L / \rho_0 \tag{2}$$

$$\text{FCC and ECC} : \overline{\rho} = 3\pi \left(\frac{d}{l_1}\right)^2 \left(\frac{\sqrt{2}}{2} - \frac{5d}{6l_1}\right) \tag{3}$$

$$\text{VC} : \overline{\rho} = \frac{3\pi}{2} \left(\frac{d}{l_1}\right)^2 \left(\frac{1}{2} - \frac{d}{3l_1}\right) \tag{4}$$

where ρ_0 and ρ_L are the density of the base material and of the lattice material, respectively, d represents strut diameter, l_1 is cube length, and d/l_1 denotes the aspect ratio.

It should be mentioned that FCC and ECC share the same expression of relative density (Equation (3)). Figure 3 plots the curves showing the change in the relative density versus aspect ratio d/l_1 given by Equations (3) and (4). Computer-aided design (CAD) predictions are also given in the same figure to properly validate the responses of both equations. As can be seen from Figure 3, good consistency is observed between theoretical and CAD predictions.

2.3. Numerical Results

Lattice materials derived from topology optimisation (labelled as -TO) and crystal inspiration (labelled as -CI) are presented in Table 2. The relative density of the generated cells varies in the interval [0.1, 0.3], which is comparable to that of aerogel, alumina nanolattices, and other ultralight materials [45]. The boundary surface of the final optimised topology of the lattice is fitted smoothly by using FreeCAD software (open source); thus, the geometry model of the lattice cell is obtained. High consistency is observed in geometry between topology-optimised and crystal-inspired materials, but the solutions provided by topology optimisation are generally non-uniform in terms of strut thickness and joint shape, unlike the regular topology of crystal-inspired cells. By increasing the relative density, walls are likely to be formed between neighbouring struts (see lattice materials with $\bar{\rho} = 0.30$). It should be mentioned that the FCC in this work is similar to the octahedral lattice in [5,46]. The ECC in this work is also named as octahedron in Gross et al. [13], and the VC in this work has the same cell topology as the cubic lattice in Mei et al. [2].

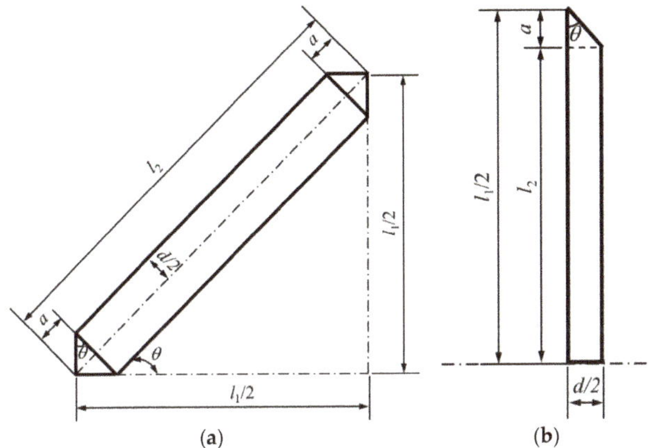

Figure 2. Individual strut geometries of: (**a**) FCC and ECC; (**b**) VC.

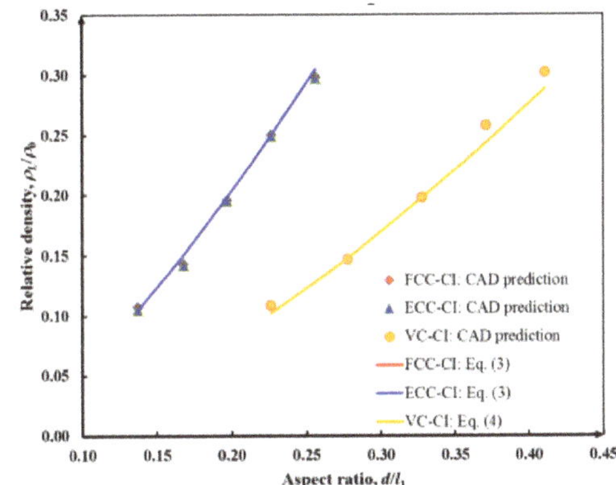

Figure 3. Comparison of the relative density predicted by theory and the CAD model.

3. Finite Element Modelling

Finite element models are built in Ls-dyna® to predict the uniaxial compressive behaviour of the lattice materials. The geometry models of lattice materials are meshed into tetrahedral elements, which are constant stress solid elements. Instead of using periodic boundary conditions on a single unit cell, a 3 × 3 × 3 cell model with dimension of 22.5 × 22.5 × 22.5 mm³ is built. The influence of the number of cells on the elastic modulus of lattice material was numerically studied by Maskery et al. [47], and they found that the converged modulus of the 3 × 3 × 3 cell diamond lattice was just 1% below the upper bound of the theoretical elastic modulus. The FE model of FCC-TO is shown in Figure 4. The lattice materials are placed between two parallel plates, which are modelled as rigid bodies by *Mat.020_Mat_Rigid [48]. The bottom plate is fixed while the top plate impacts the specimen at a constant speed (10 m/s). "*Mat.024_Mat_Piecewise_Linear_Plasticity" [48] is selected to bilinearly approximate the stress–strain curve of elastic-plastic material in Table 1. The impact force is captured by using the *Automatic_Surface_To_Surface contact algorithm [48] applied among specimens and boundaries. In this algorithm, the stiffness of contact elements is penalised by a scale factor [48]. Mesh size is determined, as a result of a sensitivity analysis, as a compromise between accuracy and low computing time. Figure 5 plots the stress–strain curves of the FCC-TO sample calculated by various element sizes, i.e., 0.10~0.15 mm, 0.125~0.200 mm, 0.15~0.25 mm, 0.20~0.30 mm, and 0.25~0.35 mm. In Figure 5, convergence can be obtained when element size is equal to 0.125~0.200 mm. Therefore, this element size is employed for computation throughout this study, which means the main parts of the lattice have an element size of 0.200 mm, while other parts (e.g., joints) are characterised by a smaller element size, i.e., 0.125 mm. The finite element model has been validated by means of experimental results obtained for a body-centred-cubic (BCC) lattice made of 316L stainless steel, taken from [49]. As shown in Figure 5b, a good agreement is observed between experimental and numerical data. It should be noted that the simulation curve in Figure 5b is filtered by using Butterworth Filter with a frequency of 20,000 Hz. The frequency is carefully selected to filter the numerical perturbation while maintaining the characteristics of the stress–strain curve.

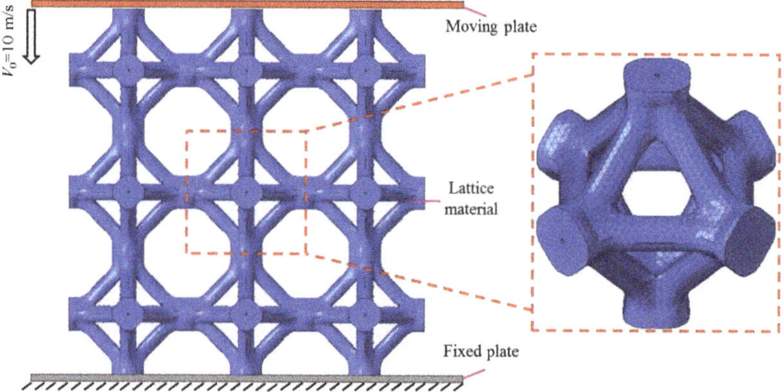

Figure 4. Finite element model of the FCC-TO with its unit cell enlarged.

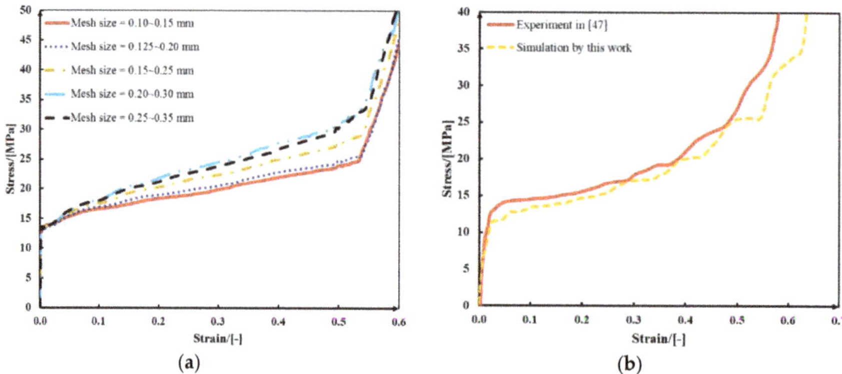

Figure 5. (**a**) Effects of element size on the stress-strain curve of the FCC-TO lattice material; (**b**) validation of the modelling method by using experimental data reproduced from [49], copyright permission: Elsevier, 2020.

4. Results and Discussion

4.1. Deformation Modes

Table 3 reports the Von Mises stress distributions in lattice materials with a relative density of 0.10 up to 50% overall deformation. Generally, each layer of lattice materials collapses almost synchronously, although the middle layer deforms relatively faster than the top and bottom layers, which was also observed in sheet-IWP latice by Al-Ketan et al. [50]. There are strong stress concentrations occurring on some supporting struts and connecting joints for each lattice material. However, the level of equivalent stress is the lowest at the horizontal rods of VC-CI because these struts are orthogonal to the loading direction. The gap between the stresses of vertical and horizontal rods results in unstable deformation in VC-CI at a low relative density, which exhibits a failure band at $\varepsilon = 0.35$. Liu et al. [51] obtained microscopy images of local stress concentrations in VC-CI where slip bands and grain boundaries were found. Comparing the topology-guided and manually generated lattices, FCC-TO and ECC-TO exhibit highly similar deformation modes to the FCC-CI and ECC-CI. However, an obvious difference is found between VC-TO and VC-CI lattices (marked in Table 3) and no shear band is observed in the VC-TO lattice. Thus, the topology-optimised VC lattice undergoes a more stable deforming process.

Table 3. Deformation features of lattice materials with $\overline{\rho} = 0.10$ at different compressive strains.

Lattice Materials	$\varepsilon = 0.015$	$\varepsilon = 0.15$	$\varepsilon = 0.35$	$\varepsilon = 0.50$
FCC-TO	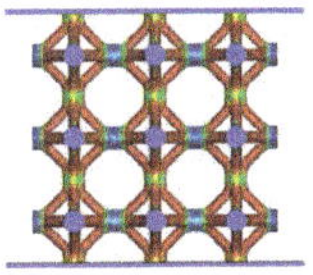	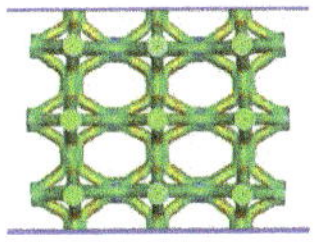		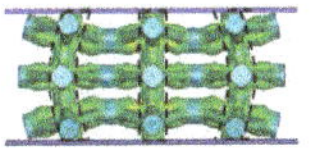
FCC-CI				
ECC-TO			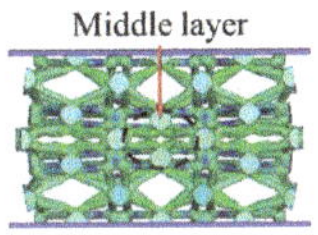 Middle layer	
ECC-CI			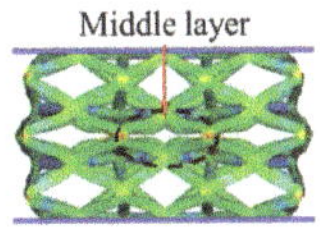 Middle layer	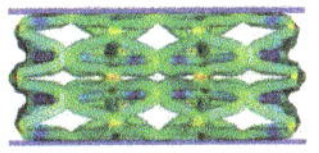
VC-TO	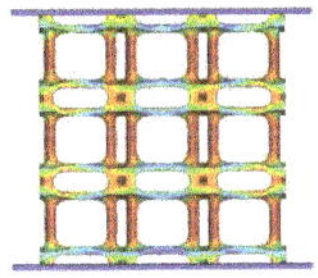	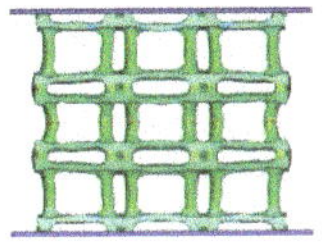	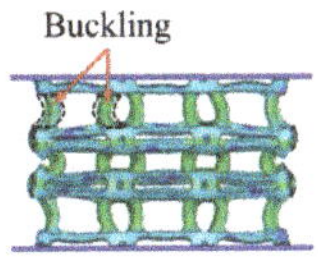 Buckling	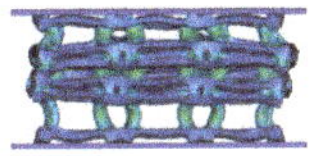
VC-CI			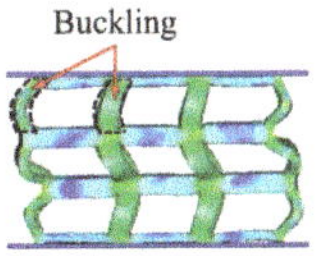 Buckling	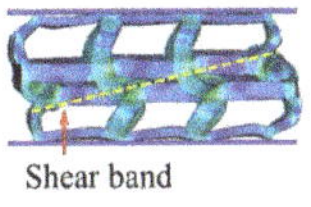 Shear band

It should be noted that the stress level related to the colour fringe of each figure in Tables 3 and 4 varies to have a better visualisation. In general, the blue colour represents the lowest stress level, while the red colour denotes the highest stress level in each figure.

Table 4. Deformation features of lattice materials with $\overline{\rho} = 0.20$ at different compressive strains.

Lattice Materials	$\varepsilon = 0.015$	$\varepsilon = 0.15$	$\varepsilon = 0.35$	$\varepsilon = 0.50$
FCC-TO	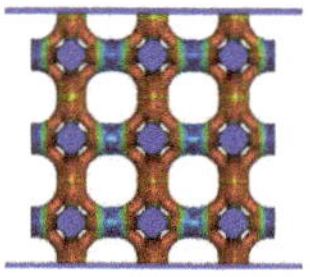	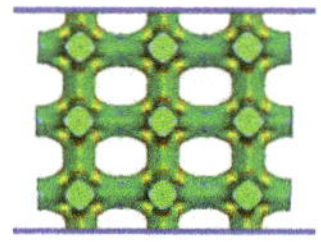	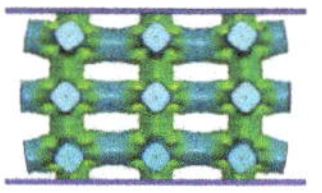	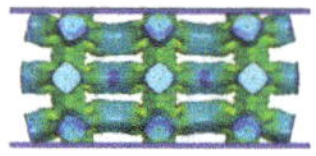
FCC-CI		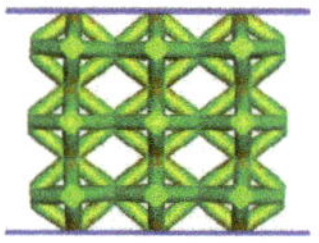	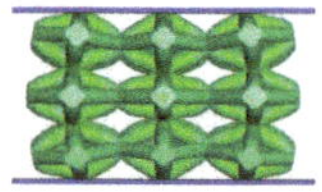	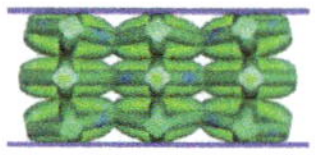
ECC-TO	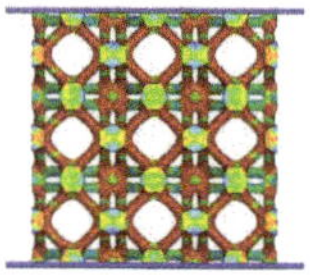			
ECC-CI				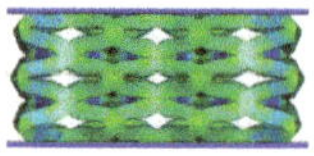
VC-TO	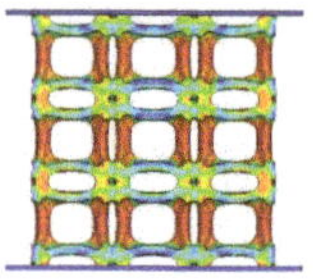	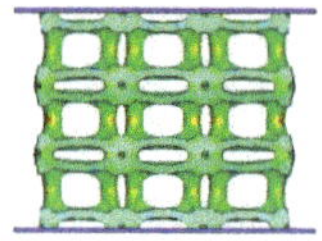	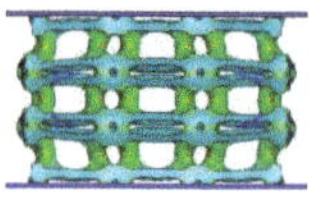	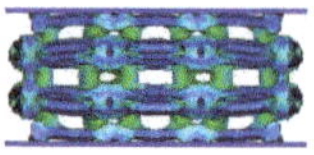
VC-CI				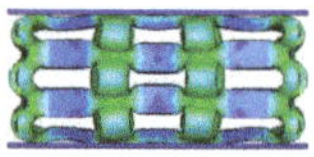

To study the effect of relative density, the deformation features of lattice materials with a relative density ranging from 0.10 to 0.30 are carefully examined. The stress distributions are similar for a relative density higher than 0.20, although the higher the relative density, the higher the Von Mises stress due to the higher stiffness. Thus, the Von Mises stress distributions in the lattice materials with relative density of 0.20 are presented in Table 4. It can be found that lattice materials deform more uniformly and collectively at a higher relative density. In addition, the shear band disappears in the VC-CI lattice when its relative density is higher than 0.20.

4.2. Stress-Strain Curves

The stress-strain (σ-ε) curves extracted from the impact simulation on the designed lattice materials with various relative densities are shown in Figure 6. Additionally, the initial compression stage with $\varepsilon < 5‰$ is enlarged. The stress and strain are calculated in accordance with the initial cross-sectional area and length of each specimen, respectively.

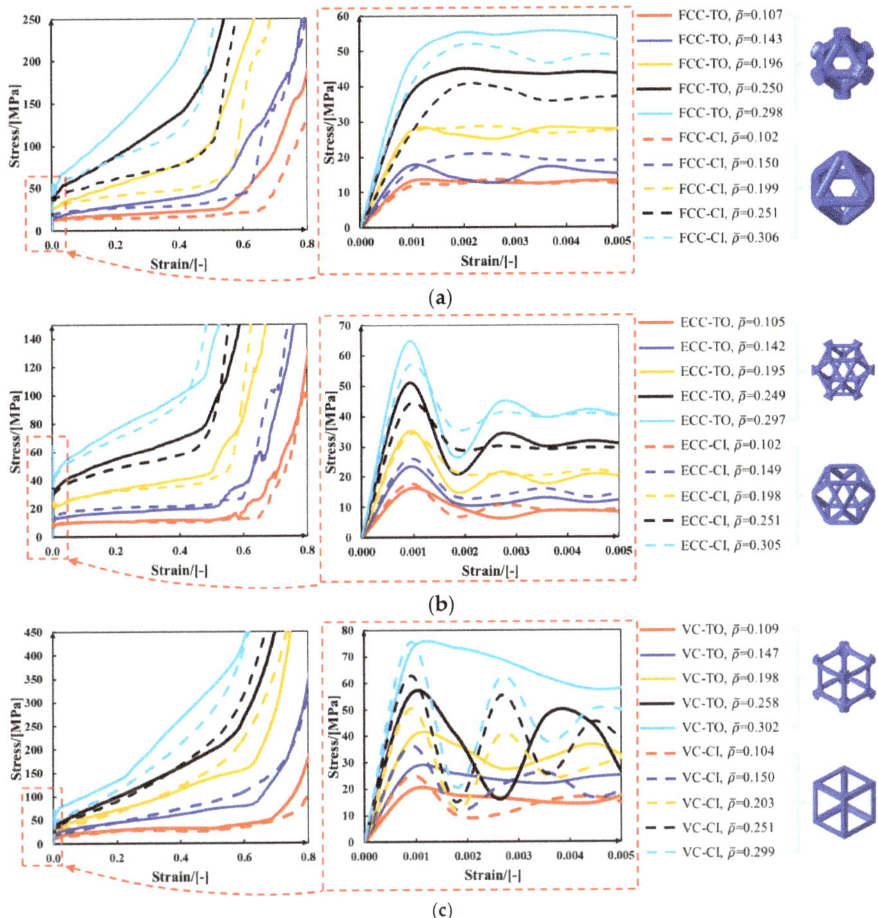

Figure 6. Stress-strain curves of lattice materials generated from two methods with various relative densities: (**a**) FCC; (**b**) ECC; (**c**) VC.

The curves are similar to those of common porous or cellular materials [52], which exhibit three ideal regimes under uniaxial compression, i.e., a pre-collapse regime (including the linear elastic stage), followed by a plateau regime with approximately constant stress and a final densification regime with steeply increasing stress. The linear elastic regime is characterized by elastic modulus (E_L), which is driven by the bending or stretching for the inclined or vertical cell struts/walls, respectively [53]. The plateau regime due to the plastic hinges at sections or joints can be measured by plateau stress (σ_{pl}). The densification regime starts from a densification strain (ε_{cd}), where the individual cell strut/wall comes into contact with each other, and exhibits dramatically increasing strength [53].

The modulus E_L is defined as the slope of the initial linear elastic region; the initial highest peak stress is defined as the strength σ_b; and the plateau stress σ_{pl} is calculated as the arithmetical mean of the stress at a strain interval between 20% and 40% according to ISO 13314: 2011 [54].

$$\sigma_{pl} = \frac{1}{\varepsilon_2 - \varepsilon_1} \int_{\varepsilon_1}^{\varepsilon_2} \sigma(\varepsilon)d\varepsilon, \tag{5}$$

where ε_1 and ε_2 equal 0.2 and 0.4, respectively.

The ε_{cd} is identified by using the energy absorption efficiency (η) method.

$$\eta(\varepsilon) = \frac{1}{\sigma(\varepsilon)} \int_0^{\varepsilon} \sigma(\varepsilon)d\varepsilon. \tag{6}$$

The onset of densification is given by using the following equation:

$$\left. \frac{d\eta(\varepsilon)}{d\varepsilon} \right|_{\varepsilon=\varepsilon_{cd}} = 0, \tag{7}$$

for which the $\eta(\varepsilon)$ reaches a maximum on the $\eta(\varepsilon) - \varepsilon$ curve. For example, the $\eta(\varepsilon) - \varepsilon$ curves of the FCC-TO lattices are depicted in Figure 7.

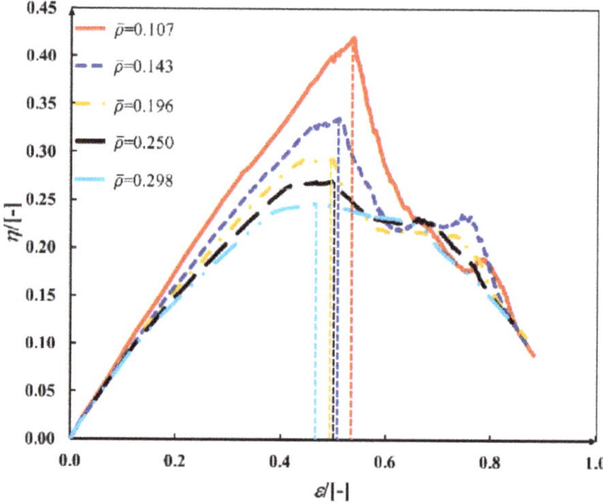

Figure 7. The $\eta(\varepsilon) - \varepsilon$ curves of FCC-TO lattices with various relative densities.

Figure 6 shows that the slopes and initial peak stresses of the σ-ε curves go up as the relative density increases and all the curves reach the maximum strength value at about 1‰~5‰ overall deformation. The plateau regime is steady at lower relative density. However, a nearly linear hardening phenomenon is observed in lattice materials with higher relative density due to an increasing stiffness and hardening effect of base material. With an increasing relative density, the onset of densification also occurs earlier, resulting in a decreasing ε_{cd} (Figure 7). Comparing the two lattice generation methods, topology-guided lattices generally produce higher σ-ε curves than manually generated structures. Specifically, the gap between TO- and CI- lattices is larger at a higher relative density, although the difference is not obvious at a lower relative density. This is because of the cell walls formed in topology-optimised lattices at high mass fraction, as shown in Table 2. In general, σ_{pl} of optimised structures is higher than that of manually designed structures, especially for FCC (see Table 5). However, three cases (i.e., ECC with $\bar{\rho} \approx 0.15$, VC with $\bar{\rho} \approx 0.15$ and 0.20) are excepted, which may be attributed to the slightly lower relative density of topology-optimised lattices comparing with the corresponding crystal-inspired lattices. For example, the $\bar{\rho}$ of VC-TO is 0.147, while that of corresponding VC-CI is 0.150 (see Figure 6).

Table 5. Plateau stress, σ_{pl}/[MPa], of the as-designed lattice materials with various relative densities.

Relative Density, $\bar{\rho}$ [-]	FCC-TO	FCC-CI	Difference	ECC-TO	ECC-CI	Difference	VC-TO	VC-CI	Difference
0.10	20.63	15.59	32.33%	11.43	10.70	6.78%	32.19	28.36	13.49%
0.15	33.69	28.08	19.98%	18.45	21.24	−13.13%	52.96	60.08	−11.84%
0.20	65.43	45.05	45.24%	36.76	34.16	7.63%	90.26	93.16	−3.12%
0.25	110.53	69.13	59.89%	61.00	52.64	15.89%	127.66	124.92	2.19%
0.30	156.75	100.09	56.60%	86.66	80.37	7.82%	190.38	158.42	20.18%

4.3. Mechanical Properties and Energy Absorption

In general, the compressive strength and elastic modulus of lattice materials would increase if the relative density increases because there is a larger amount of material withstanding the impact force. This relationship could be fitted by a power law proposed by Gibson and Ashby [55]. The elastic Modulus, E_L, and collapse strength, σ_b, scale with relative density, $\bar{\rho}$, according to the relationships:

$$\frac{\sigma_b}{\sigma_y} = C_1(\bar{\rho})^{n_1},$$

(8)

$$\frac{E_L}{E_0} = C_2(\bar{\rho})^{n_2},$$

(9)

where n_1 and n_2 represent the structural bending/stretching dominated mode. For bending-dominated structures (e.g., body-centred lattice), $n_1 = 1.5$, and $n_2 = 2$; for stretching-dominated structures (e.g., Octet-truss lattice), both n_1 and n_2 are equal to 1. C_1 and C_2 are constants related to the lattice's architecture as well as the base material properties.

The power law and simulation data are plotted in Figure 8, reflecting that the results of this study are fitted very well with the formulae. The coefficients of constant C (C_1 and C_2) and exponent n (n_1 and n_2) are tabulated in Table 6. It is noticed that the exponents, n_1 and n_2, of the as-designed lattice materials are respectively close to 1.50 and 1.70, which indicates a bending-dominated behaviour mixed with a light stretching mode. Montemayor and Greer [46] pointed out that the FCC lattice behaves as a bending-dominated structure due to the rotation between and within unit cells, although the unit cell is a stretching-dominated structure. This phenomenon can be observed in Table 3. Gross et al. [13] made it clear that the ECC lattice is also a bending-dominated structure. Compression tests were carried out on the VC-CI lattices by Mei et al. [2], where the power law with $n_1 = 1.5$ and $n_2 = 2$ was used to fit the experimental data.

Table 6. Values of the parameters of the power laws used in fitting mechanical properties and energy absorption.

Lattice Material	Collapse Strength, σ_b [MPa]		Elastic Modulus, E_L [GPa]		Toughness, U_T [MJ/m^3]		Strain Energy, W_V [MJ/m^3]	
	C_1	n_1	C_2	n_2	C_3	n_3	C_4	n_4
FCC-TO	1.048	1.548	2.369	1.707	205.554	1.773	566.801	1.782
ECC-TO	1.165	1.482	2.807	1.611	154.456	1.849	354.951	1.849
VC-TO	1.242	1.453	3.512	1.691	147.402	1.438	884.025	1.719
FCC-CI	0.830	1.418	2.106	1.705	107.533	1.502	183.828	1.241
ECC-CI	0.935	1.397	2.613	1.603	103.767	1.625	126.143	1.230
VC-CI	1.250	1.353	3.614	1.608	105.342	1.292	499.610	1.449

Comparing the six bending-dominated structures in Figure 8a,b, it is predicted that VC lattices are stronger than ECC lattices, while ECC lattices are superior to FCC lattices. Additionally, the collapse strength and elastic modulus of FCC-TO and ECC-TO are close to those of FCC-CI and ECC-CI at lower relative density. However, the topology-optimised FCC and ECC obviously outperform the corresponding crystal-inspired structures at a higher relative density. Interestingly, the inverse phenomenon is found in VC lattices where the collapse strength and elastic modulus of VC-CI are higher than those of VC-TO.

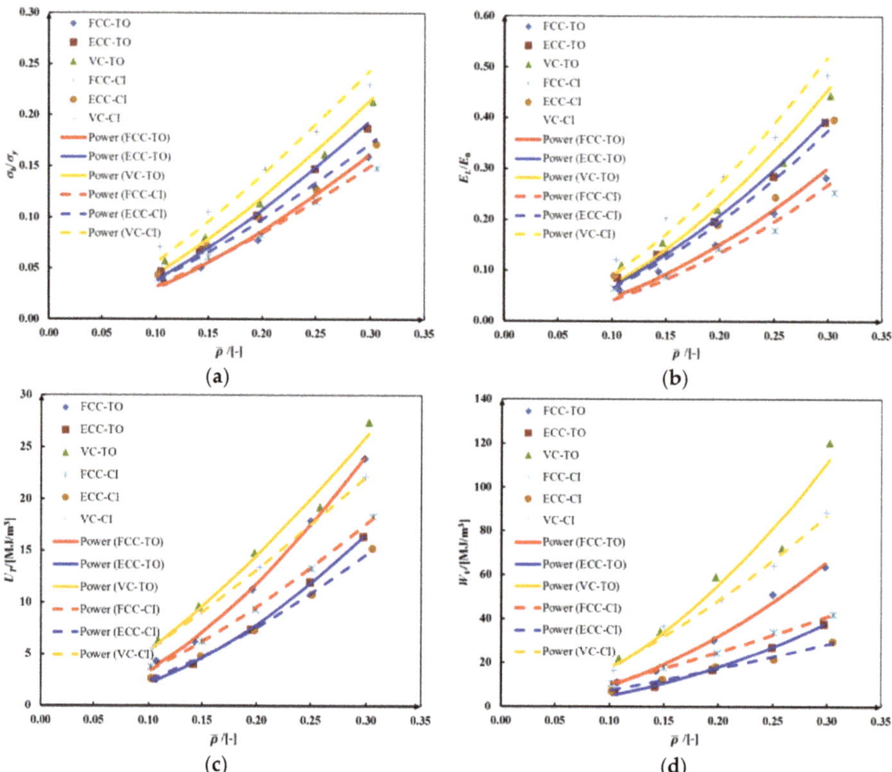

Figure 8. The relationship between the mechanical properties/energy absorption and relative density of lattice materials: (**a**) normalised collapse strength; (**b**) normalised elastic modulus; (**c**) toughness; (**d**) energy absorption per unit volume.

The energy absorption ability, namely the energy absorbed per unit volume of cellular materials, is defined by the area under the stress-strain curve up to the densification strain.

$$W_V = \int_0^{\varepsilon_{cd}} \sigma(\varepsilon)d\varepsilon \tag{10}$$

Among which, the toughness (U_T) is defined as the amount of energy per unit volume up to the strain of 0.25 [50].

$$U_T = \int_0^{\varepsilon_a} \sigma(\varepsilon)d\varepsilon, \quad \varepsilon_a = 0.25 \tag{11}$$

The U_T and W_V are also be fitted by the power law.

$$U_T = C_3(\overline{\rho})^{n_3} \tag{12}$$

$$W_V = C_4(\overline{\rho})^{n_4} \tag{13}$$

As shown in Figure 8c,d, the toughness and energy absorption of all materials rise with an increasing relative density. The relationship among three kinds of lattice materials keeps as VC > FCC > ECC, and the differences become more distinct at a higher relative density. In addition, topology-optimised lattices are characterised by performances better than those of crystal-inspired lattices and the differences also become more distinct with an increasing relative density.

Engineering application generally requires that an energy absorber ought to absorb impact energy as much as possible while maintaining a low maximum stress [55]. Cellular material with high porosity produces low plateau stress; however, the quantity of absorbed energy may also be low. In contrast, a dense material is able to absorb a large amount of energy, but a high plateau stress may exceed the stress limitation [55]. The plot of energy absorption in Figure 8d misses the information of the maximum allowable stresses. Therefore, Figure 9 shows the diagrams of FCC-TO, ECC-TO, and VC-TO, in which the W_V is plotted with respect to the stress (σ) to simplify the relationships of different compressive stages. The maximum allowable stress (σ_{max}) under a certain energy absorption ability can also be obtained. This curve helps to find a cellular material that bears the required σ_{max} by maximising the energy absorption capability [56]. An energy-efficient structure gives a high envelope. In Figure 9, VC-TO lattice is able to absorb more energy than others with the same allowable stress. For instance, if the σ_{max} is 300 MPa, the values of W_V for FCC-TO, ECC-TO, and VC-TO with $\bar{\rho} = 0.196$ are 64 MJ/m^3, 47 MJ/m^3, and 80 MJ/m^3, respectively.

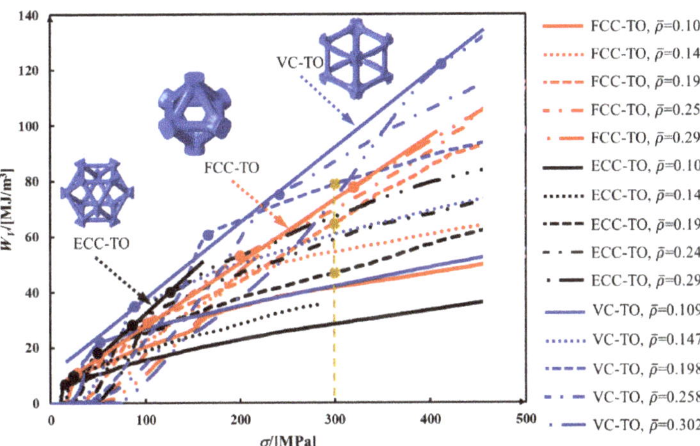

Figure 9. The energy absorption diagrams of topology-optimised lattice materials.

5. Conclusions

Three types of lattice materials, i.e., FCC, ECC, and VC, have been separately designed based on two different methods: topology optimisation and crystal inspiration. Numerical compression tests have been conducted to comparatively characterise their deformation modes, mechanical properties, and energy absorption capability. The main conclusions and contributions are summarised as follows:

a) Topology optimisation-guided lattice materials are highly similar to the corresponding crystal-inspired lattice materials, especially at a low relative density. The topology optimisation-guided lattice materials are generally non-uniform in terms of strut thickness and joints shape, while the crystal-inspired cells are uniform.

b) Formulae relating the relative density ($\bar{\rho}$) and aspect ratio (d/l_1) of crystal-inspired lattices are presented, which has been well validated by CAD predictions.

c) Comparing the topology-guided and manually generated structures, FCC-TO and ECC-TO exhibit a highly similar bending-dominated deformation mode to FCC-CI and ECC-CI, respectively.

However, differences are found between VC-TO and VC-CI lattices. Shear band is observed in VC-CI structures at a low relative density while the VC-TO lattice deforms stably.

d)   In terms of collapse strength and elastic modulus, the VC lattice is stronger than the FCC and ECC lattices because its struts are arranged along the loading direction. On the other hand, topology-generated lattices outperform the corresponding crystal-guided lattices in aspects of toughness and energy absorption per unit volume.

Author Contributions: Conceptualization, C.Y., K.X. and S.X.; methodology, C.Y. and S.X.; software, C.Y. and K.X.; validation, C.Y. and K.X.; formal analysis, C.Y. and S.X; investigation, C.Y., K.X. and S.X.; data curation, C.Y. and K.X.; writing-original draft preparation, C.Y.; writing-review and editing, K.X. and S.X; project administration, S.X.; funding acquisition, S.C.X. All authors have read and agreed to the published version of the manuscript.

Funding: The research was funded by the National Natural Science Foundation of China (No. 51775558) and the Nature Science Foundation for Excellent Youth Scholars of Hunan Province (No. 2019JJ30034).

Acknowledgments: The authors are grateful to Ping Xu and Shuguang Yao at CSU for insightful discussion.

Conflicts of Interest: The authors declare no conflict of interest.

References

1.   Xiao, Z.; Yang, Y.; Xiao, R.; Bai, Y.; Song, C.; Wang, D. Evaluation of topology-optimized lattice structures manufactured via selective laser melting. *Mater. Des.* **2018**, *143*, 27–37. [CrossRef]

2.   Mei, H.; Zhao, R.; Xia, Y.; Du, J.; Wang, X.; Cheng, L. Ultrahigh strength printed ceramic lattices. *J. Alloys Compd.* **2019**, *797*, 786–796. [CrossRef]

3.   Kadic, M.; Milton, G.; van Hecke, M.; Wegener, M. 3D metamaterials. *Nat. Rev. Phys.* **2019**, *1*, 198–210. [CrossRef]

4.   Peng, C.; Tran, P.; Nguyen-Xuan, H.; Ferreira, A.J.M. Mechanical performance and fatigue life prediction of lattice structures: Parametric computational approach. *Compos. Struct.* **2020**, *235*, 111821. [CrossRef]

5.   Helou, M.; Kara, S. Design, analysis and manufacturing of lattice structures: An overview. *Int. J. Comput. Integr. Manuf.* **2017**, *31*, 243–261. [CrossRef]

6.   Daynes, S.; Feih, S.; Lu, W.; Wei, J. Design concepts for generating optimised lattice structures aligned with strain trajectories. *Comput. Methods Appl. Mech. Eng.* **2019**, *354*, 689–705. [CrossRef]

7.   Hazeli, K.; Babamiri, B.; Indeck, J.; Minor, A.; Askari, H. Microstructure-topology relationship effects on the quasi-static and dynamic behavior of additively manufactured lattice structures. *Mater. Des.* **2019**, *176*, 107826. [CrossRef]

8.   Alaña, M.; Lopez-Arancibia, A.; Pradera-Mallabiabarrena, A.; Ruiz de Galarreta, S. Analytical model of the elastic behavior of a modified face-centered cubic lattice structure. *J. Mech. Behav. Biomed. Mater.* **2019**, *98*, 357–368. [CrossRef]

9.   Leary, M.; Mazur, M.; Elambasseril, J.; McMillan, M.; Chirent, T.; Sun, Y.; Qian, M.; Easton, M.; Brandt, M. Selective laser melting (SLM) of AlSi12Mg lattice structures. *Mater. Des.* **2016**, *98*, 344–357. [CrossRef]

10. Sienkiewicz, J.; Płatek, P.; Jiang, F.; Sun, X.; Rusinek, A. Investigations on the Mechanical Response of Gradient Lattice Structures Manufactured via SLM. *Metals* **2020**, *10*, 213. [CrossRef]

11. Gautam, R.; Idapalapati, S. Compressive Properties of Additively Manufactured Functionally Graded Kagome Lattice Structure. *Metals* **2019**, *9*, 517. [CrossRef]

12. Al-Saedi, D.; Masood, S.; Faizan-Ur-Rab, M.; Alomarah, A.; Ponnusamy, P. Mechanical properties and energy absorption capability of functionally graded F2BCC lattice fabricated by SLM. *Mater. Des.* **2018**, *144*, 32–44. [CrossRef]

13. Gross, A.; Pantidis, P.; Bertoldi, K.; Gerasimidis, S. Correlation between topology and elastic properties of imperfect truss-lattice materials. *J. Mech. Phys. Solids* **2019**, *124*, 577–598. [CrossRef]

14. Deshpande, V.; Fleck, N.; Ashby, M. Effective properties of the octet-truss lattice material. *J. Mech. Phys. Solids* **2001**, *49*, 1747–1769. [CrossRef]

15. Nguyen, D.S.; Tran, T.T.; Le, D.K.; Le, V.T. Creation of Lattice Structures for Additive Manufacturing in CAD Environment. In Proceedings of the IEEE International Conference on Industrial Engineering and Engineering Management (IEEM), Bangkok, Thailand, 16–19 December 2018; pp. 396–400.

16. Hedayati, R.; Sadighi, M.; Mohammadi-Aghdam, M.; Zadpoor, A. Analytical relationships for the mechanical properties of additively manufactured porous biomaterials based on octahedral unit cells. *Appl. Math. Model.* **2017**, *46*, 408–422. [CrossRef]

17. Bonatti, C.; Mohr, D. Mechanical performance of additively-manufactured anisotropic and isotropic smooth shell-lattice materials: Simulations & experiments. *J. Mech. Phys. Solids* **2019**, *122*, 1–26.

18. Ling, C.; Cernicchi, A.; Gilchrist, M.; Cardiff, P. Mechanical behaviour of additively-manufactured polymeric octet-truss lattice structures under quasi-static and dynamic compressive loading. *Mater. Des.* **2019**, *162*, 106–118. [CrossRef]

19. Lozanovski, B.; Downing, D.; Tran, P.; Shidid, D.; Qian, M.; Choong, P.; Brandt, M.; Leary, M. A Monte Carlo simulation-based approach to realistic modelling of additively manufactured lattice structures. *Addit. Manuf.* **2020**, *32*, 101092. [CrossRef]

20. Lozanovski, B.; Leary, M.; Tran, P.; Shidid, D.; Qian, M.; Choong, P.; Brandt, M. Computational modelling of strut defects in SLM manufactured lattice structures. *Mater. Des.* **2019**, *171*, 107671. [CrossRef]

21. Peng, C.; Tran, P. Bioinspired functionally graded gyroid sandwich panel subjected to impulsive loadings. *Compos. Part B Eng.* **2020**, *188*, 107773. [CrossRef]

22. Tran, P.; Peng, C. Triply periodic minimal surfaces sandwich structures subjected to shock impact. *J. Sandw. Struct. Mater.* **2020**, 1–30. [CrossRef]

23. Bai, L.; Zhang, J.; Chen, X.; Yi, C.; Chen, R.; Zhang, Z. Configuration optimization design of Ti6Al4V lattice structure formed by SLM. *Materials* **2018**, *11*, 1856. [CrossRef] [PubMed]

24. Hu, Z.; Gadipudi, V.; Salem, D. Topology Optimization of Lightweight Lattice Structural Composites Inspired by Cuttlefish Bone. *Appl. Compos. Mater.* **2018**, *26*, 15–27. [CrossRef]

25. Chiandussi, G. On the solution of a minimum compliance topology optimisation problem by optimality criteria without a priori volume constraint specification. *Comput. Mech.* **2005**, *38*, 77–99. [CrossRef]

26. Zhang, W.; Li, D.; Kang, P.; Guo, X.; Youn, S.-K. Explicit topology optimization using IGA-based moving morphable void (MMV) approach. *Comput. Methods Appl. Mech. Eng.* **2020**, *360*, 112685. [CrossRef]

27. Bendsoe, M.P.; Sigmund, O. *Topology Optimization: Theory, Methods, and Applications*, 2nd ed.; Springer Science & Business Media: Berlin, Germany, 2004.

28. Sigmund, O.; Maute, K. Topology optimization approaches. *Struct. Multidiscip. Optim.* **2013**, *48*, 1031–1055. [CrossRef]

29. Kentli, A. Topology optimization applications on engineering structures. In *Truss and Frames-Recent Advances and New Perspectives*; IntechOpen: London, UK, 2019.

30. Zhang, X.; Ramos, A.S.; Paulino, G.H. Material nonlinear topology optimization using the ground structure method with a discrete filtering scheme. *Struct. Multidiscip. Optim.* **2017**, *55*, 2045–2072. [CrossRef]

31. Costa, G.; Montemurro, M.; Pailhès, J. A 2D topology optimisation algorithm in NURBS framework with geometric constraints. *Int. J. Mech. Mater. Des.* **2018**, *14*, 669–696. [CrossRef]

32. Costa, G.; Montemurro, M.; Pailhès, J. Minimum length scale control in a NURBS-based SIMP method. *Comput. Methods Appl. Mech. Eng.* **2019**, *354*, 963–989. [CrossRef]

33. Costa, G.; Montemurro, M.; Pailhès, J. NURBS hyper-surfaces for 3D topology optimization problems. *Mech. Adv. Mater. Struct.* **2019**, *1582826*, 1–20. [CrossRef]

34. Lazarov, B.S.; Wang, F.; Sigmund, O. Length scale and manufacturability in density-based topology optimization. *Arch. Appl. Mech.* **2016**, *86*, 189–218. [CrossRef]

35. Geoffroy-Donders, P.; Allaire, G.; Pantz, O. 3-d topology optimization of modulated and oriented periodic microstructures by the homogenization method. *J. Comput. Phys.* **2020**, *401*, 108994. [CrossRef]

36. Da, D.; Xia, L.; Li, G.; Huang, X. Evolutionary topology optimization of continuum structures with smooth boundary representation. *Struct. Multidiscip. Optim.* **2017**, *57*, 2143–2159. [CrossRef]

37. Allaire, G.; Jouve, F.; Toader, A.-M. Structural optimization using sensitivity analysis and a level-set method. *J. Comput. Phys.* **2004**, *194*, 363–393. [CrossRef]

38. Yamada, T.; Izui, K.; Nishiwaki, S.; Takezawa, A. A topology optimization method based on the level set method incorporating a fictitious interface energy. *Comput. Methods Appl. Mech. Eng.* **2010**, *199*, 2876–2891. [CrossRef]

39. Afrousheh, M.; Marzbanrad, J.; Göhlich, D. Topology optimization of energy absorbers under crashworthiness using modified hybrid cellular automata (MHCA) algorithm. *Struct. Multidiscip. Optim.* **2019**, *60*, 1021–1034. [CrossRef]

40. Yang, C.; Xu, P.; Xie, S.; Yao, S. Mechanical performances of four lattice materials guided by topology optimisation. *Scr. Mater.* **2020**, *178*, 339–345. [CrossRef]

41. Yang, C.; Xu, P.; Yao, S.; Xie, S.; Li, Q.; Peng, Y. Optimization of honeycomb strength assignment for a composite energy-absorbing structure. *Thin-Walled Struct.* **2018**, *127*, 741–755. [CrossRef]

42. Tovar, A. Bone Remodelling as a Hybrid Cellular Automaton Optimization Process. Ph.D. Thesis, University of Notre Dame, Cotabato City, Philippines, 2004.

43. LS-TaSC™. *Theory Manual (Version 4.0) of Topology and Shape Computations*; Livermore Software Technology Corporation: Livermore, CA, USA, 2018.

44. Tro, N.J. *Chemistry: A Molecular Approach*, 2nd ed.; Prentice Hall: Upper Saddle River, NJ, USA, 2010.

45. Yuan, S.; Chua, C.; Zhou, K. 3D-Printed Mechanical Metamaterials with High Energy Absorption. *Adv. Mater. Technol.* **2019**, *4*, 1800419. [CrossRef]

46. Montemayor, L.; Greer, J. Mechanical Response of Hollow Metallic Nanolattices: Combining Structural and Material Size Effects. *J. Appl. Mech.* **2015**, *82*, 071012. [CrossRef]

47. Maskery, I.; Aremu, A.O.; Parry, L.; Wildman, R.D.; Tuck, C.J.; Ashcroft, I.A. Effective design and simulation of surface-based lattice structures featuring volume fraction and cell type grading. *Mater. Des.* **2018**, *155*, 220–232. [CrossRef]

48. Ls-dyna. *Keyword User's Manual Volume I (Version 971 R8.0)*; Livermore Software Technology Corporation: Livermore, CA, USA, 2015.

49. Zhang, L.; Feih, S.; Daynes, S.; Chang, S.; Wang, M.; Wei, J.; Lu, W. Energy absorption characteristics of metallic triply periodic minimal surface sheet structures under compressive loading. *Addit. Manuf.* **2018**, *23*, 505–515. [CrossRef]

50. Al-Ketan, O.; Rowshan, R.; Abu Al-Rub, R. Topology-mechanical property relationship of 3D printed strut, skeletal, and sheet based periodic metallic cellular materials. *Addit. Manuf.* **2018**, *19*, 167–183. [CrossRef]

51. Liu, Y.; Li, S.; Zhang, L.; Hao, Y.; Sercombe, T. Early plastic deformation behaviour and energy absorption in porous β-type biomedical titanium produced by selective laser melting. *Scr. Mater.* **2018**, *153*, 99–103. [CrossRef]

52. Sun, Y.; Li, Q.M. Dynamic compressive behaviour of cellular materials: A review of phenomenon, mechanism and modelling. *Int. J. Impact Eng.* **2018**, *112*, 74–115. [CrossRef]

53. Yang, L.; Mertens, R.; Ferrucci, M.; Yan, C.; Shi, Y.; Yang, S. Continuous graded Gyroid cellular structures fabricated by selective laser melting: Design, manufacturing and mechanical properties. *Mater. Des.* **2019**, *162*, 394–404. [CrossRef]

54. ISO 13314. *Mechanical Testing of Metals-Ductility Testing-Compression Test for Porous and Cellular Metals*; International Organization for Standardization: Geneva, Switzerland, 2011.

55. Gibson, L.J.; Ashby, M.F. *Cellular Solids: Structure and Properties*, 2nd ed.; Cambridge University Press: Cambridge, UK, 1997.

56. Yu, S.; Sun, J.; Bai, J. Investigation of functionally graded TPMS structures fabricated by additive manufacturing. *Mater. Des.* **2019**, *182*, 108021. [CrossRef]

 *metals* 

Article

Elasto-Plastic Behaviour of Transversely Isotropic Cellular Materials with Inner Gas Pressure

Zhimin Xu [1], Kangpei Meng [2], Chengxing Yang [2], Weixu Zhang [1,*], Xueling Fan [1] and Yongle Sun [2,*]

1   State Key Laboratory for Strength and Vibration of Mechanical Structures, School of Aerospace Engineering, Xi'an Jiaotong University, Xi'an 710049, China
2   School of Mechanical, Aerospace and Civil Engineering, The University of Manchester, Sackville Street, Manchester M13 9PL, UK
*   Correspondence: zhangwx@mail.xjtu.edu.cn (W.Z.); yongle.sun@manchester.ac.uk (Y.S.);
     Tel.: +86-029-82668-754 (W.Z.); +44-016-1275-1916 (Y.S.)

Received: 7 July 2019; Accepted: 14 August 2019; Published: 16 August 2019

Abstract: The fabrication process of cellular materials, such as foaming, usually leads to cells elongated in one direction, but equiaxed in a plane normal to that direction. This study is aimed at understanding the elasto-plastic behaviour of transversely isotropic cellular materials with inner gas pressure. An idealised ellipsoidal-cell face-centred-cubic foam that is filled with gas was generated and modelled to obtain the uniaxial stress–strain relationship, Poisson's ratio and multiaxial yield surface. The effects of the elongation ratio and gas pressure on the elasto-plastic properties for a relative density of 0.5 were investigated. It was found that an increase in the elongation ratio caused increases in both the elastic modulus and yield stress for uniaxial loading along the cell elongation direction, and led to a tilted multiaxial yield surface in the mean stress and Mises equivalent stress plane. Compared to isotropic spheroidal-cell foams, the size of the yield surface of the ellipsoidal-cell foam is smaller for high-stress triaxiality, but larger for low-stress triaxiality, and the yield surface rotates counter-clockwise with the Lode angle increasing. The gas pressure caused asymmetry of the uniaxial stress–strain curve (e.g., reduced tensile yield stress), and it increased the nominal plastic Poisson's ratio for compression, but had the opposite effect for tension. Furthermore, the gas pressure shifted the yield surface towards the negative mean stress axis with a distance equal to the gas pressure. The combined effects of the elongation ratio and gas pressure are complicated, particularly for the elasto-plastic properties in the plane in which the cells are equiaxed.

Keywords: foam; enclosed gas; anisotropy; elasticity; plasticity; multiaxial yielding

---

1. Introduction

Cellular materials, either natural or manmade, are unique with regard to mechanical, thermal, acoustic and electromagnetic properties, benefitting from their high porosity, which reduces the overall density, enhances the energy absorption capacity and enables the integration of multiple functions. Foams, made of metals, polymers, ceramics, etc., are typical cellular materials, and they are widely used in transport, aerospace, defence, building and biomedical industries [1,2].

In the fabrication of foams, the cell structure is determined by the foaming process, which is sensitive to the foaming agent used and the state of the base material during the foaming [3]. It is common that the cell structure is anisotropic after fabrication, particularly when gravity plays an important role in the foaming process. Foams usually consist of cells which are elongated in one direction, but equiaxed in the plane perpendicular to the elongation direction, exhibiting transversely isotropic structural characteristics [4]. However, most previous studies focused on the elasto-plastic behaviour of presumably isotropic foams [5–8], and there is still a paucity of experimental and modelling data for better understanding the anisotropic elasto-plastic behaviour of foams. Therefore, more studies on anisotropic foams are needed for both uniaxial and multiaxial loadings, which are pertinent to applications of foams as energy absorbers and cores of sandwich structures.

Closed-cell foams contain inner trapped gas in the cells after fabrication and the inner gas pressure can considerably affect the macroscopic elasto-plastic properties [1]. It is also of fundamental significance and scientific interest to study the effect of gas pressure on the elasto-plastic behaviour of closed-cell foams. For static loading, Ozgur et al. [9] analysed the effect of gas pressure on the elastic properties of cellular materials using 2D finite element (FE) models, in which hexagonal, rectangular and circular cells were considered. Öchsner and Mishuris [10] developed a 3D simple cubic cell model and numerically analysed the gas effects on the stress–strain relationship, Poisson's ratio, Young's modulus and yield surface. Zhang et al. [11] established a theoretical model using second-order moment of stress with consideration of inner gas pressure, and they overcame the limitation of an earlier theoretic model [12] and clarified the gas effect. Based on the Gurson yield function [13], Guo et al. [14–17] developed a thick-walled spherical unit cell model and systematically studied the yield behaviour of metal foams with inner gas pressure. For dynamic loading, Sun and Li [18] investigated the effects of gas pressure on the dynamic strength and deformation of cellular materials, and their findings have been applied to explain experimental observations on dynamic compressive behaviour of closed-cell foams [2]. However, these previous studies all focused on isotropic cellular materials. There is still a lack of modelling studies on the elasto-plastic behaviour of anisotropic cellular materials with inner gas pressure.

In this study, an idealised transversely isotropic cellular material with inner gas pressure, i.e., gas-filled ellipsoidal-cell face-centred-cubic (FCC) foam, is studied numerically. The ellipsoidal cell is elongated in one direction, but equiaxed in the plane normal to the elongation direction. Different elongation ratios are considered to investigate the effects of anisotropy on static elasto-plastic properties (e.g., uniaxial stress–strain relationship, Poisson's ratio and multiaxial yield surface). The effects of gas pressure on the anisotropic elasto-plastic properties are also investigated.

2. Material and Methods

The typical cell structure of a transversely isotropic closed-cell foam is shown in Figure 1a, along with the inner gas pressure. To facilitate modelling without losing key physics, one representative volume (RV) unit is considered, as shown in Figure 1b. The RV simplification of geometry implies that the cells are periodically arranged in 3D space, as well as cell deformation, in contrast to the random distribution of irregular cells and cell deformation in an actual closed-cell foam that is normally seen. Nevertheless, provided that the closed-cell foam possesses transversely isotropic macro-properties, a mechanical model based on RV geometry is deemed reliable to capture qualitative behaviour and gain general insights. The enhanced computational efficiency obtained by the geometric simplification enables the analysis of sufficient loading cases in modelling to investigate both the uniaxial and multiaxial elasto-plastic behaviour of the closed-cell foam. A similar RV approach has been widely used for the analysis of the elasto-plastic behaviour of cellular materials [10,19,20].

The RV unit is idealised to be an FCC unit with ellipsoidal cells, as shown in Figure 1b. The ellipsoid is elongated in the y direction (Figure 1c) and mathematically described as

$$\frac{(x^2 + z^2)}{a_1{}^2} + \frac{y^2}{a_2{}^2} \leq 1 \tag{1}$$

where a_1 and a_2 are the axial radii in the isotropic plane (i.e., x-z plane) and along the elongation direction (i.e., y direction), respectively. The elongation ratio is thus defined as $R = a_2/a_1$.

The relative density of a cellular material is

$$f = \rho/\rho_s \tag{2}$$

where ρ and ρ_s are the densities of the cellular material and the base material, respectively. For the closed-cell FCC foam, the relative density can also be determined from geometric parameters, viz.

$$f = 1 - \frac{16\pi a_1{}^2 a_2}{3L^3} \tag{3}$$

Here, L is the size of the cubic unit, which is kept constant. We adopted $f = 0.5$ in this study. It should be noted that, for given R, L, and f, the values of a_1 and a_2 can be uniquely determined.

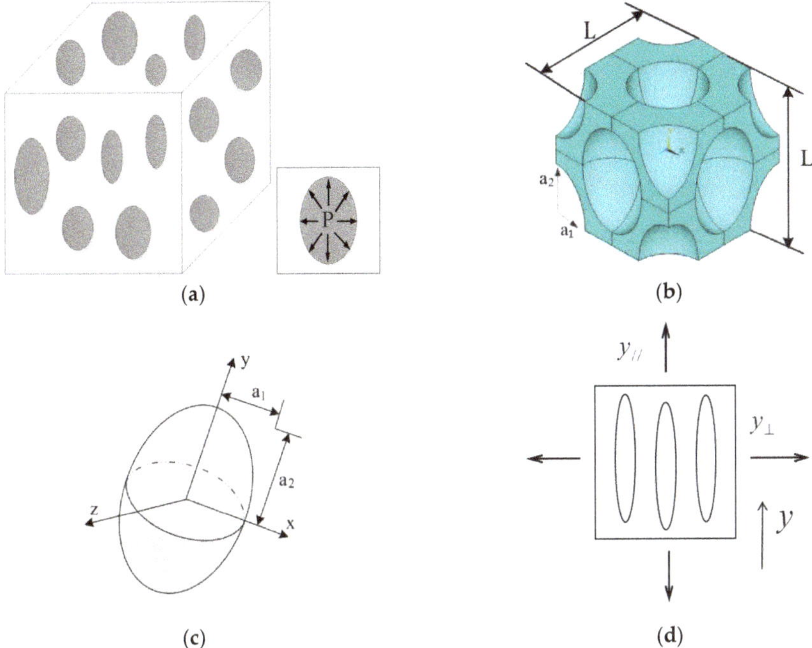

Figure 1. (**a**) Typical cell structure of transversely isotropic closed-cell foam with inner gas pressure; (**b**) representative volume (RV) unit of ellipsoidal-cell face-centred-cubic (FCC) foam; (**c**) orientation of the ellipsoidal cell, which is equiaxed in the x-z plane, but elongated in the y direction; (**d**) definition of directions parallel and perpendicular to the elongation direction (i.e., y-direction).

The multiaxial plastic behaviour of cellular materials is complicated. It has been demonstrated that the yield surface of an isotropic foam is dependent on both the first and second stress invariants [2,6]. For transversely isotropic cellular materials, the yield surface may be more complicated and the locus of yield points may not be unique if only the first and second stress invariants are used in characterisation. Therefore, a full description of the stress state should be provided. In general, three parameters are needed to determine each point in the principal stress space, and here, the mean stress, Mises equivalent stress and Lode angle are adopted. It should be noted that these stress parameters are solely associated with the type of loading and are independent of the constitutive behaviour of the material which is either isotropic or not isotropic. The mean stress is expressed as

$$\sigma_m = \frac{\sigma_{kk}}{3} \tag{4}$$

with Einstein's summation convention applied. The familiar Mises equivalent stress is given by

$$\sigma_e = \sqrt{3J_2} = \sqrt{\frac{3}{2}s_{ij}s_{ij}} \tag{5}$$

where J_2 is the second deviatoric stress invariant, and s_{ij} is the deviatoric stress, i.e., $s_{ij} = \sigma_{ij} - \sigma_m \delta_{ij}$, ($i$, $j = 1, 2, 3$). The stress triaxiality is thus obtained via

$$X_\Sigma = \frac{\sigma_m}{\sigma_e} \tag{6}$$

The Lode angle is defined as follows:

$$L = -\cos(3\theta) = -\left(3\frac{\sqrt[3]{\tfrac{1}{2}J_3}}{\sigma_e}\right)^3 = -\frac{27}{2}\frac{J_3}{\sigma_e^3} = -\frac{27}{2}\frac{Det(s_{ij})}{\sigma_e^3} \tag{7}$$

where L is the Lode parameter, θ is the Lode angle and J_3 is the third deviatoric stress invariant.

It is assumed that the principal stress direction coincides with the axial direction of the ellipsoidal cell, i.e., $\sigma_1 = \sigma_{11}$, $\sigma_2 = \sigma_{22}$ and $\sigma_3 = \sigma_{33}$. Accordingly, the relationship between the principal stress, mean stress, Mises equivalent stress and Lode angle is given as follows:

$$\frac{3}{2\sigma_e}\{\sigma_1, \sigma_2, \sigma_3\} = \left\{-\cos\left(\theta + \frac{\pi}{3}\right), -\cos\left(\theta - \frac{\pi}{3}\right), \cos\theta\right\} + \frac{3}{2\Sigma}\{1,\,1,\,1\} \tag{8}$$

In the ellipsoidal-cell FCC foam model, only one eighth of the RV unit is considered (Figure 2a) owing to symmetry. Multiaxial loading is applied according to Equation (8), as shown in Figure 2b, where T_1, T_2 and T_3 are axial loads in three principal directions. Any state of loading stress can be obtained by varying the ratio of the three axial loads.

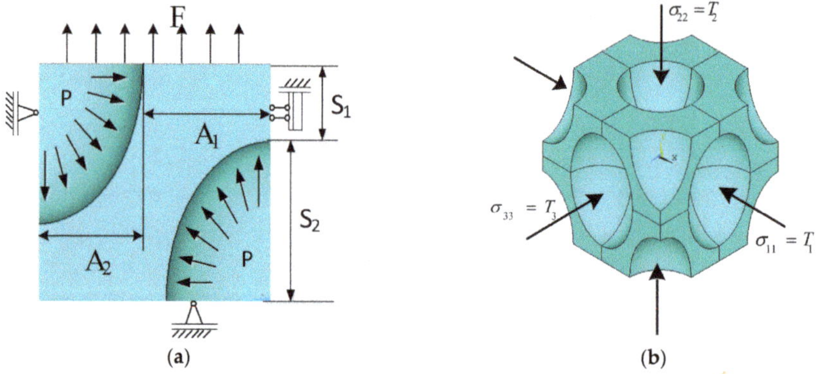

Figure 2. (**a**) Schematic of ellipsoidal-cell face-centred-cubic (FCC) foam model (only 1/8 volume is considered owing to symmetry); (**b**) application of multiaxial loading.

Details of the loading process are described below:

1) Inner gas pressure is applied and meanwhile, an additional load is applied to balance the inner gas pressure. Note that the gas pressure refers to the gauge pressure after the subtraction of ambient air pressure. Öchsner and Mishuris [10] concluded that the inner gas pressure only causes slight deformation and thus the gas pressure is assumed to be constant throughout the loading process. Taking a uniaxial loading case as an example, the area A_1 of the solid in Figure 2a is on the loading boundary, where a balancing load to the inner gas pressure P needs to be applied, in addition to the main load. The equilibrium equation is given as

$$\int_{A_2} P\,dA_2 = -\int_{A_1} F'\,dA_1 \tag{9}$$

where A_1 and A_2 are the areas of the solid and gas, respectively, as shown in Figure 2a. F' denotes the balancing load to the inner gas pressure P. The method for balancing gas pressure is the same when applying loads in other directions;

2) The relationship between triaxial loads and invariant stress parameters has been given in Equation (8). Accordingly, the loads to be applied in the three axial directions are calculated and determined for different stress states. This provides guidance on the application of multiaxial loads to maximise the attainable stress states in the principal stress space, and the multiaxial loading is applied in a proportional manner;

3) The macroscopic stress and strain is calculated. Taking uniaxial loading as an example, the macroscopic stress can be calculated using the following formula:

$$\bar{\sigma} = \frac{1}{A_1 + A_2}\left(\int_{A_1} F dA_1 + \int_{A_2} P dA_2 + \int_{A_1} F' dA_1 \right) \tag{10}$$

where F is the main load applied to the FCC foam. Taking into account Equation (9), the macroscopic stress can be expressed as

$$\bar{\sigma} = \frac{1}{A_1 + A_2}\int_{A_1} F dA_1 \tag{11}$$

The corresponding macroscopic strain $\bar{\varepsilon}$ can be expressed as

$$\bar{\varepsilon} = \bar{\varepsilon}_F - \bar{\varepsilon}_P - \bar{\varepsilon}_{F'} \tag{12}$$

where $\bar{\varepsilon}_F$, $\bar{\varepsilon}_P$ and $\bar{\varepsilon}_{F'}$ are the strains corresponding to the main load F, inner gas pressure P and the balancing load F' to the gas pressure, respectively. Similarly, macroscopic stresses and strains in other directions can be obtained. Finally, the macroscopic Mises equivalent stress and equivalent strain can be calculated via

$$\bar{\sigma}_e = \sqrt{\frac{3}{2}\bar{S}_{ij}\bar{S}_{ij}} \tag{13}$$

$$\bar{\varepsilon}_e = \sqrt{\frac{2}{3}\bar{e}_{ij}\bar{e}_{ij}} \tag{14}$$

where $\bar{S}_{ij}$ and $\bar{e}_{ij}$ are the macroscopic deviatoric stress and strain tensors, respectively.

General purpose FE software ANSYS® was employed in numerical modelling. Linear solid elements (ANSYS designation SOLID185) were used and the FE mesh consisted of 61,091 elements, as shown in Figure 3. A mesh sensitivity analysis was performed, which showed that the numerical results of stress and strain were unchanged when further refining the mesh.

It is assumed that the base material conforms to Hooke's elasticity law and von Mises plasticity theory, i.e.,

$$\sigma = \begin{cases} E_s\varepsilon, & \text{for } \varepsilon \leq \varepsilon_0 \\ \sigma_s + E_p(\varepsilon - \varepsilon_0), & \text{for } \varepsilon > \varepsilon_0 \end{cases} \tag{15}$$

where E_s and E_p are the elastic modulus and plastic hardening modulus, respectively; σ_s and ε_0 are the initial yield stress and yield strain, respectively. For a qualitative study on cellular materials in general, the following parameters are assumed: E_s = 3.5 GPa, E_p = 0.1 GPa, σ_s = 80 MPa and Poisson's ratio = 0.33.

The initial yield stress of the ellipsoidal-cell FCC foam is defined as the intersection of the extrapolated linear elastic portion and plastic portion in the macroscopic stress–strain curve, as proposed by Deshpande and Fleck [21]. For multiaxial loading, the Mises equivalent stress–strain curve and mean stress–strain curve are used to determine the initial yield surface. The Poisson's ratio of the ellipsoidal-cell FCC foam is defined as the absolute value of the ratio of macroscopic strains perpendicular and parallel to the loading direction, respectively.

Figure 3. Finite element (FE) mesh of ellipsoidal-cell face-centred-cubic (FCC) foam model (only 1/8 volume is considered owing to symmetry).

3. Results and Discussion

3.1. Uniaxial Stress-Strain Relationship

Figure 4 shows the uniaxial stress–strain curves of the FCC foam loaded along the ellipsoidal-cell elongation direction, when different values of gas pressure and elongation ratio are considered. For a typical elongation ratio ($R = 2$), increasing the gas pressure does not change the linear elastic portion of the stress–strain curve, but the plastic portion is significantly affected, as shown in Figure 4a. Without inner gas pressure, the stress–strain curve is antisymmetric between tension and compression, but such anti-symmetry is violated by the presence of gas pressure. When the gas pressure increases, the tensile yield stress markedly decreases, while the compressive yield stress only slightly decreases (in magnitude). For spheroidal-cell foam, Xu et al. [20] found a similar effect of gas pressure on tensile yield stress, but a slight increase in compressive yield stress due to gas pressure, in contrast to the observation in Figure 4a. It is surmised that the gas pressure can exacerbate the deformation concentration at the end of the long axis when compressive load is applied along the long axis, thereby promoting macroscopic yielding at a lower stress level for the ellipsoidal-cell foam, compared to the compressive behaviour of spheroidal-cell foam [20]. Nevertheless, for both spheroidal-cell and ellipsoidal-cell foams, the effect of gas pressure on the uniaxial stress–strain curve is more pronounced in tension than in compression.

Figure 4b shows the uniaxial stress–strain curves of the gas-filled ellipsoidal-cell FCC foam with different elongation ratios. For a given value of gas pressure ($P = 15$ MPa), increasing the elongation ratio leads to an increase in yield stress for both tension and compression, to a lesser extent for the latter. The elastic modulus (i.e., slope of the linear elastic portion) also increases. The above observation demonstrates a strong effect of the elongation ratio on the uniaxial elasto-plastic behaviour of the gas-filled FCC foam loaded along the cell elongation direction. The increases in both yield stress and elastic modulus can be attributed to the fact that a higher elongation ratio implies more material is distributed to bear the load applied in the cell elongation direction.

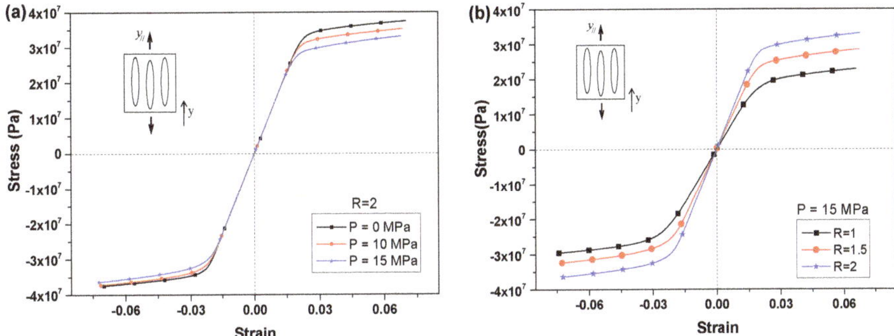

Figure 4. Uniaxial stress–strain curves of the ellipsoidal-cell face-centred-cubic (FCC) foam loaded along the cell elongation direction for different values of inner gas pressure (**a**) and elongation ratio (**b**).

Figure 5 shows the uniaxial stress–strain curves of the FCC foam when the loading direction is perpendicular to the cell elongation direction. Figure 5a shows the gas effect when the elongation ratio is two. Again, the gas pressure hardly affects the elastic modulus and the effect of gas pressure on the tensile stress–strain curve is similar to that in the parallel loading case (Figure 4a). However, the gas pressure enhances the magnitude of compressive yield stress, unlike the decreasing trend found in the parallel loading case (Figure 4a). It is expected that the gas pressure can contribute to an enhanced resistance of the FCC foam to compressive load, since the gas pressure can be directly additive to the compressive load-bearing capacity, except that undesirable deformation potentially induced by gas pressure could cause an adverse effect (e.g., slight decrease in compressive yield stress, as observed in Figure 4a). Figure 5b shows the effect of the elongation ratio on the uniaxial stress–strain curve of the gas-filled ellipsoidal-cell FCC foam. It appears that increasing the elongation ratio does not affect the elastic modulus, but causes a reduction in tensile yield stress, in contrast to the increasing trend found in the parallel coating case (Figure 4b). The effect of the elongation ratio on the compressive yield stress is similar between the perpendicular and parallel loading cases.

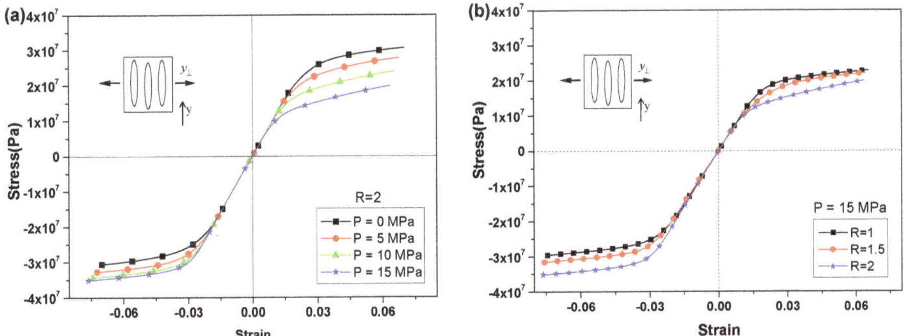

Figure 5. Uniaxial stress–strain curves of the ellipsoidal-cell face-centred-cubic (FCC) foam loaded perpendicularly to the cell elongation direction for different values of inner gas pressure (**a**) and elongation ratio (**b**).

Comparing the results shown in Figures 4 and 5, one can conclude that (1) given an elongation ratio greater than one, the elastic modulus and tensile yield stress are larger in the long axial direction than in the short axial direction, whether inner gas pressure is present or not; however, the compressive yield stress is affected by gas pressure in opposite ways between the parallel loading and perpendicular loading, although when gas pressure is absent, the compressive yield stress is higher in the long axial direction than in the short axial direction; (2) given the gas pressure, the elongation ratio plays a more significant role in the uniaxial elasto-plastic behaviour under parallel loading than under perpendicular loading; (3) in general, the gas pressure hardly affects the elastic modulus, and it has less effect on the uniaxial plastic behaviour under parallel loading than under perpendicular loading.

3.2. Poisson's Ratio

Figure 6a,b shows the Poisson's ratio of the FCC foam uniaxially loaded parallelly and perpendicularly to, respectively, the cell elongation direction. It should be noted that the Poisson's ratio of isotropic spheroidal-cell foam ($R = 1$) is independent of the loading direction. By contrast, for ellipsoidal-cell foam, the Poisson's ratio does not vary in the equiaxed-cell plane (x-z plane) for the parallel loading (Figure 6a), but it differs between the long and short axes for the perpendicular loading (Figure 6b). The prominent feature is that when the gas pressure is absent, the Poisson's ratio is symmetric between tension and compression and such symmetry is independent of the elongation ratio and loading direction. When the elongation ratio increases, the Poisson's ratio becomes larger in the parallel loading case (Figure 6a), but smaller in the perpendicular loading case (Figure 6b). The gas pressure does not affect the Poisson's ratio in the elastic stage, but its effect is significant in the plastic stage, wherein the gas pressure decreases the Poisson's ratio for tension, but increases the Poisson's ratio for compression. As a result, the Poisson's ratio becomes asymmetrical between tension and compression. The effect of gas pressure on Poisson's ratio can be explained as follows. For uniaxial tension, the gas pressure and tensile loading are aligned, which favours the deformation along the loading direction and thus reduces the Poisson's ratio. Conversely, the gas pressure counteracts the compressive load and thereby increases the Poisson's ratio under uniaxial compression. Such trends hold for different elongation ratios. A similar gas effect on Poisson's ratio was also reported by Xu et al. [20] and Öchsner and Mishuris [10] for isotropic cellular materials.

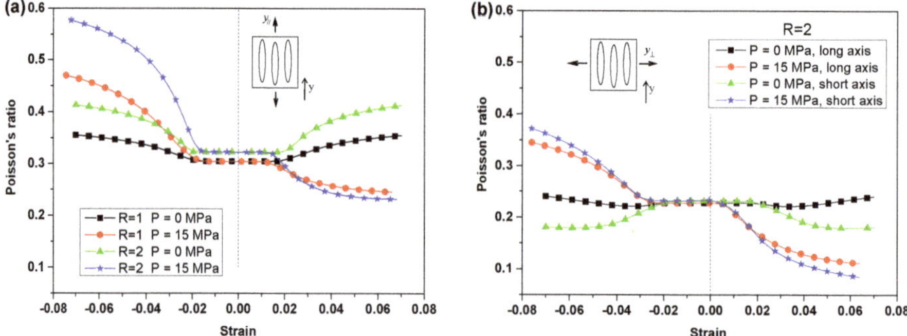

Figure 6. Poisson's ratio of the ellipsoidal-cell face-centred-cubic (FCC) foam loaded parallelly (**a**) and perpendicularly to (**b**) the cell elongation direction. Note that for perpendicular loading, the Poisson's ratio differs between the short and long axes.

3.3. Multiaxial Yield Surface

Figure 7 shows the initial yield surfaces of the FCC foam under multiaxial loading at different Lode angles. For a Lode angle of 0°, when the elongation ratio is one (i.e., spheroidal-cell foam) and the gas pressure is zero, the yield surface is an ellipse, being symmetrical about the σ_e axis in the $\sigma_e - \sigma_m$ plane, but such symmetry is violated when the elongation ratio is two, leading to a yield surface manifested as a tilted ellipse in the $\sigma_e - \sigma_m$ plane. It is clearly seen that the yield surface of the transversely isotropic ellipsoidal-cell FCC foam is distinct from the yield surface of the isotropic spheroidal-cell FCC foam. Particularly, compared to the spheroidal-cell foam, the ellipsoidal-cell foam is more susceptible to yielding when approaching a hydrostatic stress state (i.e., yield stress is lower when $X_\Sigma \to \infty$), but it is more resistant to yielding when approaching a shear stress state (i.e., yield stress is higher when $X_\Sigma \to 0$). The Lode angle does not affect the yield surface of the spheroidal-cell foam ($R = 1$), but it does affect the orientation of the tilted yield surface of the ellipsoidal-cell foam ($R = 2$), i.e., the yield surface rotates counter-clockwise when the Lode angle is increasing.

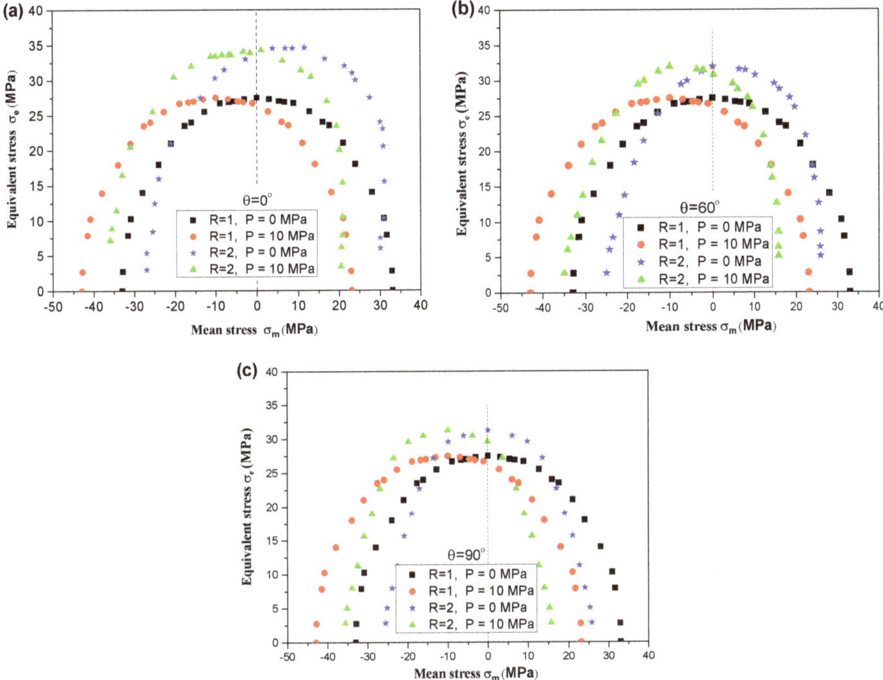

Figure 7. Initial multiaxial yield surfaces of the ellipsoidal-cell face-centred-cubic (FCC) foam for Lode angles of 0° (**a**), 60° (**b**) and 90° (**c**).

The effect of gas pressure on the yield surface is straightforward, i.e., the gas pressure shifts the yield surface towards the negative mean stress axis with a distance equal to the gas pressure in the $\sigma_e - \sigma_m$ plane. Such an effect is independent of the Lode angle and elongation ratio, which is consistent with the previous studies by Xu et al. [20] and Zhang et al. [11].

4. Conclusions

A qualitative study on the elasto-plastic properties of transversely isotropic cellular materials with inner gas pressure is presented, which is focused on a gas-filled ellipsoidal-cell FCC foam with a relative density of 0.5. The findings are summarised as follows:

(1) The elasto-plastic behaviour of the gas-filled ellipsoidal-cell FCC foam is dependent on the loading direction. The effect of the elongation ratio is most pronounced for uniaxial loading along the cell elongation direction. Increasing the elongation ratio leads to increases in the elastic modulus, yield stress and Poisson's ratio, due to more load-bearing material being distributed in the elongation direction. For perpendicular loading, increasing the elongation ratio does not affect the elastic modulus, but reduces the tensile yield stress and Poisson's ratio, and increases the compressive yield stress. In general, the elastic modulus, tensile yield stress and Poisson's ratio are higher for parallel loading than for perpendicular loading. For multiaxial loading, the initial yield surface of the ellipsoidal-cell FCC foam is a tilted ellipse in the $\sigma_e - \sigma_m$ plane and it rotates counter-clockwise when the Lode angle is increasing;

(2) The inner gas pressure causes asymmetry of the uniaxial stress–strain curve of the FCC foam. It reduces the tensile yield stress, but it has the opposite effects on the compressive yield stress for

loadings parallel and perpendicular to the cell elongation direction, i.e., slight reduction for the former, while a considerable increase for the latter. The effect of gas pressure on the uniaxial stress–strain curve is more pronounced in tension than in compression. Poisson's ratio is independent of gas pressure in the elastic regime, but it increases in the plastic compression regime and decreases in the plastic tension regime, due to the effect of gas pressure. Furthermore, the gas pressure shifts the tilted multiaxial yield surface of the ellipsoidal-cell foam towards the negative mean stress axis with a stress value identical to the gas pressure value.

Author Contributions: Conceptualization, W.Z. and Y.S.; methodology, Z.X. and X.F.; software, Z.X. and X.F.; validation, K.M. and C.Y.; formal analysis, Z.X. and Y.S.; investigation, Z.X., K.M. and C.Y.; resources, W.Z., X.F. and Y.S.; data curation, K.M. and C.Y.; writing—original draft preparation, Z.X. and Y.S.; writing—review and editing, K.M., C.Y., X.F. and W.Z.; visualization, Z.X.; supervision, W.Z.; project administration, X.F.; funding acquisition, W.Z. and Y.S.

Funding: This research was funded by the China State Key Laboratory for Strength and Vibration of Mechanical Structures through Open Project (grant number SV2018-KF-37) and the National Natural Science Foundation of China (grant numbers 11772246, 11472203, 11172227, U1530259).

Acknowledgments: The authors are grateful to Fulin Shang at XJTU for insightful discussion.

Conflicts of Interest: The authors declare no conflict of interest.

References

1. Gibson, L.J.; Ashby, M.F. *Cellular Solids: Structure and Properties*, 2nd ed.; Cambridge University Press: Cambridge, UK, 1997.
2. Sun, Y.; Li, Q.M. Dynamic compressive behaviour of cellular materials: A review of phenomenon, mechanism and modelling. *Int. J. Impact Eng.* **2018**, *112*, 74–115. [CrossRef]
3. Banhart, J. Manufacture, characterisation and application of cellular metals and metal foams. *Prog. Mater. Sci.* **2001**, *46*, 559–632. [CrossRef]
4. Huber, A.T.; Gibson, L.J. Anisotropy of foams. *J. Mater. Sci.* **1988**, *23*, 3031–3040. [CrossRef]
5. Sun, Y.; Amirrasouli, B.; Razavi, S.B.; Li, Q.M.; Lowe, T.; Withers, P.J. The variation in elastic modulus throughout the compression of foam materials. *Acta Mater.* **2016**, *110*, 161–174. [CrossRef]
6. Deshpande, V.S.; Fleck, N.A. Isotropic constitutive models for metallic foams. *J. Mech. Phys. Solids* **2000**, *48*, 1253–1283. [CrossRef]
7. Abrate, S. Criteria for yielding or failure of cellular materials. *J. Sandw. Struct. Mater.* **2008**, *10*, 5–51. [CrossRef]
8. Zhang, W.X.; Wang, T.J. Effect of surface energy on the yield strength of nanoporous materials. *Appl. Phys. Lett.* **2007**, *90*, 063104. [CrossRef]
9. Ozgur, M.; Mullen, R.L.; Welsch, G. Analysis of closed cell metal composites. *Acta Mater.* **1996**, *44*, 2115–2126. [CrossRef]
10. Öchsner, A.; Mishuris, G. Modelling of the multiaxial elasto-plastic behaviour of porous metals with internal gas pressure. *Finite Elem. Anal. Des.* **2009**, *45*, 104–112. [CrossRef]
11. Zhang, W.; Xu, Z.; Wang, T.J.; Chen, X. Effect of inner gas pressure on the elastoplastic behavior of porous materials: A second-order moment micromechanics model. *Int. J. Plast.* **2009**, *25*, 1231–1252. [CrossRef]
12. Kitazono, K.; Sato, E.; Kuribayashi, K. Application of mean field approximation to elastic-plastic behaviour for closed-cell metal foams. *Acta Mater.* **2003**, *51*, 4823–4836. [CrossRef]
13. Gurson, A.L. Continuum Theory of Ductile Rupture by Void Nucleation and Growth: Part I—Yield Criteria and Flow Rules for Porous Ductile Media. *J. Eng. Mater. Technol.* **1977**, *99*, 2–15. [CrossRef]
14. Guo, T.F.; Cheng, L. Modeling vapor pressure effects on void rupture and crack growth resistance. *Acta Mater.* **2002**, *50*, 3487–3500. [CrossRef]
15. Guo, T.F.; Cheng, L. Vapor pressure and void size effects on failure of a constrained ductile film. *J. Mech. Phys. Solids* **2003**, *51*, 993–1014. [CrossRef]
16. Cheng, L.; Guo, T.F. Void interaction and coalescence in polymeric materials. *Int. J. Solids Struct.* **2007**, *44*, 1787–1808. [CrossRef]
17. Guo, T.F.; Faleskog, J.; Shih, C.F. Continuum modeling of a porous solid with pressure-sensitive dilatant matrix. *J. Mech. Phys. Solids* **2008**, *56*, 2188–2212. [CrossRef]

18. Sun, Y.; Li, Q.M. Effect of entrapped gas on the dynamic compressive behaviour of cellular solids. *Int. J. Solids Struct.* **2015**, *63*, 50–67. [CrossRef]

19. McElwain, D.L.S.; Roberts, A.P.; Wilkins, A.H. Yield criterion of porous materials subjected to complex stress states. *Acta Mater.* **2006**, *54*, 1995–2002. [CrossRef]

20. Xu, Z.M.; Zhang, W.X.; Wang, T.J. Deformation of closed-cell foams incorporating the effect of inner gas pressure. *Int. J. Appl. Mech.* **2010**, *02*, 489–513. [CrossRef]

21. Deshpande, V.S.; Fleck, N.A. Multi-axial yield behaviour of polymer foams. *Acta Mater.* **2001**, *49*, 1859–1866. [CrossRef]

Article

Optimization and Validation of Sound Absorption Performance of 10-Layer Gradient Compressed Porous Metal

Fei Yang [1], Xinmin Shen [1,2,*], Panfeng Bai [1], Xiaonan Zhang [1], Zhizhong Li [1,3] and Qin Yin [1]

[1] Department of Mechanical Engineering, College of Field Engineering, Army Engineering University, No. 1 Haifu Street, Nanjing 210007, China; 19962061916@163.com (F.Y.); baipanfeng1990@foxmail.com (P.B.); zxn8206@163.com (X.Z.); lizz0607@163.com (Z.L.); dafengyinqin@126.com (Q.Y.)

[2] State Key Laboratory of Ultra-Precision Machining Technology, Department of Industrial and Systems Engineering, The Hong Kong Polytechnic University, Kowloon 999077, Hong Kong, China

[3] State Key Laboratory of Disaster Prevention & Mitigation of Explosion & Impact, College of Defense Engineering, Army Engineering University, No. 1 Haifu Street, Nanjing 210007, China

* Correspondence: xmshen@polyu.edu.hk; Tel.: +86-025-8082-1451

Received: 2 May 2019; Accepted: 16 May 2019; Published: 21 May 2019

Abstract: Sound absorption performance of a porous metal can be improved by compression and optimal permutation, which is favorable to promote its application in noise reduction. The 10-layer gradient compressed porous metal was proposed to obtain optimal sound absorption performance. A theoretical model of the sound absorption coefficient of the multilayer gradient compressed porous metal was constructed according to the Johnson-Champoux-Allard model. Optimal parameters for the best sound absorption performance of the 10-layer gradient compressed porous metal were achieved by a cuckoo search algorithm with the varied constraint conditions. Preliminary verification of the optimal sound absorber was conducted by the finite element simulation, and further experimental validation was obtained through the standing wave tube measurement. Consistencies among the theoretical data, the simulation data, and the experimental data proved accuracies of the theoretical sound absorption model, the cuckoo search optimization algorithm, and the finite element simulation method. For the investigated frequency ranges of 100–1000 Hz, 100–2000 Hz, 100–4000 Hz, and 100–6000 Hz, actual average sound absorption coefficients of optimal 10-layer gradient compressed porous metal were 0.3325, 0.5412, 0.7461, and 0.7617, respectively, which exhibited the larger sound absorption coefficients relative to those of the original porous metals and uniform 10-layer compressed porous metal with the same thickness of 20 mm.

Keywords: gradient compressed porous metal; sound absorption performance; optimal parameters; theoretical modeling; cuckoo search algorithm; finite element simulation; experimental validation

---

1. Introduction

Noise pollution is one of the major environmental problems all over the world [1]. Especially in the urban area, the increasing noise pollution is harmful to persons, animals, precise instruments, and structural buildings, and the harmful level is determined by its frequency, intensity, and time [2]. Therefore, noise reduction is one of the research focuses in the field of environmental protection [3], and development of a sound absorber with fine sound absorption performance is considered an effective method to achieving noise reduction [4,5]. Normally, sound absorption performance of the sound absorber with certain thickness is evaluated by the average sound absorption coefficient in the given frequency range, which indicates that the desired sound absorber must achieve a higher average sound absorption coefficient and utilize smaller total thickness simultaneously. Moreover, taking the practical applications into account, besides the outstanding sound absorption performance, the desired sound absorber should have the additional advantages of low fabrication cost, excellent fire resistance, fine environmental friendliness, convenient installability, and easy maintainability [6,7]. Therefore, these factors should be taken into consideration in developing the sound absorber for practical application.

The proposed sound absorber in this study is expected to be used for noise reduction in workshops. Although many sound absorbers made of wood or cork have been developed and can exhibit fine sound absorption performance, they will be favorable to use in the workshops when their fire resistance, mechanical strength, and maintainability are improved in future, because the materials used in the workshops should obey the stricter requirements. Among the present common sound absorbing materials, porous metal is considered as one potential candidate to fabricate the sound absorber for mass production and large-scale application in workshops, and many sound absorbers that consist of porous metal have been developed [8–14]. Ru et al. [8] prepared the porous copper by the resin curing and foaming method, which exhibited a greater sound absorption capacity than the lotus-type copper. A porous polycarbonate material sample was developed by Liu et al. [9] through the additive manufacturing method, and the results indicated that increasing the angle of the slanted pores decreased the sound absorption coefficient when the porosity was kept as constant. Ning and Zhao [10] studied sound absorption characteristics of the multilayer porous materials backed with the air gap, which could achieve better sound absorption behavior for an increase of the acoustic resistance. Sound absorption performance of the porous metals was enhanced by Otaru [11] through using X-ray tomography images to reliably characterize pore structure-related parameters of the high density porous metallic structures, and the flow simulations were used to deduce parameters that determine the acoustical properties, which could assist in minimizing the operating cost and the time involved in sound absorption measurement. Chen et al. [12] presented a method for calculating and optimizing the sound absorption coefficient of the multi-layered porous fibrous metals in the low frequency range, and the numerical examples demonstrated that the optimization model was very applicable and efficient. The various nickel foam-based multilayer sound absorbing structures were developed by Cheng et al. [13], which could achieve the optimal sound absorption coefficient of 0.4 in the 1000–1600 Hz for the composite structure of five-layer foams with a backed 5 mm-thick cavum. Ao et al. [14] conducted the sound absorption characteristics and the structure optimization of porous stainless steel fibrous felt materials, and it indicated that the sample with excellent sound absorption performance could be prepared by adjusting the match between the mean pore size and the fibrous diameter for different frequencies. This research [8–14] proved that sound absorption behavior of the porous metal was determined by its structural parameters, and it could be improved by using some optimization methods, which indicated that the suitable optimization was a crucial procedure to develop a novel sound absorber made of the porous metal.

It was proved by Bai et al. [15] that the sound absorption efficiency of the porous metal could be improved through compression, and the average sound absorption coefficient of the compressed and the microperforated porous metal panel absorber in the 100–6000 Hz reached 0.5969 with a backed cavity of 20 mm [16]. Meanwhile, Yang et al. [17] proposed a high-efficiency and thin-thickness acoustic absorber by compression and assembly of the porous metals, which could obtain an excellent average sound absorption coefficient of 0.6033 in the 100–6000 Hz with a total thickness of 11 mm. Therefore, a 10-layer gradient compressed porous metal was optimized to obtain an excellent sound absorption performance in this research. Firstly, the structural parameters of each compressed layer of porous metal were derived according to their definitions, and the theoretical model of the sound absorption coefficient of the 10-layer gradient compressed porous metal was constructed by the transfer matrix method [18,19] based on the Johnson-Champoux-Allard model [20,21]. Secondly, parameters of the proposed sound absorbers were optimized for best average sound absorption coefficients in varied frequency ranges by the cuckoo search algorithm [22,23], which was proved effective in optimizing parameters of the sound absorbing structures [24]. Thirdly, finite element simulation models of the obtained optimal sound absorbers were founded in the virtual acoustic laboratory [25], which was considered an effective assistant method to preliminarily verify the sound absorption performance of the composite structures [19,24]. Afterwards, the desired 10-layer gradient compressed porous metals were fabricated according to the optimal parameters by compression and assembly. Finally, the sound absorption coefficients of the sound absorbers were tested by the standing wave tube measurement [26,27], which was treated as effective validation for the theoretical sound absorption model, the cuckoo search optimization algorithm, and the finite element simulation method.

2. Theoretical Modeling

The 10-layer gradient compressed porous metal consisted of 10 single compressed porous metals, thus it was necessary to derive the structural parameters for each layer. The major structural parameters were thickness, porosity, and static flow resistivity. Based on their definitions, when the compression ratio of the single compressed porous metal was η (defined as the ratio of the reduced thickness by the compression to the initial thickness), the corresponding thickness d, the porosity ϕ, and the static flow resistivity σ could be calculated by Equations (1), (2), and (3), respectively [17]. Here, d_0, ϕ_0, and σ_0 were thickness, porosity, and static flow resistivity of initial porous metal before compression.

$$d = d_0 \cdot (1 - \eta) \tag{1}$$

$$\phi = \frac{\phi_0 - \eta}{1 - \eta} \tag{2}$$

$$\sigma == \frac{\phi_0}{(\phi_0 - \eta)(1 - \eta)} \sigma_0 \tag{3}$$

Based on the Johnson-Champoux-Allard model [20,21], complex effective density $\rho(\omega)$ and complex effective bulk modulus $K(\omega)$ of the single compressed porous metal could be obtained according to the derived structural parameters, as shown in Equations (4) and (5), respectively. Here, ω was the sound angular frequency, which could be calculated by Equation (6); ρ was the density of the air with normal temperature, 1.21 Kg/m^3; γ was the specific heat ratio of the air, 1.40; P_0 was the standard static pressure of the air, $1.013 \cdot 10^5$ Pa; N_u was the Nusselt number, 4.36; P_r was the Prandtl number, 0.71 [15–21]. Meanwhile, j was the symbol of the imaginary number.

$$\rho(\omega) = \rho\left[1 + \left(3^2 + \frac{4\omega\rho}{\sigma\phi}\right)^{-0.5} - j\frac{\sigma\phi}{\omega\rho}\left(1 + \frac{\omega\rho}{4\sigma\phi}\right)^{0.5}\right] \tag{4}$$

$$K(\omega) = \gamma P_0\left[\gamma - (\gamma - 1)\left[1 - N_u\left(j\frac{8\omega\rho P_r}{\sigma\phi} + N_u\right)^{-1}\right]\right]^{-1} \tag{5}$$

$$\omega = 2\pi f \tag{6}$$

Moreover, for the single layer compressed porous metal, the wave number k in it and its corresponding characteristic impedance Z could be calculated through Equations (7) and (8), respectively. Furthermore, transfer matrix P of the single layer compressed porous metal could be calculated by Equation (9) [17–19].

$$k = \omega\sqrt{\frac{\rho(\omega)}{K(\omega)}} \tag{7}$$

$$Z = \sqrt{\rho(\omega)K(\omega)} \tag{8}$$

$$P = \begin{bmatrix} \cos(kd) & jZ\sin(kd) \\ jZ^{-1}\sin(kd) & \cos(kd) \end{bmatrix} \tag{9}$$

The utilized porous metal samples were all purchased from YiYang Foam metal New material Co., Ltd., Yiyang, China, and their initial thicknesses, porosities, and static flow resistivities were the same, which indicated that the structural parameters of the compressed porous metals were determined by the compression ratio η. Therefore, supposing the transfer matrix of the ith layer single compression porous metal with the compression ratio of η_i was P_i, the total transfer matrix T of the 10-layer gradient compressed porous metal could be calculated by Equation (10) based on the transfer matrix method [18,19], and the corresponding sound absorption coefficient α could further be obtained by Equation (11). In Equation (11), $\mathrm{Re}()$ and $\mathrm{Im}()$ corresponded to the real part and the imaginary part of one complex number, respectively. In this way, the theoretical sound absorption model of the 10-layer gradient compressed porous metal was constructed, which provided the theoretical basis for further parameter optimization.

$$T = \begin{bmatrix} T_{11} & T_{12} \\ T_{21} & T_{22} \end{bmatrix} = \prod_{i=1}^{10} P_i \tag{10}$$

$$\alpha = \frac{4\mathrm{Re}\left(\frac{T_{11}}{T_{21}} \cdot \frac{1}{\rho c}\right)}{\left[1 + \mathrm{Re}\left(\frac{T_{11}}{T_{21}} \cdot \frac{1}{\rho c}\right)\right]^2 + \left[\mathrm{Im}\left(\frac{T_{11}}{T_{21}} \cdot \frac{1}{\rho c}\right)\right]^2} \tag{11}$$

3. Parameter Optimization

As mentioned above, all the utilized porous metal samples were purchased from YiYang Foam metal New material Co., Ltd., Yiyang, China, and their initial thicknesses, porosities, and static flow resistivities were 5 mm (measured by Vernier caliper), 0.95 (measured according to its definition), and 10,200 Pa·s·m^{-2} (measured based on the water tank method), respectively, which were marked by the manufacturer and further validated through the experimental measurements in this study. Therefore, for the proposed 10-layer gradient compressed porous metal in this research, its sound absorption performance was completely determined by the 10 compression ratios, and parameter optimization was realized by using the cuckoo search algorithm [22,23], which was proved effective and practical to optimize parameters of the sound absorbing structures [24]. The proposed sound absorber was desired for use in workshops, and the available space to install it was limited to 20 mm, which indicated that the summation of the 10 compression ratios needed to be smaller than six (summation of thicknesses of the 10 initial porous metals was 50 mm), as shown in Equation (12). Meanwhile, from Equation (2), it could be observed that each compression ratio η_i needed to be smaller than the initial porosity ϕ_0. Furthermore, it was validated through the compression experiments that it was difficult to further compress the utilized porous metal sample when the compression ratio exceeded 0.9, because the structure of the compressed sample was more and more dense. Therefore, the additional constraint condition of the limit for each compression ratio was given and is shown in Equation (13).

$$\sum_{i=1}^{10} \eta_i \geq 6 \tag{12}$$

$$\eta_i \leq 0.9 \quad i = 1, 2, \ldots, 10 \tag{13}$$

For the various equipment and machines installed in the different workshops, frequency ranges of their noise differed. Taking practical applications into consideration, the investigated frequency ranges were 100–1000 Hz, 100–2000 Hz, 100–4000 Hz, and 100–6000 Hz. In the optimization process, maximization of the average sound absorption coefficient in the corresponding frequency range was treated as the optimization target, as shown in Equation (14). Here, $f_{\min}$ and $f_{\max}$ were the upper limit and the lower limit of the investigated frequency range.

$$\max(average(\alpha(f)), f \in [f_{\min}, f_{\max}]) \tag{14}$$

As is well known, the cuckoo search algorithm may obtain a local optimal solution instead of a global optimal solution. There are two factors that affect the optimization results, which include initial values of the parameters and the optimization time. Therefore, in order to improve the reliability of the optimization results, each optimization process continued as long as possible, and it was stopped until the obtained optimal parameters had no changes for more than 1 h. Meanwhile, in order to further improve accuracy of the optimization results, different initial values of the parameters were selected for each optimization process. Consistency of the optimal parameters for each investigated sound absorber was obtained, which validated the reliability and the accuracy of the cuckoo search algorithm. According to the cuckoo search algorithm [22–24], compression ratios of each layer in the optimal 10-layer gradient compressed porous metals with the different investigated frequency ranges were obtained and are summarized in Table 1. The total thickness of each optimal sound absorber was 20 mm. The corresponding average sound absorption coefficients were 0.3805, 0.5678, 0.7503, and 0.8231 when the investigated frequency ranges were 100–1000 Hz, 100–2000 Hz, 100–4000 Hz, and 100–6000 Hz, respectively. By contrast, the average sound absorption coefficients of the original porous metal with the same thickness of 20 mm were 0.1639, 0.3070, 0.4981, and 0.6050 when the frequency ranges were 100–1000 Hz, 100–2000 Hz, 100–4000 Hz, and 100–6000 Hz, respectively. Meanwhile, when the 10 compression ratios in the 10-layer gradient compressed porous metals were equal to 0.6, it was named by the uniform compressed porous metal as a special case, which was also treated as the contrast. The average sound absorption coefficients of the uniform compressed porous metal were 0.2593, 0.5015, 0.7168, and 0.7980 when the investigated frequency ranges were 100–1000 Hz, 100–2000 Hz, 100–4000 Hz, and 100–6000 Hz, respectively. It could be observed that the sound absorption performance in each investigated frequency range was significantly improved by the 10-layer gradient compressed porous metals relative to the original porous metal with the same thickness of 20 mm, and the optimal 10-layer gradient compressed porous metals exhibited better sound absorption performance than the uniform compressed porous metal.

Table 1. Summary of compression ratios of each layer in the optimal 10-layer gradient compressed porous metals with the different investigated frequency ranges.

Layer Sequences	Investigated Frequency Ranges			
	100–1000 Hz	**100–2000 Hz**	**100–4000 Hz**	**100–6000 Hz**
1st layer	7.11%	37.17%	85.03%	32.16%
2nd layer	90.00%	90.00%	32.65%	74.59%
3rd layer	90.00%	46.36%	28.15%	32.52%
4th layer	78.15%	42.65%	32.74%	35.33%
5th layer	43.31%	45.28%	42.15%	45.27%
6th layer	43.83%	50.51%	53.95%	56.51%
7th layer	47.02%	57.32%	66.56%	70.10%
8th layer	51.91%	65.54%	78.76%	79.40%
9th layer	58.67%	75.17%	90.00%	84.13%
10th layer	90.00%	90.00%	90.00%	90.00%

Sound absorption coefficients of the optimal 10-layer gradient compressed porous metals were obtained through taking the optimal parameters in Table 1 into the constructed theoretical sound absorption models in Equations (1) to (11), and their distributions are shown in Figure 1. Meanwhile, the sound absorption coefficients of the original porous metal and those of the uniform compressed porous metal with the same thickness of 20 mm were also calculated according to the Johnson-Champoux-Allard model [20,21] and are shown in Figure 1 for contrast. It could further prove the improvement of the sound absorption performance by these obtained optimal 10-layer gradient compressed porous metals. Meanwhile, it could be found that the sound absorption coefficients were reduced in the 100–1000 Hz when the investigated frequency range was enlarged, because the corresponding sound absorption peak shifted to the high-frequency direction.

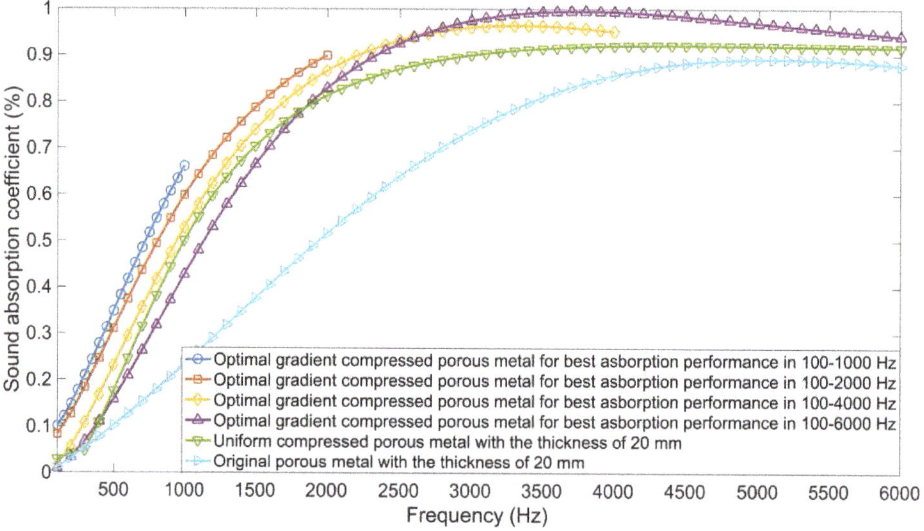

Figure 1. Comparisons of sound absorption coefficients of the optimal 10-layer gradient compressed porous metals and those of the original porous metal and uniform compressed porous metal.

4. Finite Element Simulation

Finite element simulation models were considered as an effective assistant method to preliminarily verify the sound absorption performance of the composite structures [19,24], and the obtained optimal sound absorbers were simulated in a virtual acoustic laboratory [25]. The constructed finite element simulation model for the 10-layer gradient compressed porous metal is shown in Figure 2. Meanwhile, the finite element simulation model for the original porous metal and uniform compressed porous metal with a thickness of 20 mm is shown in Figure 3, which was treated as the contrast. The used model for the porous metal in this study was the Delany-Bazley-Miki model [28,29], and the structural parameters we needed to set were thickness, porosity, and static flow resistivity, which could be obtained by Equations (1), (2), and (3) according to the optimal compression ratios in Table 1 and the parameters of the initial porous metal. Plane wave was imported in the acoustic source inlet, which was the incident sound source in the finite element simulation model. Sound pressures of the incident wave and those of the reflected wave at the two microphones were measured, and the sound absorption coefficient a_s could be calculated through Equation (15). Here, H_{12}, H_i, and H_r were transfer functions of the total sound field, the incident wave, and the reflected wave, respectively; k_0 was the wave number; x_l was the distance between microphone 1 and the surface of the detected sample. The length of the standing wave tube was 300 mm, and the distance between the two microphones was 40 mm [19]. The mesh grid was set to 5 mm in these constructed finite element simulation models, as shown in Figures 2 and 3. Meanwhile, in order to validate the influence of the mesh grid to the simulation results, the corresponding finite element simulation models with mesh grids of 0.5 mm, 1 mm, and 3 mm were also constructed. The influence of the mesh grid was evaluated by average deviations of the sound absorption coefficients at the corresponding frequency points among these simulation data obtained by the various finite element simulation models. It could be found that the calculated average deviations were smaller than 1%, which indicated that the influence of the mesh grid was insignificant in this study. Therefore, in order to improve the computing speed and promote the simulation efficiency, the mesh grid of 5 mm was selected in the finite element simulation models in this study, as shown in Figures 2 and 3.

$$\alpha_s = 1 - \left| \frac{H_{12} - H_i}{H_r - H_{12}} e^{j2k_0 x_l} \right|^2 \tag{15}$$

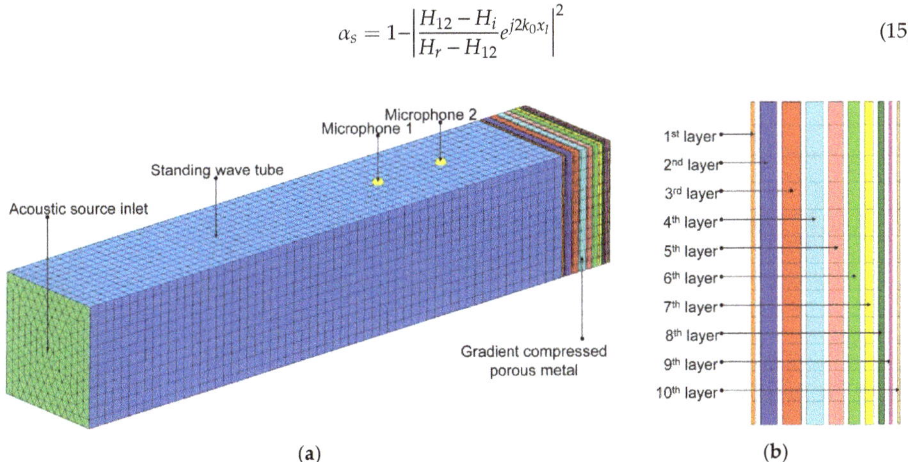

Figure 2. The constructed finite element simulation model for 10-layer gradient compressed porous metal. (**a**) Frame diagram of the system of the standing wave tube measurement; (**b**) schematic drawing compositions of the 10-layer gradient compressed porous metal.

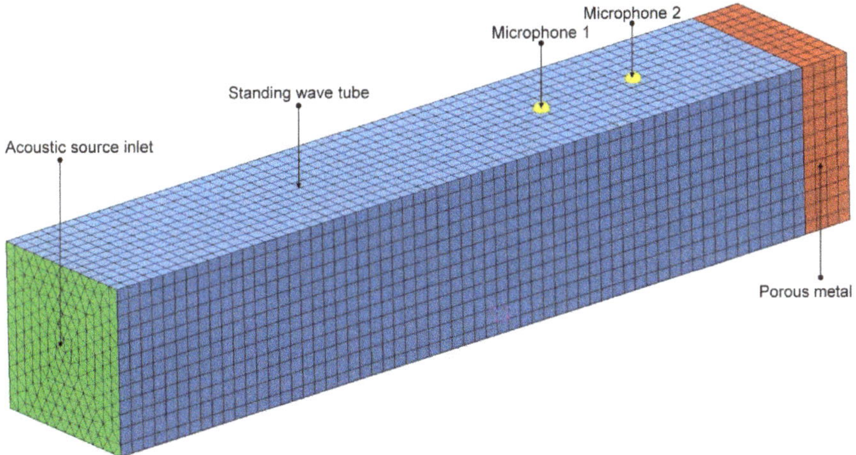

Figure 3. The constructed finite element simulation model for the original porous metal and the uniform compressed porous metal.

Comparisons of the sound absorption coefficients of the investigated sound absorber in theory as well as those in simulations are shown in Figure 4. It could be observed that, for the four optimal 10-layer gradient compressed porous metals, the original porous metal with the thickness of 20 mm, and the uniform compressed porous metal with the thickness of 20 mm, simulation results were consistent with theoretical results, which could preliminarily verify the effective improvement of the sound absorption performance by the optimal 10-layer gradient compressed porous metals. The existing differences between theoretical data and simulation data were primarily generated by the different acoustic models used for the porous metal. The theoretical sound absorption models were constructed based on the Johnson-Champoux-Allard model in this study [20,21], and the acoustic model for porous metal in the finite element simulation model was the Delany-Bazley-Miki model [28,29]. It is well known that single-piece fabrication of the porous metal with different structural parameters is high-cost and time-consuming, and the measurement of sound absorption coefficients of the sound absorber also takes a great deal of time. Therefore, finite element simulation could be treated as an effective supplement for further experimental validation [19,24].

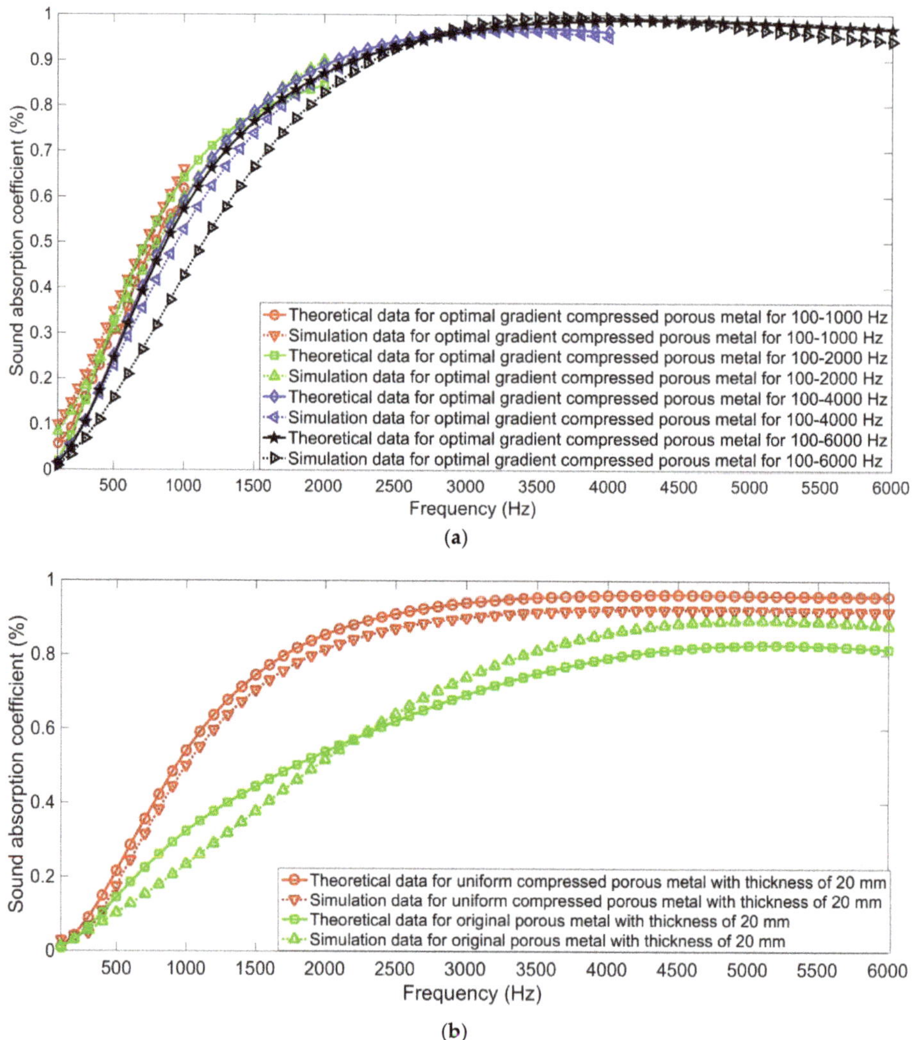

Figure 4. Comparisons of the sound absorption coefficients in theory and those in simulation. (**a**) The optimal 10-layer gradient compressed porous metals for the varied frequency ranges; (**b**) the original porous metal and uniform compressed porous metal with same thicknesses of 20 mm.

5. Standing Wave Tube Measurement

The initial porous metal was porous copper, which was purchased from YiYang Foam metal New material Co., Ltd., Yiyang, China. According to the obtained compression ratios of each layer in the optimal 10-layer gradient compressed porous metals in Table 1, the initial porous metals were compressed by the CTM2050 universal testing machine (Wuxi City Bleecker Trading Co., Ltd., Wuxi, China), as shown in Figure 5 [15–17]. Afterwards, the 10 single compressed porous metals were assembled by the glue water according to the orders in Table 1. The neighboring layers were glued by the glue water at three boundary points, and the bonding area was limited to smaller than 1 mm^2, which aimed to reduce the influence of the glue water on the sound absorption performance of the 10-layer gradient compressed porous metals. In this method, the four optimal 10-layer gradient compressed porous metals were fabricated.

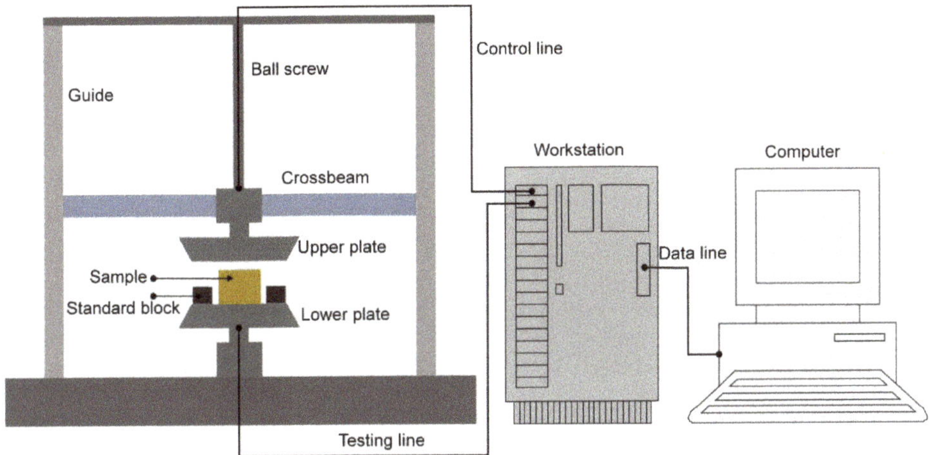

Figure 5. Schematic diagram of the used CTM2050 universal testing machine for fabrication of the optimal 10-layer gradient compressed porous metals.

The fabricated optimal 10-layer gradient compressed porous metals are shown in Figure 6, and their sound absorption coefficients were measured by the AWA6128A detector (Hangzhou Aihua Instruments Co., Ltd., Hangzhou, China), as shown in Figure 7 [15–17,27]. According to the operating manual of the AWA6128A detector, measurement of the sound absorption coefficient in the 100–2000 Hz range required the sample with a diameter of 96 mm, and that in the 2000–6000 Hz required the sample with a diameter of 30 mm. Therefore, it was found that, for the investigated frequency ranges of 100–1000 Hz and 100–2000 Hz, there was one sample with a diameter of 96 mm in Figure 6a,b, respectively. Meanwhile, for the investigated frequency ranges of 100–4000 Hz and 100–6000 Hz, there were additional samples with the diameter of 30 mm in Figure 6c,d, respectively. In order to reduce the random error in the measuring process, the four 10-layer gradient compressed porous metals, the original porous metal with the thickness of 20 mm, and the uniform compressed porous metal with the thickness of 20 mm were measured 10 times, and the final actual sound absorption coefficients were the average of the 10 experimental measurements for each sound absorber. It was found that the standard deviation for each sound absorption coefficient data was smaller than 0.6%, which indicated that the experimental uncertainty in this research was negligible, and the experimental data obtained by the standing wave tube measurement were accurate and believable.

Comparisons of theoretical data, simulation data, and experimental data of the sound absorption coefficient of the four optimal 10-layer gradient compressed porous metals, those of the uniform compressed porous metal with a thickness of 20 mm, and those of the original porous metal with a thickness of 20 mm are shown in Figure 8. When the investigated frequency ranges were 100–1000 Hz, 100–2000 Hz, 100–4000 Hz, and 100–6000 Hz, the average deviations between theoretical data and experimental data for the optimal 10-layer gradient compressed porous metals were 0.0520, 0.0438, 0.0304, and 0.0624, respectively, and the corresponding deviations between the simulation data and the experimental data were 0.0315, 0.0294, 0.0296, and 0.0356, respectively. For the uniform compressed porous metal with a thickness of 20 mm, the average deviation between theoretical data and experimental data was 0.0596, and that between simulation data and experimental data was 0.0393. For the original porous metal with a thickness of 20 mm, the average deviation between theoretical data and experimental data was 0.0825, and that between simulation data and experimental data was 0.1069. Consistencies among the theoretical data, the simulation data, and the experimental data proved accuracies of the theoretical sound absorption model, the cuckoo search optimization algorithm, and the finite element simulation method, which provided a novel method to develop gradient compressed porous metals.

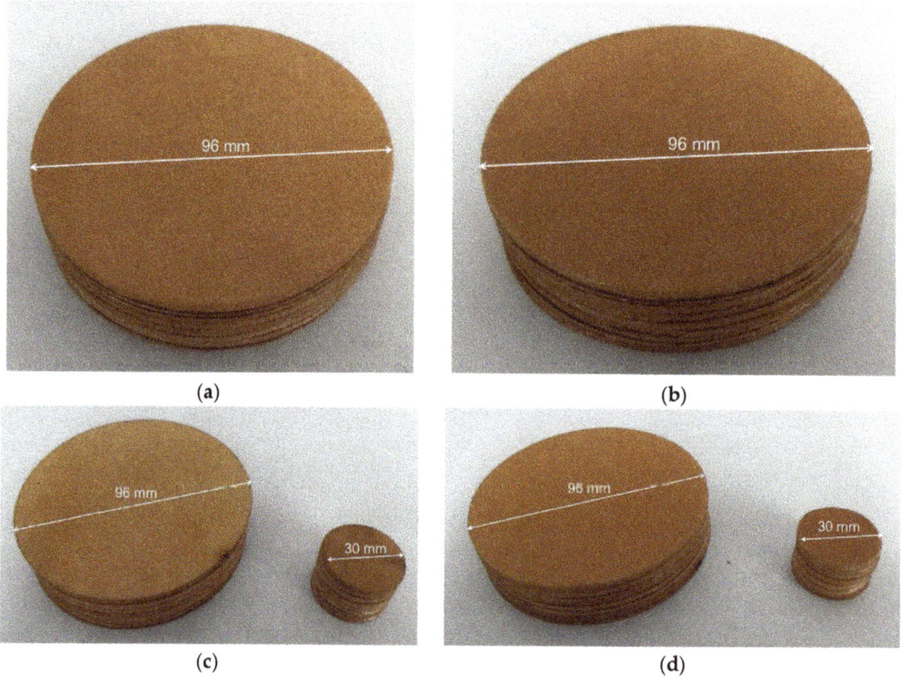

Figure 6. The prepared 10-layer gradient compressed porous metals for the varied frequency ranges. (**a**) 100–1000 Hz; (**b**) 100–2000 Hz; (**c**) 100–4000 Hz; (**d**) 100–6000 Hz.

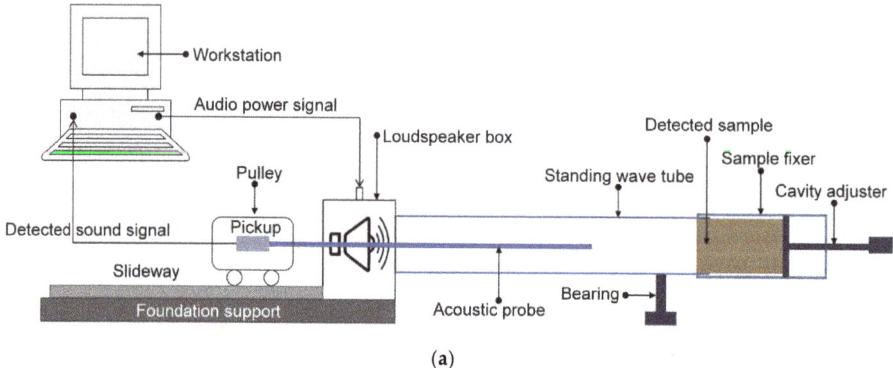

Figure 7. *Cont.*

(b)

Figure 7. The AWA6128A detector for the standing wave tube measurement. (**a**) Schematic diagram; (**b**) actual picture.

Comparisons of the average sound absorption coefficients of the four optimal 10-layer gradient compressed porous metals, those of the uniform compressed porous metal with a thickness of 20 mm, and those of the original porous metal with a thickness of 20 mm are shown in Table 2. The actual average sound absorption coefficients of the optimal 10-layer gradient compressed porous metals were 0.3325, 0.5412, 0.7461, and 0.7617 when the investigated frequency ranges were 100–1000 Hz, 100–2000 Hz, 100–4000 Hz, and 100–6000 Hz, respectively, which exhibited higher sound absorption coefficients than those of the uniform compressed porous metal and those of the original porous metal with the same thickness of 20 mm. The improvement was evaluated by the growth ratio of the average sound absorption coefficient, as shown in Equation (16). Here, $average(\alpha_G(f))$, $average(\alpha_U(f))$, and $average(\alpha_O(f))$ were the actual average sound absorption coefficients of the optimal 10-layer gradient compressed porous metal, that of the uniform compressed porous metal with a thickness of 20 mm, and that of the original porous metal with a thickness of 20 mm, respectively. Relative to the original porous metal, it could be calculated by Equation (16) that the growth ratios were 117.5%, 118.5%, 76.3%, and 44.3% when the investigated frequency ranges were 100–1000 Hz, 100–2000 Hz, 100–4000 Hz, and 100–6000 Hz, respectively. Meanwhile, for the uniform compressed porous metal, the growth ratios were 58.5%, 19.4%, 9.2%, and 3.1% when the investigated frequency ranges were 100–1000 Hz, 100–2000 Hz, 100–4000 Hz, and 100–6000 Hz, respectively. It could be found that the improvement of the sound absorption coefficient in the low-frequency range was more remarkable than that in the high-frequency range, and the optimal 10-layer gradient compressed porous metal could make up for the shortage of the original porous metal and broaden its sound absorption broadband. Meanwhile, relative to the uniform compressed porous metal, the sound absorption performance of the 10-layer gradient compressed porous metal could be further improved by parameter optimization.

$$\begin{cases} \varepsilon_1 = \dfrac{average(\alpha_G(f)) - average(\alpha_O(f))}{average(\alpha_O(f))} \\[2mm] \varepsilon_2 = \dfrac{average(\alpha_U(f)) - average(\alpha_G(f))}{average(\alpha_G(f))} \end{cases} \tag{16}$$

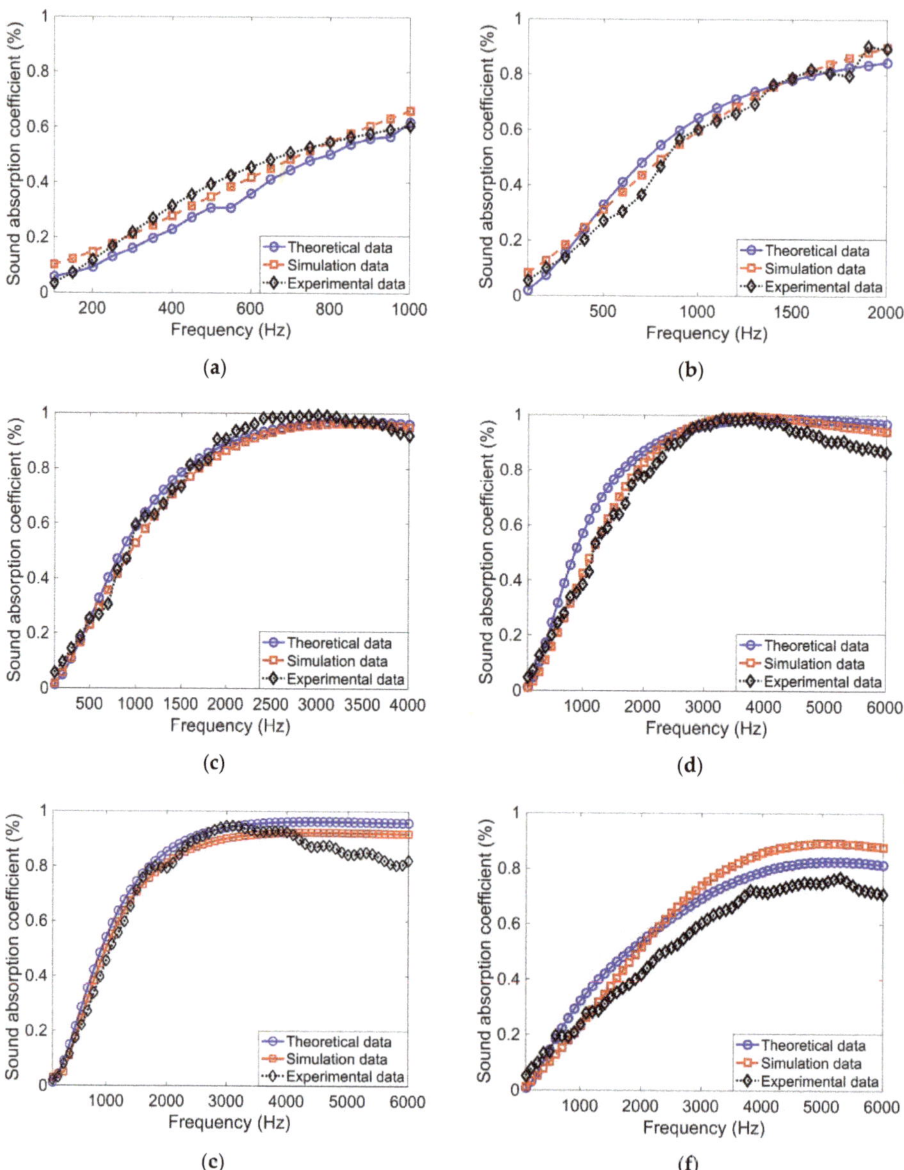

Figure 8. Comparisons of the theoretical data, the simulation data, and the experimental data of the sound absorption coefficient of the investigated sound absorbers. (**a**) Optimal 10-layer gradient compressed porous metals for 100–1000 Hz; (**b**) optimal 10-layer gradient compressed porous metals for 100–2000 Hz; (**c**) optimal 10-layer gradient compressed porous metals for 100–4000 Hz; (**d**) optimal 10-layer gradient compressed porous metals for 100–6000 Hz; (**e**) uniform compressed porous metal with the thickness of 20 mm; (**f**) original porous metal with the thickness of 20 mm.

Table 2. Comparisons of average sound absorption coefficients of the optimal 10-layer gradient compressed porous metal and those of original porous metal and the uniform compressed porous metal.

Investigated Frequency Range	Type of Material	Average Sound Absorption Coefficient		
		In Actual	In Theory	In Simulation
100–1000 Hz	Original porous metal	0.1529	0.1639	0.1179
	Uniform compressed porous metal	0.2098	0.2593	0.2292
	Gradient compressed porous metal	0.3325	0.3805	0.3804
100–2000 Hz	Original porous metal	0.2477	0.3070	0.2540
	Uniform compressed porous metal	0.4531	0.5015	0.4664
	Gradient compressed porous metal	0.5412	0.5678	0.5645
100–4000 Hz	Original porous metal	0.4233	0.4981	0.4917
	Uniform compressed porous metal	0.6831	0.7168	0.6793
	Gradient compressed porous metal	0.7461	0.7503	0.7244
100–6000 Hz	Original porous metal	0.5280	0.6050	0.6227
	Uniform compressed porous metal	0.7391	0.7980	0.7597
	Gradient compressed porous metal	0.7617	0.8231	0.7870

6. Structural Analysis

Distributions of the compression ratio of the optimal 10-layer gradient compressed porous metals are shown in Figure 9. It could be observed that, for each optimal sound absorber, there was a layer with a high compression ratio in the beginning two layers. Afterwards, the compression ratio decreased quickly in the next several layers, and then it increased gradually until the last layer. These were common structural characteristics for the four optimal 10-layer gradient compressed porous metals, which influenced their sound absorption performance.

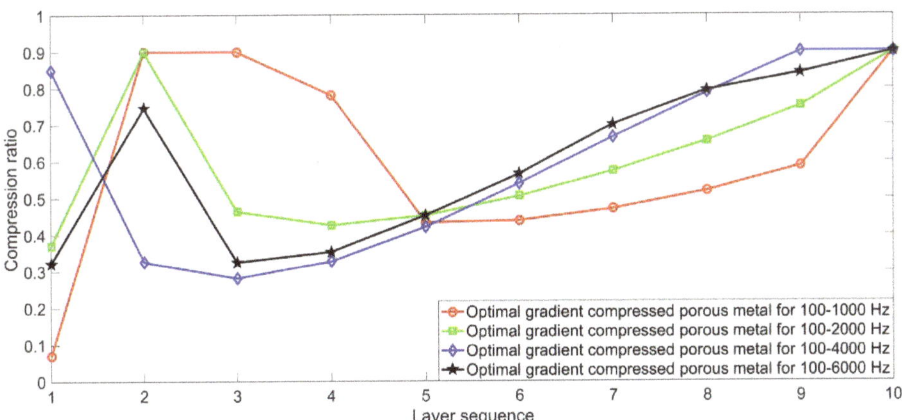

Figure 9. Distributions of compression ratio of optimal 10-layer gradient compressed porous metals.

As mentioned above, the sound absorption performance of porous metal was primarily determined by its structural parameters, and these structural parameters were controlled by the compression ratios in the optimal 10-layer gradient compressed porous metals. Therefore, structural parameters of the optimal 10-layer gradient compressed porous metals, which included thickness, porosity, and static flow resistivity, were calculated by Equations (1), (2), and (3), respectively [17], as shown in Table 3. Distributions of the structural parameters of the optimal 10-layer gradient compressed porous metals along the thickness direction are shown in Figure 10. It was found that there was a layer with low porosity and high static flow resistivity in the first several layers. After the gradient decrease of the porosity and the gradient increase of the static flow resistivity in the middle several layers, there were more layers with low porosity and high static flow resistivity in the last several layers. It was presumed that the varied pore structures in the optimal 10-layer gradient compressed porous metals were effective in absorbing the sound in different frequency ranges [17], and the two layers with low porosity and high static flow resistivity enhanced the sound absorption effect. In the future, interactions between the material and the sound wave will be further quantitatively investigated and theoretically modeled, which is favorable to reveal the sound absorption mechanism of the gradient compressed porous metal and to explain its sound absorption characteristics with different frequencies and parameters.

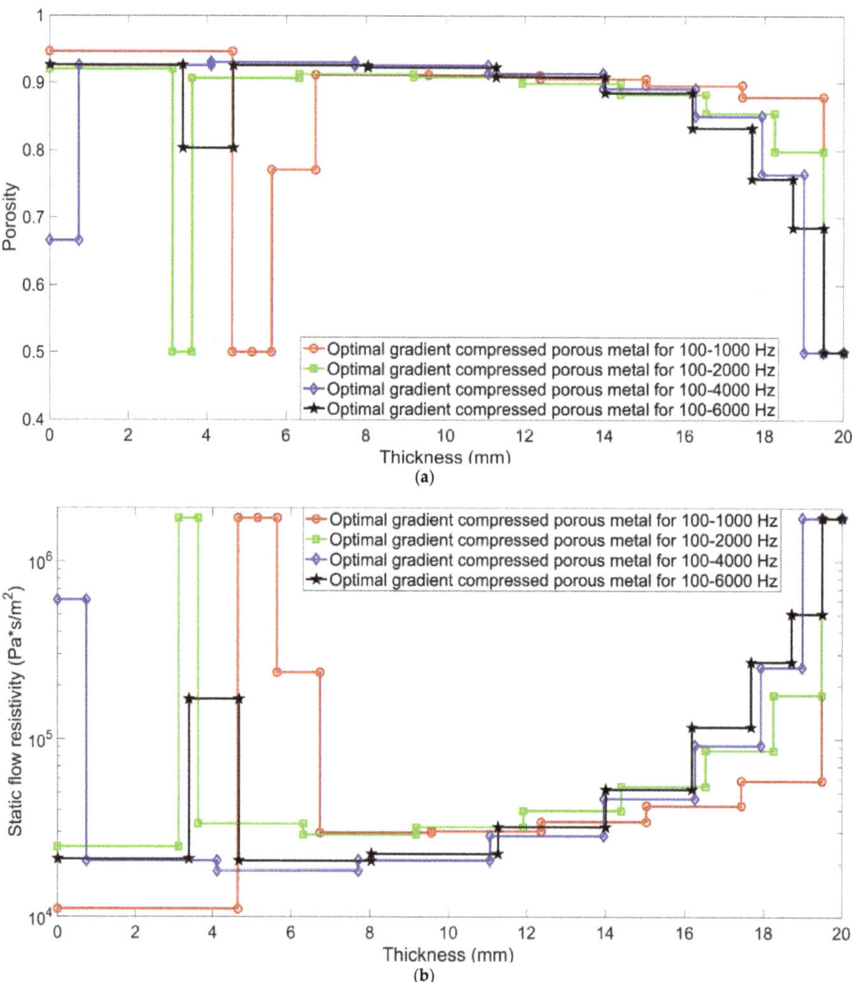

Figure 10. Distribution of structural parameters of the optimal 10-layer gradient compressed porous metals along the thickness direction. (**a**) Porosity; (**b**) static flow resistivity.

Table 3. Structural parameters of the optimal 10-layer gradient compressed porous metals.

Parameters	Frequency Range (Hz)	1st Layer	2nd Layer	3rd Layer	4th Layer	5th Layer	6th Layer	7th Layer	8th Layer	9th Layer	10th Layer
Thickness (mm)	100–1000	4.645	0.500	0.500	1.093	2.835	2.809	2.649	2.404	2.066	0.500
	100–2000	3.141	0.500	2.682	2.868	2.736	2.475	2.134	1.723	1.242	0.500
	100–4000	0.748	3.367	3.592	3.363	2.893	2.302	1.672	1.062	0.500	0.500
	100–6000	3.392	1.270	3.374	3.233	2.737	2.175	1.495	1.030	0.794	0.500
Porosity	100–1000	0.946	0.500	0.500	0.771	0.912	0.911	0.906	0.896	0.879	0.500
	100–2000	0.920	0.500	0.907	0.913	0.909	0.899	0.883	0.855	0.799	0.500
	100–4000	0.666	0.926	0.930	0.926	0.914	0.891	0.850	0.765	0.500	0.500
	100–6000	0.926	0.803	0.926	0.923	0.909	0.885	0.833	0.757	0.685	0.500
Static flow resistivity (10^5 Pa·s·m^{-2})	100–1000	1.108	174.8	174.8	23.73	2.983	3.041	3.438	4.218	5.821	174.8
	100–2000	2.490	174.8	3.350	2.911	3.213	3.969	5.434	8.611	17.75	174.8
	100–4000	60.64	2.081	1.820	2.087	2.858	4.624	9.192	25.35	174.8	174.8
	100–6000	2.122	16.86	2.073	2.265	3.211	5.221	11.74	27.19	50.63	174.8

7. Conclusions

Optimization and validation of the sound absorption performance of 10-layer gradient compressed porous metals were conducted in this study. Through theoretical modeling, parameter optimization, finite element simulation, sample preparation, standing wave tube measurement, and structural analysis, the following conclusions were obtained.

(1) According to the constructed theoretical sound absorption model, parameter optimizations of the 10-layer gradient compressed porous metals were realized by the cuckoo search algorithm, and the optimal actual average sound absorption coefficients were 0.3325, 0.5412, 0.7461, and 0.7617 when the investigated frequency ranges were 100–1000 Hz, 100–2000 Hz, 100–4000 Hz, and 100–6000 Hz, respectively, which proved that 10-layer gradient compressed porous metals could obtain excellent sound absorption performance relative to the original porous metal and the uniform compressed porous metal with the same thickness of 20 mm.

(2) Sound absorption performance of the optimal 10-layer gradient compressed porous metals, the uniform compressed porous metal, and the original porous metal were preliminarily verified by the finite element simulation method and further validated through the standing wave tube measurement. Consistencies among the theoretical data, the simulation data, and the experimental data proved accuracies and reliability of the theoretical sound absorption model, the cuckoo search optimization algorithm, and the finite element simulation method.

(3) Structural analysis of the optimal 10-layer gradient compressed porous metal was realized through calculation and discussion of the structural parameters, which included the thickness, the porosity, and the static flow resistivity. The varied pore structure in the optimal 10-layer gradient compressed porous metals was effective in absorbing the sound wave with different frequency ranges, and the two layers with low porosity and high static flow resistivity enhanced the sound absorption effect.

The developed optimal 10-layer gradient compressed porous metals exhibited outstanding sound absorption performance in the varied frequency ranges, which is favorable for promoting their practical applications in the reduction of industrial noise and the protection of an acoustic environment.

Author Contributions: Conceptualization, X.S.; Software, Z.L.; Validation, F.Y. and X.Z.; Formal analysis, F.Y. and P.B.; Investigation, Q.Y.; Data curation, F.Y. and P.B.; Writing—original draft preparation, F.Y. and X.S.; Writing—review and editing, X.S. and P.B.; Supervision, X.S.; Funding acquisition, X.S.

Funding: This research was funded by National Natural Science Foundation of China, grant number 51505498; Natural Science Foundation of Jiangsu Province, grant number BK20150714; National Key R & D Program of China, grant number 2016YFC0802900; Hong Kong Scholars Program, grant number XJ2017025.

Acknowledgments: The authors wish to express their sincere thanks to Hangzhou Aihong instruments Co., Ltd., China for the support of the AWA6128A detector.

Conflicts of Interest: The authors declare no conflict of interest.

References

1. Buxton, R.T.; McKenna, M.F.; Mennitt, D.; Fristrup, K.; Crooks, K.; Angeloni, L.; Wittemyer, G. Noise pollution is pervasive in U.S. protected areas. *Science* **2017**, *356*, 531–533. [CrossRef] [PubMed]
2. Nieuwenhuijsen, M.J. Influence of urban and transport planning and the city environment on cardiovascular disease. *Nat. Rev. Cardiol.* **2018**, *15*, 432–438. [CrossRef] [PubMed]
3. Vogiatzis, K.; Vanhonacker, P. Noise reduction in urban LRT networks by combining track based solutions. *Sci. Total Environ.* **2016**, *568*, 1344–1354. [CrossRef] [PubMed]
4. Yang, M.; Sheng, P. Sound Absorption Structures: From Porous Media to Acoustic Metamaterials. *Annu. Rev. Mater. Res.* **2017**, *47*, 83–114. [CrossRef]
5. Liu, P.S.; Qing, H.B.; Hou, H.L. Primary investigation on sound absorption performance of highly porous titanium foams. *Mater. Des.* **2015**, *85*, 275–281. [CrossRef]
6. Wang, J.Z.; Ao, Q.B.; Ma, J.; Kang, X.T.; Wu, C.; Tang, H.P.; Song, W.D. Sound absorption performance of porous metal fiber materials with different structures. *Appl. Acoust.* **2019**, *145*, 431–438. [CrossRef]
7. Jin, W.; Liu, J.A.; Wang, Z.L.; Wang, Y.H.; Cao, Z.; Liu, Y.H.; Zhu, X.Y. Sound absorption characteristics of aluminum foams treated by plasma electrolytic oxidation. *Materials* **2015**, *8*, 7511–7518. [CrossRef]
8. Ru, J.M.; Bo, K.; Liu, Y.G.; Wang, X.L.; Fan, T.X.; Zhang, D. Microstructure and sound absorption of porous copper prepared by resin curing and foaming method. *Mater. Lett.* **2015**, *139*, 318–321. [CrossRef]
9. Liu, Z.Q.; Zhan, J.X.; Fard, M.; Davy, J.L. Acoustic properties of a porous polycarbonate material produced by additive manufacturing. *Mater. Lett.* **2016**, *181*, 296–299. [CrossRef]
10. Ning, J.F.; Zhao, G.P. Sound absorption characteristics of multilayer porous metal materials backed with an air gap. *J. Vib. Control* **2016**, *22*, 2861–2872.
11. Otaru, A.J. Enhancing the sound absorption performance of porous metals using tomography images. *Appl. Acoust.* **2019**, *143*, 183–189. [CrossRef]
12. Chen, W.J.; Liu, S.T.; Tong, L.Y.; Li, S. Design of multi-layered porous fibrous metals for optimal sound absorption in the low frequency range. *Theor. Appl. Mech. Lett.* **2016**, *6*, 42–48. [CrossRef]
13. Cheng, W.; Duan, C.Y.; Liu, P.S.; Lu, M. Sound absorption performance of various nickel foam-base multi-layer structures in range of low frequency. *Trans. Nonferr. Metal. Soc.* **2017**, *27*, 1989–1995. [CrossRef]
14. Ao, Q.B.; Wang, J.Z.; Tang, H.P.; Zhi, H.; Ma, J.; Bao, T.F. Sound absorption characteristics and structure optimization of porous metal fibrous materials. *Rare Metal Mat. Eng.* **2015**, *44*, 2646–2650.
15. Bai, P.F.; Shen, X.M.; Zhang, X.N.; Yang, X.C.; Yin, Q.; Liu, A.X. Influences of compression ratio on sound absorption performance of porous nickel-iron alloy. *Metals* **2018**, *8*, 539. [CrossRef]
16. Bai, P.F.; Yang, X.C.; Shen, X.M.; Zhang, X.N.; Li, Z.Z.; Yin, Q.; Jiang, G.L.; Yang, F. Sound absorption performance of the acoustic absorber fabricated by compression and microperforation of the porous metal. *Mater. Des.* **2019**, *167*, 107637. [CrossRef]
17. Yang, X.C.; Shen, X.M.; Bai, P.F.; He, X.H.; Zhang, X.N.; Li, Z.Z.; Chen, L.; Yin, Q. Preparation and characterization of gradient compressed porous metal for high-efficiency and thin-thickness acoustic absorber. *Materials* **2019**, *12*, 1413. [CrossRef]
18. Verdière, K.; Panneton, R.; Elkoun, S.; Dupont, T.; Leclaire, P. Transfer matrix method applied to the parallel assembly of sound absorbing materials. *J. Acoust. Soc. Am.* **2013**, *134*, 4648–4658. [CrossRef]
19. Shen, X.M.; Bai, P.F.; Yang, X.C.; Zhang, X.N.; To, S. Low-frequency sound absorption by optimal combination structure of porous metal and microperforated panel. *Appl. Sci.* **2019**, *9*, 1507. [CrossRef]
20. Kino, N. Further investigations of empirical improvements to the Johnson–Champoux–Allard model. *Appl. Acoust.* **2015**, *96*, 153–170. [CrossRef]
21. Yang, X.C.; Peng, K.; Shen, X.M.; Zhang, X.N.; Bai, P.F.; Xu, P.J. Geometrical and dimensional optimization of sound absorbing porous copper with cavity. *Mater. Des.* **2017**, *131*, 297–306. [CrossRef]
22. Yang, X.S.; Deb, S. Engineering Optimisation by Cuckoo Search. *Int. J. Math. Model. Numer. Optim.* **2010**, *1*, 330–343. [CrossRef]
23. Yang, X.S.; Deb, S. Cuckoo search: recent advances and applications. *Neural Comput. Appl.* **2014**, *24*, 169–174. [CrossRef]
24. Yang, X.C.; Bai, P.F.; Shen, X.M.; To, S.; Chen, L.; Zhang, X.N.; Yin, Q. Optimal design and experimental validation of sound absorbing multilayer microperforated panel with constraint conditions. *Appl. Acoust.* **2019**, *146*, 334–344. [CrossRef]

25. Liu, J.; Chen, T.T.; Zhang, Y.H.; Wen, G.L.; Qing, Q.X.; Wang, H.X.; Sedaghati, R.; Xie, Y.M. On sound insulation of pyramidal lattice sandwich structure. *Compos. Struct.* **2019**, *208*, 385–394. [CrossRef]
26. Bujoreanu, C.; Nedeff, F.; Benchea, M.; Agop, M. Experimental and theoretical considerations on sound absorption performance of waste materials including the effect of backing plates. *Appl. Acoust.* **2017**, *119*, 88–93. [CrossRef]
27. Yang, X.C.; Bai, P.F.; Shen, X.M.; Zhang, X.N.; Zhu, J.W.; Yin, Q.; Peng, K. Theoretical modeling and experimental validation of sound absorbing coefficient of porous iron. *J. Porous Media* **2019**, *22*, 225–241. [CrossRef]
28. Delany, M.E.; Bazley, E.N. Acoustical properties of fibrous absorbent materials. *Appl. Acoust.* **1970**, *3*, 105–116. [CrossRef]
29. Miki, Y. Acoustical properties of porous materials-Modifications of Delany-Bazley models. *J. Acoust. Soc. Jpn. (E)* **1990**, *11*, 19–24. [CrossRef]

Article

Aluminum Foam-Filled Steel Tube Fabricated from Aluminum Burrs of Die-Castings by Friction Stir Back Extrusion

Yoshihiko Hangai [1],*, Ryusei Kobayashi [1], Ryosuke Suzuki [1], Masaaki Matsubara [1] and Nobuhiro Yoshikawa [2]

[1] Faculty of Science and Technology, Gunma University, Kiryu 376-8515, Japan; t13302045@gunma-u.ac.jp (R.K.); r_suzuki@gunma-u.ac.jp (R.S.); m.matsubara@gunma-u.ac.jp (M.M.)

[2] Institute of Industrial Science, The University of Tokyo, Tokyo 153-8505, Japan; nobyoshi@iis.u-tokyo.ac.jp

* Correspondence: hanhan@gunma-u.ac.jp

Received: 30 December 2018; Accepted: 18 January 2019; Published: 24 January 2019

Abstract: A mixture of Al burrs of Al high-pressure die-castings and a blowing agent powder was used to fabricate Al foam-filled steel tubes by friction stir back extrusion (FSBE). It was shown that the mixture can be sufficiently consolidated to form an Al precursor that is coated on the inner surface of a steel tube by the plastic flow generated during FSBE. Namely, a precursor coated steel tube can be fabricated from Al burrs by FSBE. By heat treatment of the precursor coated steel tube, an Al foam-filled steel tube can be fabricated. Al foam was sufficiently filled in the steel tube, and the porosity was almost homogeneously distributed in the entire sample. In compression tests of the samples, the Al foam-filled steel tube fabricated from Al burrs exhibited similar compression properties to an Al foam-filled steel tube fabricated from the bulk Al precursor. Consequently, it was shown that an Al foam-filled steel tube cost-effectively fabricated from Al burrs by FSBE compares favorably with an Al foam-filled steel tube fabricated from the bulk Al precursor.

Keywords: cellular materials; composites; friction welding; foam; recycle

---

1. Introduction

Composite structures consisting of a dense metallic tube filled with aluminum (Al) foam have attracted interest in various industrial fields owing to their light-weight and good energy absorption properties [1,2]. Generally, Al foam-filled metallic tubes have been fabricated by simply placing Al foam into a metallic tube with a gap between the Al foam and the metallic tube, which adversely affects the mechanical properties of the tube [3,4]. Hamada et al. and Toksoy et al. demonstrated that bonding between the Al foam and Al tube with adhesives resulted in superior absorption properties to those without any bonding [3,5]. Bonaccorsi et al. fabricated an Al foam-filled steel tube with metal bonding between the Al foam and steel tube by foaming a precursor in the steel tube [6]. The precursor was a solid Al homogeneously containing blowing agent [7,8]. Duarte et al. fabricated an Al foam-filled Al tube with metal bonding between the Al foam and Al tube by foaming a precursor in the Al tube [9].

Friction stir back extrusion (FSBE) has been developed for coating wear-resistant Al-25%Si on the surface of an Al-Si-Cu alloy AC2B engine cylinder [10], and for fabricating Al tubes and copper tubes with a fine grain structure from bulk starting materials [11,12]. Hangai et al. developed a fabrication process for Al foam-filled steel tubes [13,14] and Al foam-filled Al tubes [15,16] by FSBE. In this process, a bulk Al precursor was placed into a metal tube; then a rotating tool was plunged onto the precursor. The generated friction heat between the rotating tool and precursor softened the precursor, then the plastic flow of the precursor occurred, causing it to fill the space between the rotating tool and the metal tube, resulting in the coating of the precursor on the surface of the metal tube. The plastic flow induced the exposure of the new surface of the precursor and metal tube, realizing strict contact between them. Heat treatment of the precursor resulted in the foaming of the precursor, which filled the tube and bonded to the tube with metallic bonding. A bilayer tube consisting of an outer Al foam and an inner Al tube can also be fabricated by a similar process [17].

A reduction in cost and the high recyclability of Al foam-filled metal tubes can be expected by using Al chip waste from industrial production [18,19]. Consolidation of Al chips has been attempted by hot extrusion [18,19], compression torsion processing [20,21], friction extrusion [22], friction consolidation [23–25], screw extrusion [26], and FSBE [27]. Among these processes, FSBE can fabricate an Al tube directly from Al chips and realize strict contact with an outer metal tube.

In this study, a mixture of Al burrs of Al high-pressure die-castings and a blowing agent powder was used to fabricate Al foam-filled steel tubes by FSBE. It was expected that a foamable precursor would be fabricated by consolidating the mixture by the plastic flow generated during FSBE, along with the coating of the precursor on the inner surface of the steel tube. By heat treatment of the precursor coated steel tubes, Al foam-filled steel tubes were obtained. The pore structures of the inner Al foam were nondestructively observed to determine whether sufficient foaming occurred, and the pores were found to be homogeneously distributed in the whole specimen. Thereafter, compression tests of the obtained Al foam-filled steel tubes were conducted to determine whether similar mechanical properties to those of an Al foam-filled tube fabricated from the bulk Al precursor could be obtained.

2. Materials and Methods

2.1. Fabrication of Al Foam-Filled Steel Tube

Figure 1 shows Al burrs used to fabricate the Al foam of the Al foam-filled steel tube. The Al burrs were from Al-Si-Cu alloy ADC12 (equivalent to A383.0 Al alloy) die-castings. Figure 1a shows a scanning electron microscope (JEOL Ltd., Tokyo, Japan) image of the ADC12 burrs in powder form. As received ADC12 burrs were sieved with a mesh size of 180 μm to remove the impurities and for easy mixing with the blowing agent powder. Figure 1b shows the ADC12 burrs in round-plate form. Relatively thick (approximately 0.2 mm thickness) ADC12 burrs were cut into a circular shape with a diameter of 18 mm.

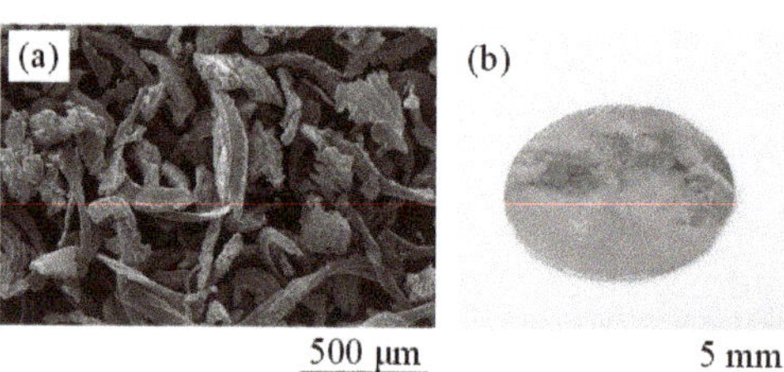

Figure 1. ADC12 burrs (**a**) in powder form and (**b**) in round-plate form.

Figure 2 shows the fabrication process of the Al foam-filled steel tube by FSBE. First, as shown in Figure 2a, ADC12 burrs in powder form and a blowing agent powder were mixed. TiH$_2$ powder (<45 μm) was used as the blowing agent. The amount of TiH$_2$ was one mass% relative to the mass of ADC12 burrs in accordance with reference [13]. As shown in Figure 2b, the mixture was placed in a SUS304 stainless-steel tube fixed by a steel jig. The steel tube had a diameter of 20 mm, a thickness of 0.2 mm, and 40 mm height. Then, five ADC12 burrs in round-plate form were laminated on top of the mixture in the steel tube to prevent the mixture from scattering during FSBE. As shown in Figure 2c, a rotating tool was plunged from above until the tip of the tool reached 3 mm above the bottom of the steel tube using friction stir welding machine (Hitachi Setsubi Engineering Co., Ltd., Hitachi, Japan). The rotation speed and plunging rates were 1500 rpm and 4 mm/min, respectively. The tool was made of tool steel with a diameter of 16 mm and had a flat bottom. The mixture was consolidated by loading the tool and the plastic flow of the mixture, resulting in the formation of a precursor and the coating on the inner surface of the steel tube. The temperature during the plunging of the tool was obtained by a K-type thermocouple placed at half the height of the steel tube and 2 mm from its surface. As shown in Figure 2d, the upper and lower parts of the obtained precursor coated steel tube were machined by wire electrical discharge machining, then the entire tube was foamed by heat treatment in an electric furnace previously held at 675 °C for 6.5–7.0 min. Then, as shown in Figure 2e, an Al foam-filled steel tube was obtained. The upper and lower parts of the obtained Al foam-filled steel tube were machined by wire electrical discharge machining to obtain a compression test specimen with 20 mm diameter and 20 mm height. Fourteen compression test specimens were obtained. In addition, similar compression test specimens were fabricated from a bulk Al precursor, which was fabricated by the friction stir welding route [28,29], by the same process as above using FSBE.

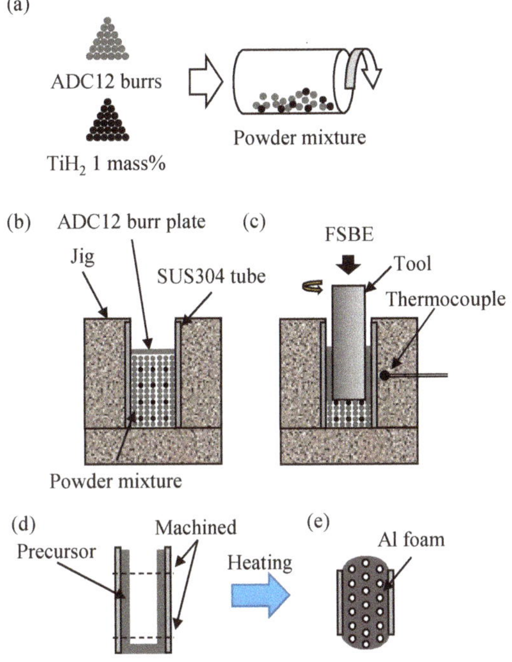

Figure 2. Schematic illustration of fabrication process of Al foam-filled steel tube by FSBE.

2.2. Evaluation of Pore Structures

The pore structures of the compression test specimens were nondestructively observed by a microfocus X-ray CT system (Shimadzu Corporation, Kyoto, Japan) at room temperature in accordance with reference [30]. The X-ray source was tungsten. A cone-beam CT system, where only one rotation of the specimen is required to obtain a set of cross-sectional X-ray CT images with the slice pitch equal to the length of one pixel in the CT image, was employed. The number of slices was about 450. The voltage and current of the X-ray tube were 80 kV and 30 µA, respectively. The equivalent length of a pixel was approximately 74 µm.

The porosity (volume fraction of pores) p (%) of the Al foam part of each compression test specimen was evaluated as follows. First, the density of the Al foam part ρ_f was evaluated as

$$\rho_f = \frac{m_{\text{FFT}} - m_{\text{tube}}}{V_{\text{FFT}} - V_{\text{tube}}},$$

where m_{FFT} and V_{FFT} are the mass and volume of the compression test specimen, respectively, which were obtained from the measured weight and dimensions of the specimen. m_{tube} and V_{tube} are the mass and volume of the steel tube part of the compression test specimen, respectively. Then, p was evaluated as

$$p = \frac{\rho_i - \rho_f}{\rho_i} \times 100,$$

where ρ_i is the density of the precursor without the tube before foaming, for which the density of ADC12 Al alloy [31] was used.

2.3. Compression Tests

Compression tests of the obtained Al foam-filled steel tubes were conducted using a universal testing machine (Shimadzu Corporation, Kyoto, Japan). The compression rate was 4 mm/min in accordance with reference [32]. The compression stress was obtained by dividing the compression load by the area of the circle equivalent to the outer diameter of the Al foam-filled steel tube.

3. Results and Discussion

3.1. Time–Temperature Relationship during FSBE

Figure 3 shows typical relationships obtained between torque N and time t and between temperature T and time t. $t = 0$ was defined as the time when the tool first came in contact with the ADC12 burrs in the round-plate form. As the indentation of the tool increased, the ADC12 burrs were densified, and plastic flow occurred. It has been reported that the filling of the Al between the rotating tool and tube occurs from the upper part, which is softened by the friction heat generated between the rotating tool and the Al precursor [15]. It is considered that plastic flow occurred in the vicinity of the tool where friction heat was generated. The torque and temperature gradually increased in this stage. At around $t = 5$ min, the torque increased rapidly. The temperature exhibited a similar tendency except with a small delay due to the time for the generated heat to be transferred to the thermocouple. In this stage, densification of the ADC12 burrs had finished, and only plastic flow occurred, resulting in a rapid increase in the resistance force and temperature. Similar tendencies were observed for all other samples, and the maximum temperatures of all the samples were between 350 and 415 °C. Also, similar tendencies were observed for the samples using the bulk Al precursor. These temperatures are much lower than the solidus temperature of ADC12 (515 °C [31]), at which the foaming of the precursor begins [33,34]. It has been demonstrated that the decomposition of TiH_2 gradually occurs between 390 and 920 °C [35], 427 and 850 °C [36], and 440 and 800 °C [37]. In the FSBE process, it is assumed that little decomposition of TiH_2 occurred because the temperature of the precursor reached these reported decomposition temperatures for only a limited time. Therefore, almost all the TiH_2 remained without decomposition.

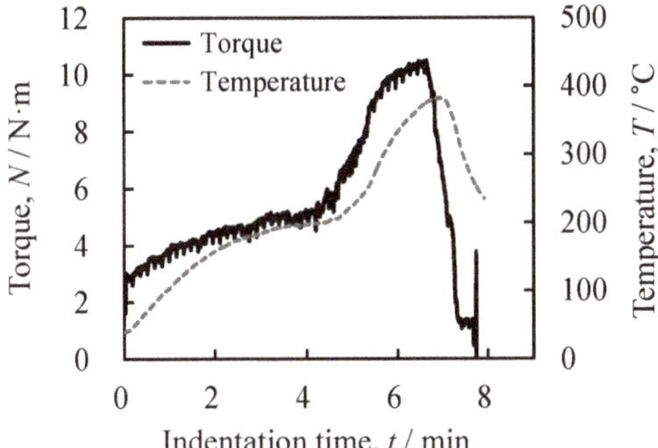

Figure 3. Torque N– indentation time t and temperature T–t relationships during FSBE.

3.2. Obtained Al Foam-Filled Steel Tube

Figure 4a shows the as-fabricated precursor coated tube. It was shown that the ADC12 burrs were sufficiently consolidated except in the upper part of the tube. The upper part of the tube was subjected to a less plastic flow during FSBE, resulting in insufficient consolidation of the ADC12 burrs. These parts were machined before foaming as shown in Figure 4b. Figure 4c shows a cross-section of the machined precursor before foaming, and Figure 4d shows an enlarged image of the bonding region of Figure 4c. It was shown that the ADC12 burrs were sufficiently consolidated to form the precursor and were coated on the surface of the tube without a gap between the tube and precursor by the load of the tool and the intense plastic flow during the FSBE.

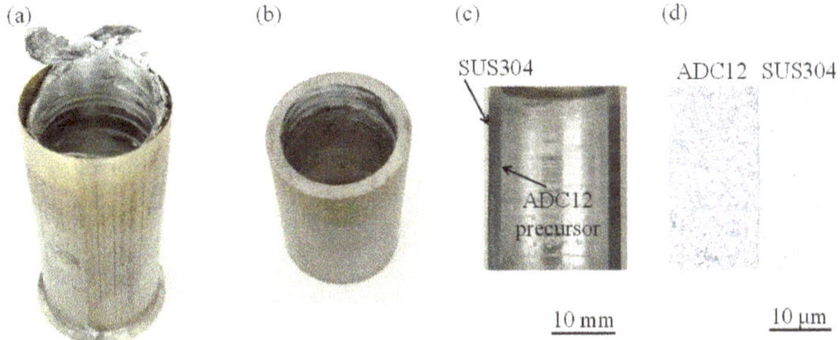

Figure 4. (**a**) As-fabricated precursor-coated steel tube. (**b**) machined precursor-coated steel tube before foaming and (**c**) its cross section. (**d**) enlarged image of bonding region of (**c**).

Figure 5a,b show the as-foamed Al foam-filled steel tubes after foaming for 6.5 min and 7.0 min, Figure 5c,d show the Al foam-filled steel tube compression test specimens obtained by machining the samples shown in Figure 5a,b, and Figure 5e,f show their cross-sectional X-ray CT images, respectively. Black and white regions in the X-ray CT images indicate pores and Al, respectively. Although a few large pores were observed, it was shown that the coated precursor was sufficiently foamed and filled the tubes. The porosities of the Al foam part of the compression test specimens shown in Figure 5c,d were $p = 79.3\%$ and $p = 90.2\%$, respectively.

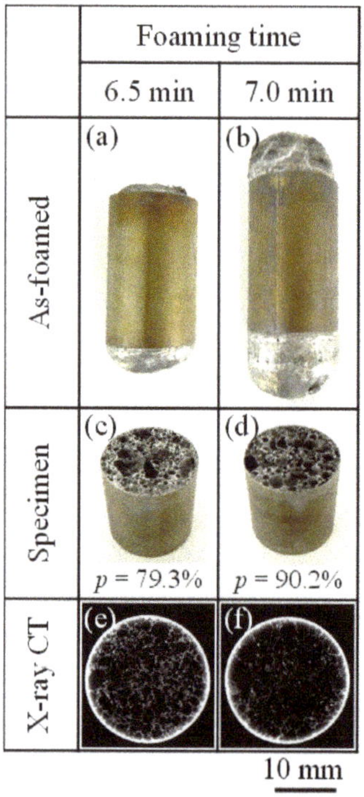

Figure 5. (**a**,**b**) As-foamed Al foam-filled steel tubes after foaming for 6.5 min and 7.0 min, respectively. (**c**,**d**) Al foam-filled steel tube compression test specimens machined from (**a**,**b**), respectively. (**e**,**f**) Their cross-sectional X-ray CT images.

Figure 6 shows the relationship between the porosity and foaming time of all the compression test specimens. The figure indicates that vigorous foaming occurred between 6.5 min and 7.0 min. In previous studies on the foaming of ADC12 precursor [33,34], it was demonstrated that the precursor started to foam once the precursor temperature exceeded the solidus temperature and gradually foamed between the solidus and liquidus temperatures. Then, vigorous foaming occurred once the precursor temperature exceeded the liquidus temperature. Although some variation among the porosities appeared at a foaming time of 6.5 min owing to the difficulty of obtaining the same foaming behavior for each specimen using an electric furnace, a foaming time of around 7.0 min induced sufficient foaming for the precursor temperature of all the samples to exceed the liquidus temperature.

Figure 7 shows the porosity distributions of the compression test specimens shown in Figure 5c,d in the height direction for foaming times of 6.5 min ($p = 79.3\%$) and 7.0 min ($p = 90.2\%$), respectively, which were evaluated using X-ray CT images such as shown in Figure 5e,f by image processing software. Although there were a few large pores, the Al foam sufficiently filled the tube with an almost homogeneous porosity distribution throughout the entire specimen, especially for the sample foamed for 7.0 min, which exhibited sufficient foaming of the precursor. The variation of the porosity and the existence of large pores are considered to be due to the insufficient distribution of the foaming agent compared with that in the bulk precursor because it was difficult to homogeneously mix the ADC12 burr and TiH_2 powders with different particle sizes. Also, slight segregation occurred during the insertion of the mixture into the tube and during the FSBE. This may be prevented by using ADC12 burr particles with a similar size to that of TiH_2.

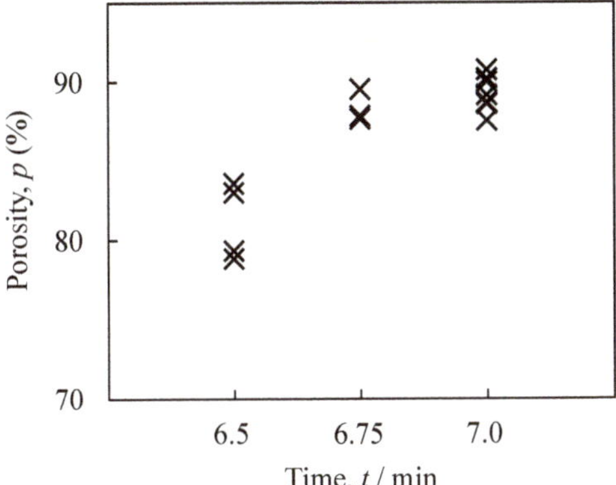

Figure 6. Relationship between porosity and foaming time of obtained Al foam-filled steel tube compression test specimens.

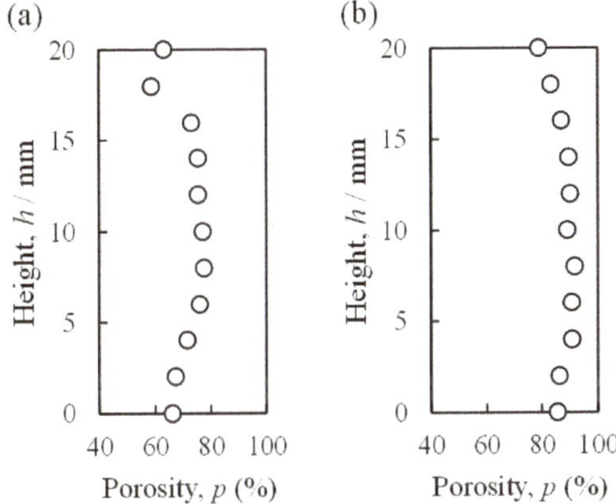

Figure 7. Porosity distributions in the height direction of Al foam-filled steel tube compression test specimens with (**a**) p = 79.3% and (**b**) p = 90.2% corresponding to Figure 5c,d, respectively.

3.3. Compression Properties

Figure 8 shows the deformation behavior of the fabricated Al foam-filled steel tubes with p = 79.3% and p = 90.2%, respectively, which are shown in Figure 5c,d during the compression test, along with that of the tube fabricated from the bulk Al precursor with p = 84.5% by FSBE in accordance with Ref. [13]. The three black lines on the surface of the samples perpendicular to the compression direction were marks made to easily observe the deformation behavior. It was found that the deformation occurred layer by layer. Little difference was observed in the deformation behavior between the Al foams with different porosities, although the number of buckling folds were different. This difference is because the lower-porosity Al foam expanded in the width direction as the deformation increased. In contrast, the higher-porosity Al foam deformed with little expansion in the width direction. The deformation behavior was similar for the other specimens fabricated in this study, including those fabricated from the bulk Al precursor.

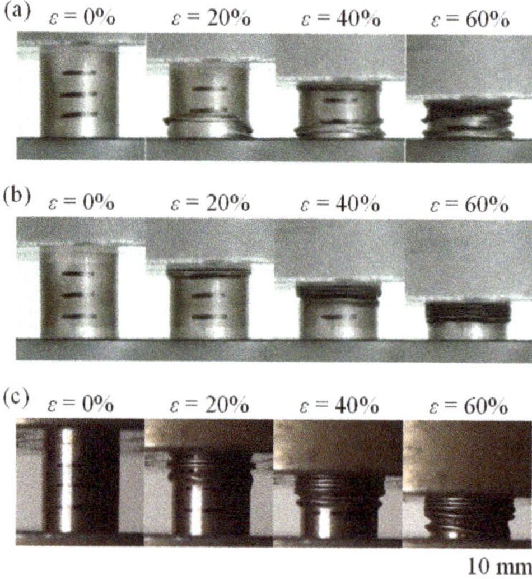

Figure 8. Deformation behavior during compression test of the fabricated Al foam-filled steel tube with (**a**) p = 79.3%, corresponding to Figure 5b,c. (**b**) p = 90.2%, corresponding to Figure 5d,c Al foam-filled steel tube fabricated using bulk Al precursor with (**c**) p = 84.5%.

Figure 9 shows the stress σ – strain ε curves of the fabricated Al foam-filled steel tubes with p = 79.3% and p = 90.2%; along with that of the tube fabricated from the bulk Al precursor with p = 84.5%, whose deformation behavior is shown in Figure 8. An elastic region in the early stage of the σ–ε curves, where σ rapidly increased to an initial maximum stress was observed. Then, a plateau region exhibiting an appropriately constant or slightly increasing σ was observed as ε increased. Then, a densification region where σ rapidly increased was observed. The increases and decreases in σ in the plateau region were observed when the folds were generated. These σ–ε curves were also observed for other specimens fabricated in this study regardless of the porosity.

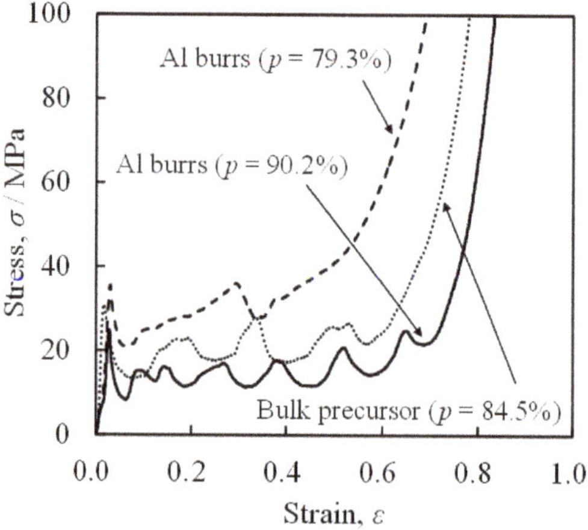

Figure 9. Stress–strain curves for the specimens in Figure 8.

Figure 10 shows the plateau stress σ_{pl}– porosity p and energy absorption per unit volume E_V– porosity p relationships for all the samples including those fabricated from the bulk Al precursor. σ_{pl} was defined as the average σ for ε = 0.2–0.3 in the σ–ε curves and E_V was defined as the area under the σ-ε curves for ε = 0–0.5 in accordance with Ref. 32. It was found that σ_{pl} and E_V decreased as p increased. This tendency was also observed for the samples fabricated from the bulk Al precursor. Consequently, it was shown that Al foam-filled steel tubes can be fabricated from ADC12 burrs that exhibit almost the same compression properties as Al foam-filled steel tubes fabricated from the bulk Al precursor when the porosity of the Al foam part is the same.

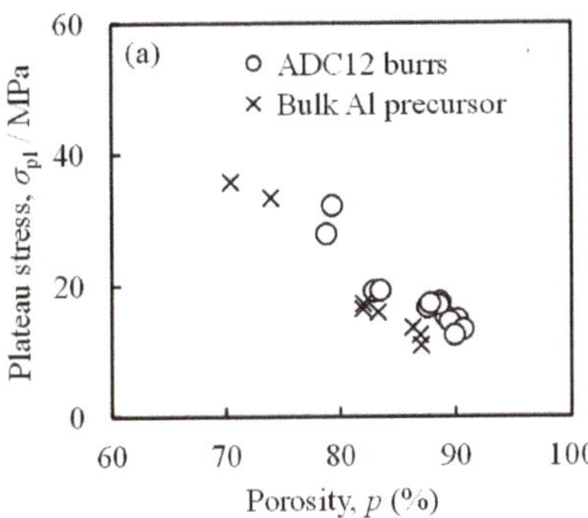

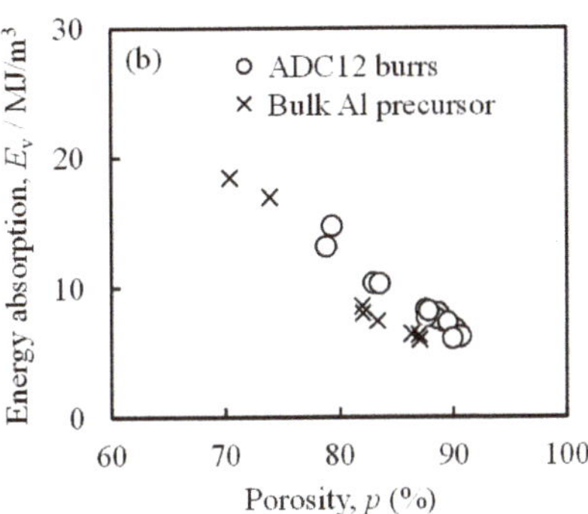

Figure 10. Relationship between (**a**) plateau stress σ_{pl} and (**b**) energy absorption per unit volume E_V and porosity p for the obtained Al foam-filled steel tubes fabricated from ADC12 burrs along with those fabricated from the bulk Al precursor.

4. Conclusions

In this study, a mixture of ADC12 burrs of Al high-pressure die-castings and a blowing agent powder was used to fabricate Al foam-filled steel tubes by FSBE. The experimental results led to the following conclusions.

(1) A mixture of ADC12 burrs and a blowing agent can be sufficiently consolidated to form a precursor that is coated on the inner surface of a steel tube by FSBE. Namely, a precursor-coated steel tube can be fabricated from ADC12 burrs by FSBE.

(2) An Al foam-filled steel tube can be fabricated by heat treatment of the precursor coated steel tube. Al foam was sufficiently filled in the steel tube, and the porosity was almost homogeneously distributed in the entire sample.

(3) The Al foam-filled steel tube fabricated from ADC12 burrs exhibited similar compression properties to an Al foam-filled steel tube fabricated from the bulk Al precursor.

Consequently, it was shown that an Al foam-filled steel tube can be cost-effectively fabricated from Al burrs by FSBE and compares favorably with an Al foam-filled steel tube fabricated from the bulk Al precursor.

Author Contributions: Y.H. elaborated the research idea and wrote this manuscript. R.K. and R.S. performed the experimental work and contributed to the discussion of the experimental results. M.M. and N.Y. performed literature review and contributed to the discussion of the experimental results.

Conflicts of Interest: The authors declare no conflict of interest.

References

1. Banhart, J. Manufacture, characterisation and application of cellular metals and metal foams. *Prog. Mater. Sci.* **2001**, *46*, 559–632. [CrossRef]
2. García-Moreno, F. Commercial applications of metal foams: Their properties and production. *Materials* **2016**, *9*, 85. [CrossRef] [PubMed]

3. Hamada, T.; Kanahashi, H.; Kanetake, N. Axial crushing performance of porous aluminum filled members. *J. Jpn. Inst. Met. Mater.* **2009**, *73*, 453–461. [CrossRef]

4. Duarte, I.; Vesenjak, M.; Krstulović-Opara, L. Dynamic and quasi-static bending behaviour of thin-walled aluminium tubes filled with aluminium foam. *Compos. Struct.* **2014**, *109*, 48–56. [CrossRef]

5. Toksoy, A.K.; Tanoglu, M.; Guden, M.; Hall, I.W. Effect of adhesive on the strengthening of aluminum foam-filled circular tubes. *J. Mater. Sci.* **2004**, *39*, 1503–1506. [CrossRef]

6. Bonaccorsi, L.; Proverbio, E.; Raffaele, N. Effect of the interface bonding on the mechanical response of aluminium foam reinforced steel tubes. *J. Mater. Sci.* **2010**, *45*, 1514–1522. [CrossRef]

7. Baumgartner, F.; Duarte, I.; Banhart, J. Industrialization of powder compact foaming process. *Adv. Eng. Mater.* **2000**, *2*, 168–174. [CrossRef]

8. Hangai, Y.; Utsunomiya, T.; Hasegawa, M. Effect of tool rotating rate on foaming properties of porous aluminum fabricated by using friction stir processing. *J. Mater. Process. Technol.* **2010**, *210*, 288–292. [CrossRef]

9. Duarte, I.; Vesenjak, M.; Krstulovic-Opara, L.; Ren, Z.R. Static and dynamic axial crush performance of in-situ foam-filled tubes. *Compos. Struct.* **2015**, *124*, 128–139. [CrossRef]

10. Shinoda, T. Applications of friction technology on cast materials. *J. JFS* **2011**, *83*, 695–701.

11. Abu-Farha, F. A preliminary study on the feasibility of friction stir back extrusion. *Scr. Mater.* **2012**, *66*, 615–618. [CrossRef]

12. Dinaharan, I.; Sathiskumar, R.; Vijay, S.J.; Murugan, N. Microstructural Characterization of pure copper tubes produced by a novel method friction stir back extrusion. *Procedia Mater. Sci.* **2014**, *5*, 1502–1508. [CrossRef]

13. Hangai, Y.; Saito, M.; Utsunomiya, T.; Kitahara, S.; Kuwazuru, O.; Yoshikawa, N. Fabrication of aluminum foam-filled thin wall steel tube by friction welding and its compression properties. *Materials* **2014**, *7*, 6796–6810. [CrossRef]

14. Hangai, Y.; Nakano, Y.; Utsunomiya, T.; Kuwazuru, O.; Yoshikawa, N. Drop weight impact behavior of Al-Si-Cu alloy foam-filled thin-walled steel pipe fabricated by friction stir back extrusion. *J. Mater. Eng. Perform.* **2017**, *26*, 894–900. [CrossRef]

15. Hangai, Y.; Nakano, Y.; Koyama, S.; Kuwazuru, O.; Kitahara, S.; Yoshikawa, N. Fabrication of aluminum tubes filled with aluminum alloy foam by friction welding. *Materials* **2015**, *8*, 7180–7190. [CrossRef] [PubMed]

16. Hangai, Y.; Otazawa, S.; Utsunomiya, T. Aluminum alloy foam-filled aluminum tube fabricated by friction stir back extrusion and its compression properties. *Compos. Struct.* **2018**, *183*, 416–422. [CrossRef]

17. Hangai, Y.; Otazawa, S.; Utsunomiya, T.; Suzuki, R.; Koyama, S.; Matsubara, M.; Yoshikawa, N. Fabrication of bilayer tube consisting of outer aluminum foam tube and inner dense aluminum tube by friction stir back extrusion. *Mater. Today Commun.* **2018**, *15*, 36–42. [CrossRef]

18. Gronostajski, J.Z.; Kaczmar, J.W.; Marciniak, H.; Matuszak, A. Direct recycling of aluminium chips into extruded products. *J. Mater. Process. Technol.* **1997**, *64*, 149–156. [CrossRef]

19. Tekkaya, A.E.; Schikorra, M.; Becker, D.; Biermann, D.; Hammer, N.; Pantke, K. Hot profile extrusion of AA-6060 aluminum chips. *J. Mater. Process. Technol.* **2009**, *209*, 3343–3350. [CrossRef]

20. Kanetake, N.; Kobashi, M.; Tsuda, S. Foaming behavior of aluminum precursor produced from machined chip waste. *Adv. Eng. Mater.* **2008**, *10*, 840–844. [CrossRef]

21. Takahashi, T.; Kume, Y.; Kobashi, M.; Kanetake, N. Solid state recycling of aluminum machined chip wastes by compressive torsion processing. *J. Japan Inst. Light Met.* **2009**, *59*, 354–358. [CrossRef]

22. Tang, W.; Reynolds, A.P. Production of wire via friction extrusion of aluminum alloy machining chips. *J. Mater. Process. Technol.* **2010**, *210*, 2231–2237. [CrossRef]

23. Tang, W.; Reynolds, A.P. *Friction Consolidation of Aluminum Chips*; John Wiley & Sons: Chichester, UK, 2011; pp. 289–298.

24. Hagihara, M.; Katoh, K.; Maeda, M.; Nomoto, M. Effect of chip shape on frictional consolidated 2017 aluminum alloy chips. In Proceedings of the 129th Conference of Japan Institute of Light Metals, Tsudanuma, Japan, 21 November 2015; pp. 215–216.

25. Li, X.; Baffari, D.; Reynolds, A.P. Friction stir consolidation of aluminum machining chips. *Int. J. Adv. Manuf. Technol.* **2018**, *94*, 2031–2042. [CrossRef]

26. Wideroe, F.; Welo, T. Using contrast material techniques to determine metal flow in screw extrusion of aluminium. *J. Mater. Process. Technol.* **2013**, *213*, 1007–1018. [CrossRef]

27. Hangai, Y.; Kobayashi, R.; Suzuki, R.; Matsubara, M.; Yoshikawa, N. Fabrication of aluminum pipe from aluminum chips by friction stir back extrusion. *J. Jpn. Inst. Met. Mater.* **2018**, *82*, 33–38. [CrossRef]
28. Hangai, Y.; Ozeki, Y.; Utsunomiya, T. Foaming conditions of porous aluminum in fabrication of ADC12 aluminum alloy die castings by friction stir processing. *Mater. Trans.* **2009**, *50*, 2154–2159. [CrossRef]
29. Hangai, Y.; Takada, K.; Endo, R.; Fujii, H.; Aoki, Y.; Utsunomiya, T. Foaming of aluminum foam precursor during friction stir welding. *J. Mater. Process. Technol.* **2018**, *259*, 109–115. [CrossRef]
30. Hangai, Y.; Takahashi, K.; Yamaguchi, R.; Utsunomiya, T.; Kitahara, S.; Kuwazuru, O.; Yoshikawa, N. Nondestructive observation of pore structure deformation behavior of functionally graded aluminum foam by X-ray computed tomography. *Mater. Sci. Eng. A* **2012**, *556*, 678–684. [CrossRef]
31. The-Japan-Institute-of-Light-Metals. *Structures and Properties of Aluminum*; The Japan Institute of Light Metals: Tokyo, Japan, 1991.
32. JIS-H-7902. *Method for Compressive Test of Porous Metals*; Japanese Standards Association: Tokyo, Japan, 2016.
33. Hangai, Y.; Amagai, K.; Tsurumi, N.; Omachi, K.; Shimizu, K.; Akimoto, K.; Utsunomiya, T.; Yoshikawa, N. Forming of aluminum foam using light-transmitting material as die during foaming by optical heating. *Mater. Trans.* **2018**, *59*, 1854–1859. [CrossRef]
34. Hangai, Y.; Amagai, K.; Omachi, K.; Tsurumi, N.; Utsunomiya, T.; Yoshikawa, N. Forming of aluminum foam using steel mesh as die during foaming of precursor by optical heating. *Opt. Laser Technol.* **2018**, *108*, 496–501. [CrossRef]
35. Matijasevic-Lux, B.; Banhart, J.; Fiechter, S.; Görke, O.; Wanderka, N. Modification of titanium hydride for improved aluminium foam manufacture. *Acta Mater.* **2006**, *54*, 1887–1900. [CrossRef]
36. Illeková, E.; Harnúšková, J.; Florek, R.; Simančík, F.; Maťko, I.; Švec, P. Peculiarities of TiH$_2$ decomposition. *J. Therm. Anal. Calorim.* **2011**, *105*, 583–590. [CrossRef]
37. Peng, Q.; Yang, B.; Friedrich, B. Porous Titanium Parts Fabricated by Sintering of TiH$_2$ and Ti Powder Mixtures. *J. Mater. Eng. Perform.* **2018**, *27*, 228–242. [CrossRef]

Review

Brief Review on Experimental and Computational Techniques for Characterization of Cellular Metals

Isabel Duarte [1,*], **Thomas Fiedler** [2], **Lovre Krstulović-Opara** [3] **and Matej Vesenjak** [4]

[1] Department of Mechanical Engineering, TEMA, University of Aveiro Campus Universitário de Santiago, 3810-193 Aveiro, Portugal
[2] Centre for Mass and Thermal Transport in Engineering Materials, School of Engineering, The University of Newcastle, Callaghan, NSW 2308, Australia; Thomas.Fiedler@newcastle.edu.au
[3] Faculty of Electrical Engineering, Mechanical Engineering and Naval Architecture, University of Split, 21000 Split, Croatia; Lovre.Krstulovic-Opara@fesb.hr
[4] Faculty of Mechanical Engineering, University of Maribor, 2000 Maribor, Slovenia; matej.vesenjak@um.si
* Correspondence: isabel.duarte@ua.pt; Tel.: +350-234-370-830

Received: 23 April 2020; Accepted: 26 May 2020; Published: 30 May 2020

Abstract: The paper presents a brief review of the main experimental and numerical techniques and standards to investigate and quantify the structural, mechanical, thermal, and acoustic properties of cellular metals. The potential of non-destructive techniques, such as X-ray computed tomography and infrared thermography are also presented.

Keywords: cellular metals; X-ray computed tomography; infrared thermography; mechanical characterization; thermal characterization; acoustic characterization

1. Introduction

Cellular metals are considered to be one of the world's most versatile lightweight multifunctional materials for a wide range of commercial, industrial, and military applications [1–4]. Cellular metals are non-flammable and recyclable, extremely tough, and can plastically deform and absorb energy. They can survive high temperatures in comparison to cellular polymers (e.g., polyurethane foams), which are widely used as crash/impact energy absorbers for vehicles due to their high strength-to-weight and stiffness-to-weight ratios and high energy absorption capacity [4].

Over recent decades, many products and solutions based on cellular metals [5–18] have been developed combined with other light materials like advanced high strength steels, light metals [19,20], polymers [21–27] and carbon nanostructures (e.g., carbon fibers) [28]. For example, cellular solids are used and/or tested as filler [29–32] and cores [33,34] of thin-walled structures and sandwich panels for buildings, industrial machines, cars, trains, and aircrafts. Their use contributes to an immediate weight reduction and material savings of the components, but also to perform multiple functions due to their 3D porous cellular structures (open-or-closed-cells) [35]. The open- and closed-cell structures are used as functional materials (e.g., electrodes, heat exchangers, and biomedical implants) and structural materials (e.g., impact/crash energy absorbers in vehicles, antivibration dampers), respectively [1,2]. The field of cellular materials has significantly increased over the last decades, which is reflected in the growth of published literature (Figure 1). Several cellular metals have been emerging due to two main reasons. The first reason is the need to create cellular structures with high structural efficiency and capacity to absorb both impact and damping noise and vibration. The second reason is the need to create periodic cellular structures with an easily reproducible unit cell to achieve predefined performances, establishing the process–structure–property relationships. Recent developments include composite and nanocomposite metal foams [28], lattices [36,37], 3D printed structures [38], syntactic foams [26,29,39], hollow sphere structures [27], auxetic materials [40–42], and hybrid structures [22–25,43] (e.g., filled hollow structures and sandwich panels). The latter often leads to the introduction of new technologies (e.g., 3D additive manufacturing) and products.

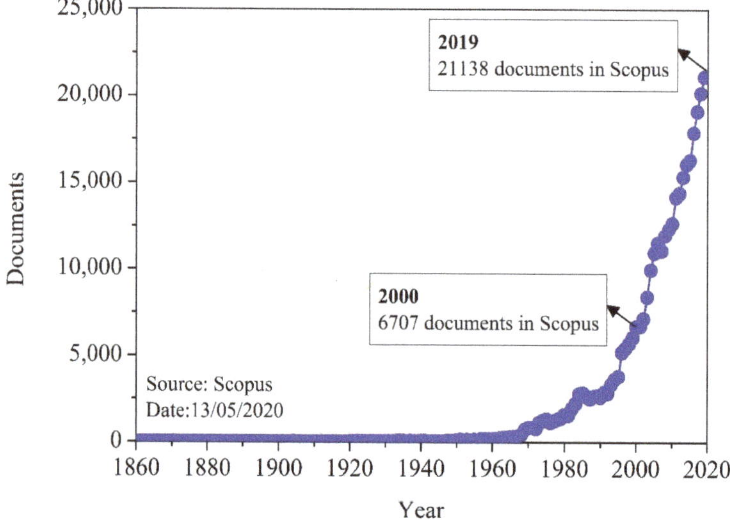

Figure 1. Number of published documents per year in the field of cellular materials (Source: Scopus).

Given the importance and diversity of these cellular materials, it is essential that the researchers use the same measurement science tools and techniques to assess their performance for comparing the different cellular materials. Most of the experimental tests are based on standards. Nevertheless, there is the need to improve and upgrade current testing techniques and methods in some cases.

2. Geometrical and Structural Characterization

Cellular metals are complex materials, formed by two or more phases (Figure 2). For example, two phases are observed in the open- and closed-cell metal foams, a discontinuous or continuous gaseous phase (cells or pores) and a continuous solid phase (base material or matrix). The base material can also have micro and nano reinforcement elements (e.g., alumina and carbon nanotubes) [28]. Other cellular materials like some syntactic foams can be formed by two solid phases (material from the porous particles/hollow spheres and interstitial material) and a discontinuous or continuous gaseous phase (e.g., pores from filler particle) [39].

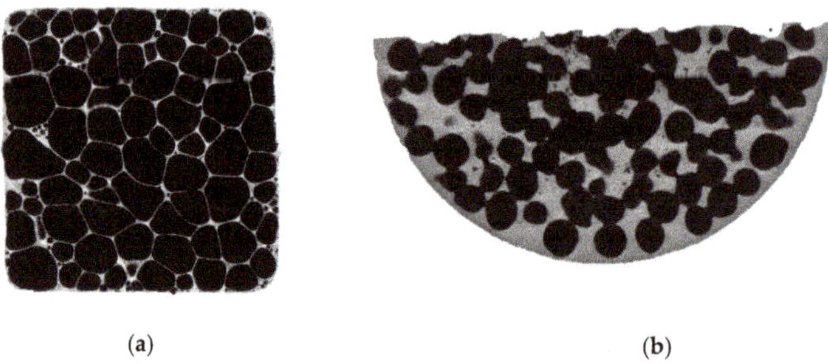

(a) (b)

Figure 2. Closed-cell foam (cross section: 20 mm × 20 mm) (**a**) and syntactic foam (diameter: 30 mm) (**b**).

Several material characterization techniques are normally used to characterize them. Microstructure, morphology formed phases/precipitates within matrix are usually analyzed by optical microscopy, scanning electron microscopy (SEM), X-ray diffraction (XRD), energy-dispersive x-ray spectroscopy (EDS), and atomic force microscopy (AFM). Formed phases, precipitates, fibers, and particles can also be investigated to some extent with nanotomography [44].

AFM is used to achieve a correlation between the surface roughness with the precipitated phase density in metallic matrix. Nano and micro indentation testers are used to determine the parameters like the stiffness and Vickers hardness of the matrix. XRD is used to identify and determine the various crystalline forms.

The morphometric parameters of the cellular metals such as a mean cell size and cell size distribution, fraction of solid, connectivity between cells, cell orientation, and porosity can be measured by a non-destructive technique, X-ray microcomputed tomography (μCT). This technique has revealed a power tool to characterize the cellular structure of these materials [45–47], but also to study their deformation behavior [48–50] and to develop numerical models to predict the mechanical, thermal and acoustic behavior. There are many different μCT scanners on the market, each having different capabilities, namely the resolution, maximum energy level, different filter options, stage set up and internal chamber area.

The equipment is composed of an X-ray source, a motor controlling rotating stage (where the specimen is mounted) and a detector (Figure 3a). The specimen is mounted between the X-ray source and the detector panel and projections are taken after each degree of rotation. Flat panel is an example of a commercial detector which is composed by an array of discrete sensors arranged as a matrix. The response of each sensor is proportional to the radiation energy that reaches its surface and is captured. The capture of data from each discrete part produces a 2D projection/scanned image (Figure 3a) that is converted into a digital image and, thus, is immediately visualized on a computer.

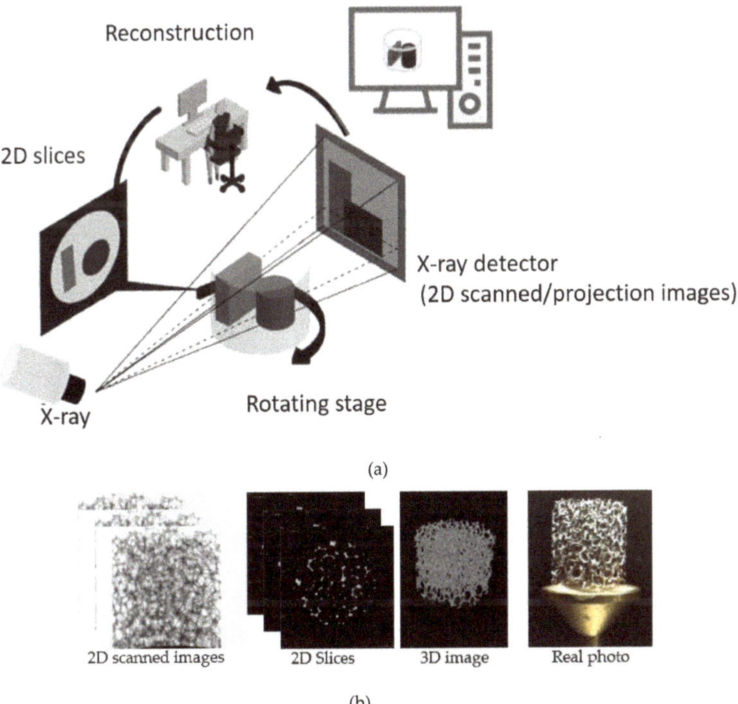

Figure 3. Scheme of the X-ray microcomputed tomography (μCT) equipment and an open-cell aluminum foam (height: 20.4 mm; diameter: 18.9 mm) (a) and showing and showing raw scans/projections, segmented images, reconstructed structure, and real photo (b).

The μCT principles are based on an X-ray source, which produces the X-ray emission that passes across the specimen and is projected on a digital detector, measuring the attenuation of the X-rays and producing a radiograph (known as scanned or projection image, Figure 3). The quality of the images mainly depends on the partial absorption and differential absorption that occurs during the scanning step. The partial absorption is associated to the absorption of some X-rays by the specimen and the transmission of others to the detector. The differential absorption refers to the absorption characteristics of the different materials to be scanned. For example, in the case that there is no differential absorption, the specimen result comes out as a uniform gray level (no contrast). Usually, it is difficult to distinguish the different materials within the specimen. Both (partial and differential) absorptions are influenced by several aspects, such as the size of the specimen and the chemical composition of the materials that are made which the X-rays have to pass. The projection images are taken incrementally over a total rotation (180° or 360°). The projection images (Figure 3a) are then processed using a computer software, producing a series of reconstructed images (2D slices) that allow to observe the internal structure of the object. Basically, the μCT data analysis is divided in four main steps, which are scanning, segmentation, reconstruction and visualization, as shown in Figure 3b. To optimize the image quality during these steps, it is necessary to adjust several parameters like magnification, incident X-ray intensity, filter type (e.g., no filter, copper, and aluminum), rotation step and acquisition time, and threshold.

There are many commercial software tools (e.g., NRecon v.1.7.3.1, Bruker, Kontich, Belgium; Octopus 8.6, Inside Matters BVBA, Aalst, Belgium and Phoenix datos, GE Sensing & Inspection Technologies GmbH, Wunstorf, Germany) that include the filtering (artefacts) and smoothing of the μCT data to enhance the quality of the images. Threshold segmentation converts a grey value image into a binary one. The resulting image then comprises two sets: one represents the background (e.g., black), the other one the object (e.g., white). This image binarization/segmentation procedure is an essential step in which the different phases/constituents using thresholding algorithms to measure the grayscale values in the μCT images are distinguished.

The reconstructed μCT images can be used for volume rendering of tomographic data, creating 3D models (Figure 4) using numerous available software tools, e.g., CTAn, CTvox, CTVol, VGStudio, and Avizo. The data visualization includes the mapping and the rendering of the results.

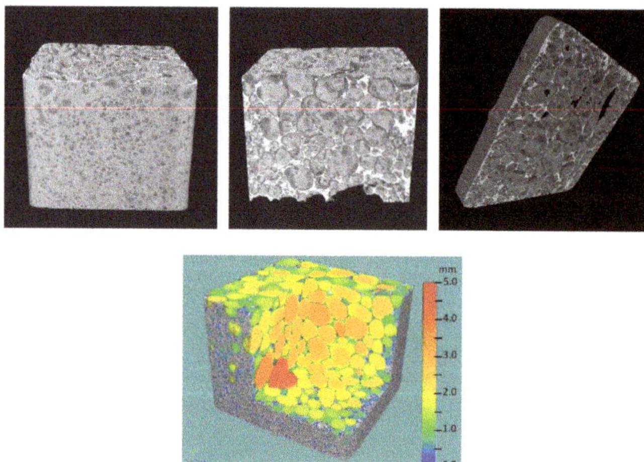

Figure 4. Volume rendering showing the cellular structure and the dense metal skin covered a closed-cell foam (width: 20 mm) with color by local structure separation, a measure of pore size.

The μCT can also be used to study the deformation and failure mode of these materials using ex-situ and in-situ tests. Ex-situ tests use specimens previously subjected to a mechanical test in a universal testing machine. In-situ imaging is performed inside μCT scanner using a material testing stage that allows the study of the deformation and failure behavior. The specimen is placed into the material testing stage (MTS) chamber and an accurate force, measured by a load cell, is applied to the specimen. The resulting deformation is measured by a precision displacement sensor. During scanning, the loading curve is displayed on the screen in real time.

Figure 5 shows a force–displacement curve of a closed-cell foam compressed with MTS up to 1540 N and 11 mm displacement, showing the collapse of pores of cellular structure.

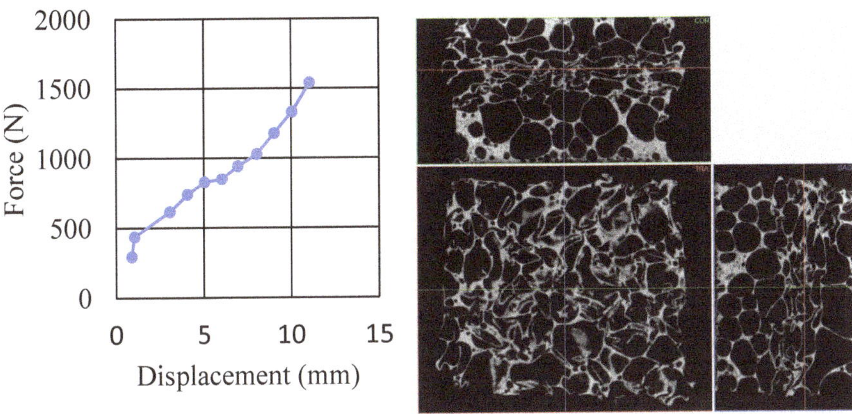

Figure 5. Force–displacement curve, showing the three orthogonal reconstructed slices (from Bruker's Data Viewer software [51]) of a closed-cell foam (cross section: 15 mm × 15 mm).

The μCT is also used together with finite element modelling [52,53] to develop the numerical models to describe and predict different process parameters and the mechanical, acoustic, and thermal behavior of the cellular materials, studying the different geometrical parameters. Finite element method (FEM) models can help to select the best microstructure for a given property (then for a given application) without the traditional, complex, time consuming, and labor-intensive experimental tests. The μCT scanned images are reconstructed using algorithms (e.g., filtering and smoothing) and software like NRecon (Bruker, Belgium). The μCT scanned images are used to create FEM meshes (Figure 6).

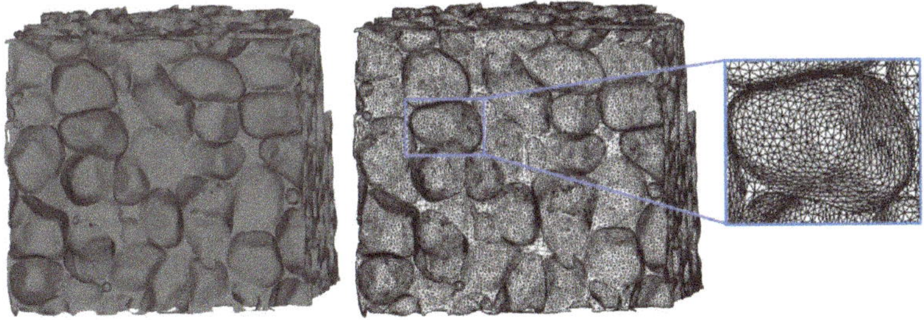

Figure 6. Finite element method (FEM) meshes of aluminum closed-cell foam (cross section: 10 mm × 10 mm). Courtesy of Diogo Heitor, Universidade de Aveiro, Portugal.

After segmentation, geometry cleanup, and refinement the model can be discretized and analyzed in a software (e.g., Abaqus, ANSYS). The geometrical properties of the mesh elements are based on the actual structure of the solid phase measured in the μCT image [54,55].

3. Mechanical Characterization

Mechanical characterization provides an insight into the mechanical response of cellular metals when subjected to various loading conditions. For this purpose, mechanical tests are being used. They allow to determine the deformation mechanics and mechanical properties of various types of cellular materials and structures. The tests can vary with respect to loading conditions (e.g., type, direction, and velocity), tested materials and expected results. Most commonly, mechanical tests are carried out by experimental or numerical (computational) methods, as the complexity of the cellular structure usually exceeds the efficiency and applicability of the analytical approach.

3.1. Experimental Methods

The most frequently applied experimental method for cellular metals is the uniaxial compressive test [56–59] because: (i) cellular metals are in applications often subjected to compressive loading and (ii) this type of test is straightforward, cost-effective and has minimum time requirements (short preparation time). From the compressive tests, the force-displacement (engineering/true stress–strain) diagrams can be obtained via load cells (or additional sensors) and the deformation behavior by visual capturing. Regardless of the type of the cellular metal, porosity or relative density (influences the elastic modulus, plastic strength, and energy absorption [60], base material, topology, and morphology (e.g., cell size effect [61], graded porosity [62]), a characteristic compressive behavior can be observed (Figure 7), which can be distributed in four main sections. Quasi-linear response (i) represents the initial stiffness of the cellular metal and is followed by transition zone (ii), where the cellular structure constituents (cell walls and struts) start to compress, bend, or stretch, resulting in local yielding. After additional loading, the stress–strain curve flattens into the stress plateau (iii). The stress is kept almost constant due to bending, buckling, stretching, and crushing of the cellular structure up to densification (iv), where the stiffness increases and approaches the base material properties at full compaction. The most important mechanical properties are initial stiffness, stress plateau σ_{pl} (in cellular metal literature also sometimes referred to as yield stress, yield strength), densification strain ε_d, and energy absorption capacity (grey area under the stress–strain curve).

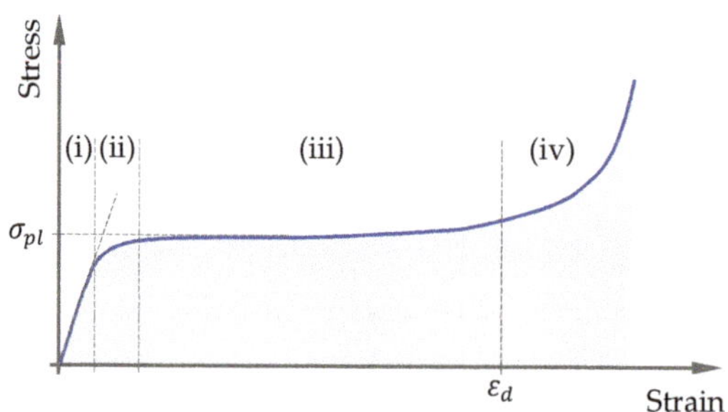

Figure 7. Characteristic compressive behavior of cellular materials.

The other types of mechanical tests involve bending [43,63], tensile [59,64], shear [59] and torsion load. The described loading types can be also combined into multiaxial loading conditions (to define e.g., multiaxial loading/stress states [59], yield surfaces [65]). For these tests, longer preparation time for specimen preparation and setting up the experimental device is usually needed.

The cellular specimens can be during the tests subjected to quasi-static or dynamic (monotonic or cyclic [66–68]) loading. According to [69], the quasi-static loading is considered for strain rates up to approx. $5\,\mathrm{s}^{-1}$. During compressive loading, three different deformation modes (in respect to critical velocities) can occur [70]: homogeneous, transitional, or shock mode. For the homogeneous mode at low loading velocities, it is common that the material starts to deform at its weakest point (negligible effect of inertia), while at the shock mode the main deformation takes place at the loading surface, regardless to the position of the weakest point in the material (substantial effect of inertia and strain rate sensitivity). In addition to dynamic servo-hydraulic testing machines other testing methods are also being used to achieve higher strain rates, e.g., drop towers, Taylor tests, Split–Hopkinson Pressure Bar (SPHB) tests [71].

The deformation mechanics and collapse mechanisms can be monitored by recording the visual data. For this purpose, cameras with a sufficient frame rate must be used, especially for the high-velocity testing. The deformation behavior can additionally be monitored and analyzed by e.g., digital image correlation (DIC) techniques (e.g., ARAMIS and GOM Correlate [72]), ex- and in-situ μCT (see Section 2), infrared thermography (see Section 4).

3.2. Numerical Methods

In the last two decades, numerical methods became a useful tool to supplement the mechanical characterization of various types of cellular metals. They can be used before experimental testing to reduce the number of specimens and experimental tests by predicting the mechanical response or after to validate computer models and reduce the number of further experimental tests. In some cases, computer simulations provide additional insight into the deformation behavior of cellular metals since they also provide data for material points inside the specimens. In solid mechanics, the finite element method (FEM) is the most used tool for analyzing the mechanical behavior of materials and structures subjected to different types of loading conditions [73]. Advanced quasi-static and dynamic numerical models of cellular structures allow to simultaneously account for material (e.g., strain rate dependent elasto-plasticity with damage and failure), geometrical (e.g., large deformation) and structural (e.g., contact and changing boundary conditions) nonlinearities. There are several ways how to model and discretize a cellular metal. The structure can be discretized in a detail (using beam [74], shell [75], or solid [76] finite elements, usually resulting in longer computational times) or as a homogenized material (most commonly as solid finite elements, resulting in shorter computational times). The geometry of the detailed computer models is either -based on computer aided design (for regular and simple cellular geometries) or μCT-based (for irregular and complex cellular structures [55]), as described in Section 2. In the case of homogenized models, the volume of the complex foam structure is simply fully discretized with solid finite elements (structured mesh) with proper constitutive relation, e.g., crushable foam [77]. If a regularity or periodicity can be found in the cellular structure, a part of the structure can be represented with a unit cell (representative volume element—RVE [78]) and proper boundary conditions, which drastically decreases the computational time. In addition to the finite element method, also some other numerical methods can be used to study the mechanical properties with highly effective accuracy and time consumption ratio, e.g., discrete element method (DEM) [79]. Numerical methods are also a very efficient tool for performing parametric studies (varying topology and morphology of the cellular structure). Furthermore, specific effects in the mechanical behavior can be easily isolated and studied. Their influence can be analyzed by changing one single or several material or geometric parameters. Additional possibilities and applicability of numerical methods represent topological optimization and current trends and development of additive manufacturing and other production techniques allowing fabrication of pre-designed cellular metals targeted for specific applications in various industries [80–84]. Although it is usually easier to predict the behavior of pre-designed structures (performing numerical simulations before the fabrication), special care should be taken to account for the additive manufacturing effects, e.g., surface roughness [85,86] (crucial also for studying the osseointegration [87]) and balling [88–90].

4. Infrared Thermography

Infrared (IR) thermography is a non-destructive testing method and already part of engineering norms and standard procedures [91–93] applicable in evaluation of dynamic deformation process of cellular materials. As cellular materials are often used as energy absorption structural elements, such components are mostly loaded in compression, where a compression yield plateau is characterizing the force-displacement, i.e., energy absorption response. It is common to use compression tests and evaluation at loading rates starting from quasi-static to dynamic and impact velocities. To understand the deformation scenario and what locally happens with cellular specimen, the strain at the specimen's surface is relevant. The passive IR thermography [91,94] is similar to the digital image correlation (DIC), both applicable to such testing as these methods are providing strain distribution during loading process. In references [95–99], it is shown how thermal image is comparable to effective strain obtained by DIC. In [95], we have shown for the first time the similarity of strain pattern and thermal image on a simple tension example, while in [96], more profound evaluation including tension and three-point bending is provided. Both methods provide strain pattern (field) on specimen surface and deformation of energy absorbing material, energy is mostly converted to the heat, thermal image reveals, where the energy is locally transformed, where yielding occurs, where plastic strain is created and how plastification zones spread through the specimen [96,97]. The thermographic approach, where DIC is combined with IR thermography has been adopted also by other authors [98,99]. IR thermography is not applicable at quasi-static to low-rate dynamic loading velocities as the heat dissipates trough material, what makes the DIC a more appropriate method. For higher loading rates DIC is requiring faster CCD cameras, where the available hardware limits the ability to obtain clear images relevant for surface pattern recognition, as reported in [96–99]. Contrary to DIC, higher heat generation with less dissipation enables better IR images with lower integration time. Importantly, IR hardware has specific acquisition rate characteristics. Most appropriate are middle-wave IR focal point array detectors (e.g., InSb detector) cooled to cryogenic temperatures [92]. Such detector enables approximately 150 fps of clear images for full resolution (when data from all detector's pixels are acquired) up to 1000 fps for reduced size image resolution (when reduced number of data from detector are used, so called window). Focal Point Array detectors of cooled middle wave (MW) IR cameras are capable to acquire data for high frame rate. To transfer all data from detector to computer, windowing (reducing active pixels) enables higher frame rates, with a drawback of recording just a portion of full frame image. For the FLIR SC 5000 MW camera (FLIR Systems Inc., Portland, OR, USA) with InSb, Focal Point Array detector was used in this research, with full size of 320×265 pixels, while minimum window size enabling faster frame rates with 160×128 pixels and 64×120 pixels. Although, there are MW cameras capable of acquiring 35,112 fps for a strip for reduced window size (64×4 pixels), it has been proven that the above mentioned camera provides for reliable observation of cellular metals dynamic deformation behavior. The method based on IR thermography is not limited to any particular type of cellular material. Various types of cellular metals can be evaluated, e.g., open- and closed-cell foams, advanced pore morphology (APM) foam elements, syntactic foams, pre-designed cellular structures [25,100]. Typical conclusions that can be drawn are, e.g., where yielding is initialized, how plastification fronts propagate, where densification occurs. This was shown in the previous research [25,100] related to thermography in experimental testing of porous materials.

It is a full field method and the temperature image can be processed as a thermal gradient image [97], giving information about direction of thermal change. Thus, from a single thermal image it can be concluded what is happening with a specimen (Figure 8). The example in Figure 8 shows the case of dynamical compression of a syntactic aluminum specimen (diameter: 26.9 mm; height: 43.2 mm). Infrared images are showing the specimen's surface, where generated heat is representing energy dissipated during compression process and, thus, the source of plastic deformation and collapse. IR camera acquisition rate was 489 Hz. IR images present specimen's deformation at a strain increment of approximately 10%.

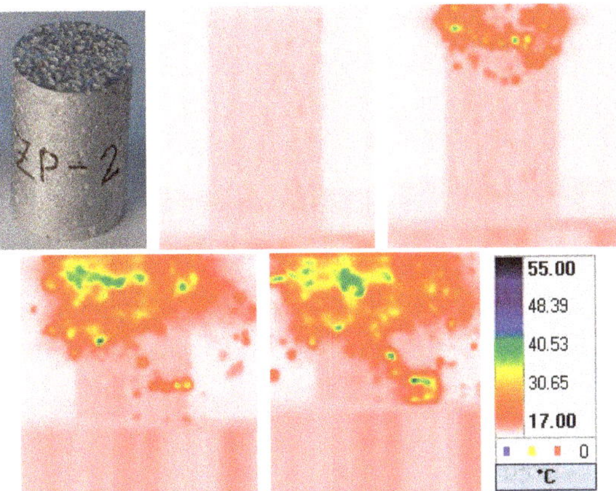

Figure 8. Photo and Infrared (IR) image of aluminum syntactic specimen compressed at the rate of 284 mm/s.

5. Thermal Characterization

Thermal characterization methods of cellular metals predominantly address their thermal conductivity and heat capacity. The heat capacity is readily determined based on foam density and the heat capacity of the metallic matrix. Hence, this overview concentrates on commonly used measurement techniques for the effective thermal conductivity of cellular metals. All methods are summarized in Table 1. Techniques can be differentiated into transient and steady-state methods. In general, transient methods are faster, permit simpler setups, and can operate within a narrow temperature band. Steady-state methods are highly accurate and permit directional measurements (thermal anisotropy) yet often require complex equipment and significant temperature gradients. A challenge that affects all experimental techniques is thermal contact resistance. Cellular metals are particularly susceptible to thermal contact resistance due to their porous surfaces and resulting small contact areas. Failure to minimize and/or consider thermal contact resistance will result in an underestimation of the effective thermal conductivity. Mitigation strategies include the application of compressive force, thermal paste, and soft metallic foils at the interphases of samples, sensors, heaters, and cooling elements.

5.1. Transient Methods

The well-established hot wire method [101,102] is not commonly used for cellular metals. In order to measure the effective thermal conductivity of cellular metals, a representative volume element (RVE) must be considered. The relatively small sample volume that is affected by hot wire measurements often does not meet this requirement. Instead, transient hot strip (THS), transient hot disc (THD), or transient hot plane (THP) methods with larger planar sensors are utilized [103–109]. In all cases, a wire is attached to a carrier material and acts simultaneously as the sensor (measuring the variation of electrical resistance) and the heat source. The planar sensor must be positioned between two samples and the measured thermal diffusivity is an average value of both samples. The evaluation of these methods is relatively complex and requires the solution of Fourier's differential equation [110]. The intermediate result is the effective thermal diffusivity, which can be converted into an effective thermal conductivity using the foam's density and specific heat capacity. Benefits of these transient methods are a minor temperature increase of only 1–5 K and the ability to measure thermal conductivities between 0.01–200 W/mK [111]. The maximum measurement temperature is limited by the carrier material and is currently about 800 °C [111].

Table 1. Commonly used methods for the determination of the effective thermal conductivity of cellular metals.

Method	Type	Relevant Standards	Measurement Data	Result	Cost	Meas. Volume	Accuracy
Hot wire	Transient	ASTM D5930-17	Electrical resistance	Thermal diffusivity	Low	Small	Good
Transient plane source (THS, THD, THP)	Transient	ISO 22007-2	Electrical resistance	Thermal diffusivity	Low	Medium	Good
Laser flash	Transient	ASTM E1461-13	Temperature	Thermal diffusivity	Medium-High	Small	Good
Guarded hot plate	Steady-state	DIN 52612-1 DIN EN 12664 ASTM C177-19	Temperature, electrical power	Thermal conductivity	Medium	Large	Very good
Heat-flow-meter	Steady-state	ASTM C 518	Temperature	Thermal conductivity	Medium	Large	Very good
Panel test technique	Steady-state	DIN EN 1094-7 ASTM C 201	Temperature, mass flow rate	Thermal conductivity	High	Large	Very good
Cut bar method	Steady-state	ASTM E1225-13	Temperature	Thermal conductivity	Medium	Large	Very good
Finite element method	Numerical	N/A	N/A	Thermal conductivity	Low-Medium	Large	Fair
Lattice Monte Carlo	Numerical	N/A	N/A	Thermal conductivity	Low-Medium	Large	Fair

Another commonly used transient method is laser flash measurement [112]. However, the complex geometry of the foam surface does not ensure a controlled energy transfer to the sample. In addition, sample sizes are often too small to achieve RVEs. Hence, this method is not usually applied for the characterization of cellular metals.

5.2. Steady-State Methods

Steady-state measurements impose a time-independent temperature gradient on a sample. Key benefits are a high accuracy, simple test evaluation and the ability to measure directional conductivities. In addition, heat flux passes through the entire sample facilitating measurements on RVEs. As a negative, a significant temperature gradient is required that complicates the measurement of the variation of thermal conductivity with temperature. Another disadvantage are the long measurement times required to ensure thermal steady-state, in particular for materials with low thermal diffusivities.

In absolute steady-state methods, the effective thermal conductivity is measured directly [113]. A commonly used method is the guarded hot plate [114–116]. The sample is positioned between a heating and a cooling element to generate a thermal gradient within the sample. After reaching steady-state, Fourier's law is used for the evaluation, i.e., $\dot{Q} = k \cdot A \cdot \Delta T / \Delta x$, where k is the effective thermal conductivity of the cellular metal, A is the sample cross section, and $\Delta T / \Delta x$ the temperature gradient. In the guarded hot plate method, the heat flux $\dot{Q}$ is determined as the energy input into the heating element at steady state. The term "guarded" describes the thermal isolation of the measurement setup to prevent convective and radiation heat losses. The Heat-Flow-Meter [115,117] resembles the guarded hot plate, however, instead of directly measuring the energy input into the heating element a heat flux sensor is positioned between the sample and the cooling element. The panel test technique [104,118–120] is another variation, where the heat flux is measured using a calorimeter that simultaneously acts as the cooling element. Compared to the guarded hot plate method, higher measurement temperatures can be achieved.

In relative steady-state methods, the thermal conductivity is measured relative to a reference material [121–124]. For increased accuracy, the thermal conductivities of the sample and the reference should be similar. In the cut bar method, the sample and either one or two reference bodies are stacked on a heating (cooling) element. The upper surface of the stack is brought into contact with a cooling (heating) element and usually a compressive force is applied to minimize thermal contact resistances within the stack. After reaching steady state, the reference bodies act as the heat flux sensors (similar to a heat-flow-meter) and Fourier's equation can be used for the evaluation.

5.3. Numerical Methods

In addition to the experimental methods outlined above, numerical analysis can be used to predict the effective thermal conductivity of metallic foams. An important step is the discretization of the typically complex cellular geometries. Unit cell models [125] and simplified model structures [126] suffice in some cases, however, improved accuracy can be obtained by deriving the numerical models from microcomputed tomography data [127,128]. Analogous to the experimental methods, transient or steady-state calculations can be performed. Due to the significantly higher computational cost of transient methods, almost exclusively steady-state analysis is selected. The two predominant numerical methods are the Finite Element Method (FEM) [125–128] and the Lattice Monte Carlo (LMC) method [129]. The result of the numerical analyses is typically a relative conductivity, i.e., the effective thermal conductivity is determined as a fraction of the matrix material. These numbers can readily be scaled for different type of matrices and temperatures. Comparison of effective thermal conductivities obtained by numerical analysis and experimental studies usually results in good agreement [129,130].

6. Acoustic Characterization

The acoustic characterization of the cellular metals is essential, since they are subjected to the noise and vibration applications (e.g., high resonance response). However, the knowledge and data are lacking for this type of materials. Few works have been published in this research topic [131–140]. The natural cork (100%) is an example of a natural cellular material widely used to minimize noise and vibration problems in mechanical systems, including industrial machines, appliances, vehicles, and buildings. The noise, vibration, and harshness (NVH) are one of the most fundamental issues for automotive industry. The NVH analysis has commonly two targets, which are to reduce and control noise and vibration for the benefits of users (occupants' comfort and environmental noise emission) and to minimize the effects of harsh forces and vibration on equipment (e.g., loosen joints and material fatigue). The vibrations are measured in two ways (frequency and amplitude) and expressed as waves per second (Hz). Vibrations that are felt are under 200 Hz, while vibrations between 20–20,000 Hz are audible by humans. Vibrations over 20,000 Hz are ultrasonic and not audible by humans.

The sound absorption coefficient (α), noise reduction coefficient (NRC), sound transmission loss, impact sound transmission and damping behavior (e.g., vibration frequencies, modes, and damping ratios) are examples of the parameters measured to characterize the acoustic properties of cellular metals.

The ability of a material to reduce sound reflections, reverberation, and echo within an enclosed space are determined based on the sound absorption coefficient and noise reduction coefficient. These are the parameters used to evaluate the sound absorption characteristics of the cellular materials and are assessed using an impedance tube according to the ASTM E1050 standard [141]. The cylindrical specimen is placed into the impedance tube (e.g., inner diameter: 37 mm) at one end, while a sound source consisting of a loudspeaker emitting a random noise is introduced at the other end of the tube [22,23]. Two microphones are placed into the tube between the sound source and the studied specimen to detect the sound pressure wave transmitted through the specimen and the wave portion that was reflected. The acoustic absorption coefficient is defined as a ratio of absorbed sound intensity in a given material and the incident sound intensity that is imposed on that material. The acoustic absorption coefficient varies from 0 (0% sound absorption) to 1 (100% sound absorption). The value 0 means no sound is absorbed, while the value 1 means the opposite. The noise reduction coefficient is calculated by defining an average of the sound absorption coefficients at the frequencies 250, 500, 1000, and 2000 Hz and rounding off the result to the nearest multiple of 0.05. The absorption and noise reduction values are also measured in a small echo chamber, either in sound transmission test or echo chamber.

The vibroacoustic behavior is measured by microphones (to measure sound pressure), accelerometers (to measure acceleration), and force/displacement/velocity transducers (to measure forces), in which the measurements are made by setting the instrument for a certain bandwidth and center frequency vibration analyzer, covering the values over that range. The ability to act as a barrier preventing airborne sound transmission from one space to another, sound transmission loss coefficient at 1/3 octave bands, will be performed according to the ASTM E90. The modal analyses are usually performed by vibration of specimens, exciting by a shaker or an impact hammer, measuring the response using a laser scanning vibrometer [140]. The time domain response and loss factor from their damped wave decay is determined using a spectral analyzer. The damping ratio of the resonance mode is determined using the half-power bandwidth method.

7. Conclusions

Different experimental and numerical characterization methods of the cellular metals are presented, i.e., different techniques to determine their structural, mechanical, thermal, and acoustic properties. The methodology, the experimental set-up, the equipment, and its principles and standards were summarized. The potential of non-destructive methods like X-ray computed tomography and infrared thermography are also presented.

X-ray microcomputed tomography is a powerful tool to characterize the complex cellular structure of cellular metals in terms of their structural morphometric parameters, to study their deformation and failure behavior and to develop numerical models to predict the mechanical, thermal, and acoustic behavior. Infrared thermography is a versatile tool well suited for the study of dynamic deformation processes of cellular metals (e.g., closed-cell foams, syntactic foams and metal hollow spheres structures) and filled profiles and sandwich panels based on cellular materials.

In general, there is a huge demand for experimental and computational characterization of physical properties of novel cellular metals. Thus, new opportunities for advances in new experimental techniques and computational modelling arise. Furthermore, computational modelling can help in designing new cellular materials (in combination with additive manufacturing) with desired properties and behavior demanded by an application.

Author Contributions: Conceptualization, I.D., T.F., L.K.-O., and M.V.; methodology, I.D., T.F., L.K.-O., and M.V.; software, I.D., T.F., L.K.-O., and M.V.; validation, I.D., T.F., L.K.-O., and M.V.; formal analysis, I.D., T.F., L.K.-O., and M.V.; investigation, I.D., T.F., L.K.-O., and M.V.; resources, I.D., T.F., L.K.-O., and M.V.; data curation, I.D., T.F., L.K.-O., and M.V.; writing—original draft preparation, I.D., T.F., L.K.-O., and M.V.; writing—review and editing, I.D., T.F., L.K.-O., and M.V.; visualization, I.D., T.F., L.K.-O., and M.V.; supervision, I.D., T.F., L.K.-O., and M.V. All authors have read and agreed to the published version of the manuscript.

Funding: This research was funded by Portuguese Science Foundation (FCT): UIDB/00481/2020 and UIDP/00481/2020-FCT and CENTRO-01–0145-FEDER-022083 (Centro 2020 program-Portugal 2020).

Acknowledgments: Supports given by the Portuguese Science Foundation (FCT), UIDB/00481/2020 and UIDP/00481/2020-FCT and CENTRO-01–0145-FEDER-022083 (Centro 2020 program-Portugal 2020, Centro Portugal Regional Operational Partnership Agreement through the European Regional Development Fund).

References

1. Banhart, J. Manufacture, characterisation and application of cellular metals and metal foams. *Prog. Mater. Sci.* **2001**, *46*, 559–632. [CrossRef]

2. Garcia-Moreno, F. Commercial Applications of Metal Foams: Their Properties and Production. *Materials* **2016**, *9*, 85. [CrossRef] [PubMed]

3. Marx, J.; Portanova, M.; Rabiei, A. A study on blast and fragment resistance of composite metal foams through experimental and modeling approaches. *Compos. Struct.* **2018**, *194*, 652–661. [CrossRef]

4. Degischer, H.-P.; Kriszt, B. *Handbook of Cellular Metals: Production, Processing, Applications*, 1st ed.; Wiley-VCH Verlag GmbH & Co. KGaA: Weinheim, Germany, 2001.

5. Banhart, J. Light-Metal Foams—History of Innovation and Technological Challenges. *Adv. Eng. Mater.* **2013**, *15*, 82–111. [CrossRef]

6. Baumgärtner, F.; Duarte, I.; Banhart, J. Industrialization of powder compact foaming process. *Adv. Eng. Mater.* **2000**, *2*, 168–174. [CrossRef]

7. Trejo Rivera, N.M.; Torres Torres, J.; Flores Valdés, A. A-242 Aluminium Alloy Foams Manufacture from is fine.the Recycling of Beverage Cans. *Metals* **2019**, *9*, 92. [CrossRef]

8. Duarte, I.; Banhart, J. A study of aluminium foam formation-kinetics and microstructure. *Acta Mater.* **2000**, *48*, 2349–2362. [CrossRef]

9. García-Moreno, F.; Radtke, L.A.; Neu, T.R.; Kamm, P.H.; Klaus, M.; Schlepütz, C.M.; Banhart, J. The Influence of Alloy Composition and Liquid Phase on Foaming of Al–Si–Mg Alloys. *Metals* **2020**, *10*, 189. [CrossRef]

10. Lehmhus, D.; Hünert, D.; Mosler, U.; Martin, U.; Weise, J. Effects of Eutectic Modification and Grain Refinement on Microstructure and Properties of PM AlSi7 Metallic Foams. *Metals* **2019**, *9*, 1241. [CrossRef]

11. Kuwahara, T.; Kaya, A.; Osaka, T.; Takamatsu, S.; Suzuki, S. Stabilization Mechanism of Semi-Solid Film Simulating the Cell Wall during Fabrication of Aluminum Foam. *Metals* **2020**, *10*, 333. [CrossRef]

12. García-Moreno, F.; Babcsan, N.; Banhart, J. X-ray radioscopy of liquid metal foams: Influence of heating profile, atmosphere and pressure. *Colloids Surf. A Physicochem. Eng. Asp.* **2005**, *263*, 290–294. [CrossRef]

13. Kuwahara, T.; Osaka, T.; Saito, M.; Suzuki, S. Compressive Properties of A2024 Alloy Foam Fabricated through a Melt Route and a Semi-Solid Route. *Metals* **2019**, *9*, 153. [CrossRef]

14. Nosko, M.; Simancík, F.; Florek, R. Reproducibility of aluminum foam properties: Effect of precursor distribution on the structural anisotropy and the collapse stress and its dispersion. *Mater. Sci. Eng. A* **2010**, *527*, 5900–5908. [CrossRef]

15. Duarte, I.; Ferreira, J.M.F. 2D quantitative analysis of metal foaming kinetics by hot-stage microscopy. *Adv. Eng. Mater.* **2014**, *16*, 33–34. [CrossRef]

16. Kuwahara, T.; Saito, M.; Osaka, T.; Suzuki, S. Effect of Primary Crystals on Pore Morphology during Semi-Solid Foaming of A2024 Alloys. *Metals* **2019**, *9*, 88. [CrossRef]

17. Mukherjee, M.; García-Moreno, F.; Jiménez, C.; Rack, A.; Banhart, J. Microporosity in aluminium foams. *Acta Mater.* **2017**, *131*, 156–168. [CrossRef]

18. Körner, C.; Singer, R.F. Processing of metal foams—Challenges and opportunities. *Adv. Eng. Mater.* **2000**, *2*, 159–165. [CrossRef]

19. Duarte, I.; Vesenjak, M.; Vide, M. Automated continuous production line of parts made of metallic foams. *Metals* **2019**, *9*, 531. [CrossRef]

20. Duarte, I.; Krstulović-Opara, L.; Vesenjak, M. Analysis of performance of in-situ carbon steel bar reinforced Al-alloy foams. *Compos. Struct.* **2016**, *152*, 432–443. [CrossRef]

21. Gopinathan, A.; Jerz, J.; Simancík, F.; Kovácik, J.; Pavlík, L. Assessment of the aluminium foam panel on PCM based thermal energy storage. *Mech. Technol. Struct. Mater.* **2019**, *2019*, 53–60.

22. Pinto, S.C.; Marques, P.A.A.P.; Vesenjak, M.; Vicente, R.; Godinho, L.; Krstulović-Opara, L.; Duarte, I. Characterization and physical properties of aluminium foam–polydimethylsiloxane nanocomposite hybrid structures. *Compos. Struct.* **2019**, *230*, 111521. [CrossRef]

23. Pinto, S.C.; Marques, P.A.A.P.; Vesenjak, M.; Vicente, R.; Godinho, L.; Krstulović-Opara, L.; Duarte, I. Mechanical, thermal, and acoustic properties of aluminum foams impregnated with epoxy/graphene oxide nanocomposites. *Metals* **2019**, *9*, 1214. [CrossRef]

24. Stöbener, K.; Rausch, G. Aluminium foam–polymer composites: Processing and characteristics. *J. Mater. Sci.* **2009**, *44*, 1506–1511. [CrossRef]

25. Duarte, I.; Krstulović-Opara, L.; Vesenjak, M.; Ren, R. Crush performance of multifunctional hybrid foams based on an aluminium alloy open-cell foam skeleton. *Polym. Test.* **2018**, *67*, 246–256. [CrossRef]

26. Taherishargh, M.; Linul, E.; Broxtermann, S.; Fiedler, T. The mechanical properties of expanded perlite-aluminium syntactic foam at elevated temperatures. *J. Alloys Compd.* **2018**, *737*, 590–596. [CrossRef]

27. Waag, U.; Schneider, L.; Löthman, P.; Stephani, G. Metallic hollow spheres—Materials for the future. *Metal. Powder Rep.* **2000**, *55*, 29–33.

28. Duarte, I.; Ferreira, J.M. Composite and Nanocomposite metal foams. *Materials* **2016**, *9*, 79. [CrossRef]

29. Movahedi, N.; Murch, G.E.; Belova, I.V.; Fiedler, T. Manufacturing and compressive properties of tube-filled metal syntactic foams. *J. Alloys Compd.* **2020**, *820*, 153465. [CrossRef]

30. Duarte, I.; Krstulović-Opara, L.; Vesenjak, M. Axial crush behaviour of the aluminium alloy in-situ foam filled tubes with very low wall thickness. *Compos. Struct.* **2018**, *192*, 184–192.

31. Hangai, Y.; Kobayashi, R.; Suzuki, R.; Matsubara, M.; Yoshikawa, N. Aluminum Foam-Filled Steel Tube Fabricated from Aluminum Burrs of Die-Castings by Friction Stir Back Extrusion. *Metals* **2019**, *9*, 124. [CrossRef]

32. Duarte, I.; Krstulović-Opara, L.; Dias-de-Oliveira, J.; Vesenjak, M. Axial crush performance of polymer-aluminium alloy hybrid foam filled tubes. *Thin Walled Struct.* **2019**, *138*, 124–136.

33. Banhart, J.; Seeliger, H.-W. Aluminium foam sandwich panels: Manufacture, metallurgy and applications. *Adv. Eng. Mater.* **2008**, *10*, 793–802. [CrossRef]

34. Baumeister, J.; Weise, J.; Hirtz, E.; Höhne, K.; Hohe, J. Applications of Aluminum Hybrid Foam Sandwiches in Battery Housings for Electric Vehicles. *Procedia Mater. Sci.* **2014**, *4*, 317–321. [CrossRef]

35. Duarte, I.; Peixinho, N.; Andrade-Campos, A.; Valente, R. Editorial—Special Issue on Cellular Materials. *Sci. Technol. Mater.* **2018**, *30*, 1–3.

36. Wadley, H.N.G. Multifunctional periodic cellular metals. *Philos. Trans. R. Soc. A* **2006**, *364*, 31–68.

37. Yang, C.; Xu, K.; Xie, S. Comparative Study on the Uniaxial Behaviour of Topology-Optimised and Crystal-Inspired Lattice Materials. *Metals* **2020**, *10*, 491. [CrossRef]

38. Hong, S.; Sanchez, C.; Du, H.; Kim, N. Fabrication of 3D Printed Metal Structures by Use of High-Viscosity Cu Paste and a Screw Extruder. *J. Electron. Mater.* **2015**, *44*, 836–841. [CrossRef]

39. Gupta, N.; Rohatgi, P.K. *Metal. Matrix Syntactic Foams: Processing, Microstructure, Properties and Applications*; DEStech Publications Inc.: Lancaster, PA, USA, 2014.

40. Novak, N.; Vesenjak, M.; Ren, Z. Auxetic Cellular Materials—A Review. *J. Mech. Eng.* **2016**, *62*, 485–493. [CrossRef]

41. Xue, Y.; Wang, W.; Han, F. Enhanced compressive mechanical properties of aluminum based auxetic lattice structures filled with polymers. *Compos. Part B Eng.* **2019**, *171*, 183–191. [CrossRef]

42. Carneiro, V.H.; Rawson, S.D.; Puga, H.; Meireles, J.; Withers, P.J. Additive manufacturing assisted investment casting: A low-cost method to fabricate periodic metallic cellular lattices. *Addit. Manuf.* **2020**, *33*, 101085. [CrossRef]

43. Vesenjak, M.; Duarte, I.; Baumeister, J.; Göhler, H.; Krstulović-Opara, L.; Ren, Z. Bending performance evaluation of aluminium alloy tubes filled with different cellular metal cores. *Compos. Struct.* **2020**, *234*, 111748. [CrossRef]

44. Möbus, G.; Inkson, B.J. Review. Nanoscale tomography in materials science. *Mater. Today* **2007**, *10*, 18–25. [CrossRef]

45. Bock, J.; Jacobi, A.M. Geometric classification of open-cell metal foams using X-ray micro-computed tomography. *Mater. Charact.* **2013**, *75*, 35–43.

46. Olurin, O.B.; Arnold, M.; Körner, C.; Singer, R.F. The investigation of morphometric parameters of aluminium foams using micro-computed tomography. *Mater. Sci. Eng. A* **2002**, *328*, 334–343. [CrossRef]

47. Wang, N.; Maire, E.; Cheng, Y.; Amani, Y.; Chen, X. Comparison of aluminium foams prepared by different methods using X-ray tomography. *Mater. Charact.* **2018**, *138*, 296–307. [CrossRef]

48. Zhang, Q.; Lee, P.D.; Singh, R.; Wu, G.; Lindley, T.C. Micro-CT characterization of structural features and deformation behavior of fly ash/aluminum syntactic foam. *Acta Mater.* **2019**, *57*, 3003–3011.

49. Kader, M.A.; Brown, A.D.; Hazell, P.J.; Robins, V.; Saadatfar, M. Geometrical and topological evolution of a closed-cell aluminium foam subject to drop-weight impact: An X-ray tomography study. *Int. J. Impact Eng.* **2020**, *139*, 103510. [CrossRef]

50. Chaturvedi, A.; Bhatkar, S.; Sarkar, P.S.; Chaturvedi, S.; Gupta, M.K. 3D Geometric modeling of aluminum-based foam using micro Computed Tomography technique. *Mater. Today Proc.* **2019**, *18*, 4151–4156.

51. Bruker. *Bruker microCT Method Note: DataViewer, An. Overview*; Bruker: Kontich, Belgium, 2016; p. 16.

52. Miedzińska, D.; Niezgoda, T.; Gieleta, R. Numerical and experimental aluminum foam microstructure testing with the use of computed tomography. *Comput. Mater. Sci.* **2012**, *64*, 90–95. [CrossRef]

53. Veyhl, C.; Belova, I.V.; Murch, G.E.; Fiedler, T. Finite element analysis of the mechanical properties of cellular aluminium based on micro-computed tomography. *Mater. Sci. Eng. A* **2011**, *528*, 4550–4555.

54. Fiedler, T.; Belova, I.V.; Murch, G.E. μ-CT-based finite element analysis on imperfections in open-celled metal foam: Mechanical properties. *Scr. Mater.* **2012**, *67*, 455–458. [CrossRef]

55. Kozma, I.; Zsoldos, I. CT-based tests and finite element simulation for failure analysis of syntactic foams. *Eng. Fail. Anal.* **2019**, *104*, 371–378. [CrossRef]

56. Japanese Standards Association. *JIS-H-7902 Compression Test. For. Porous And Cellular Metals*; Japanese Standards Association: Tokyo, Japan, 2016.

57. The International Organization for Standardization. *ISO 13314:2011 Mechanical Testing of Metals—Ductility Testing—Compression Test. for Porous and Cellular Metals*; The International Organization for Standardization: Geneva, Switzerland, 2011.

58. Hipke, T.; Lange, G.; Poss, R. *Taschenbuch für Aluminiumschäume*; Alu Media: Düsseldorf, Germany, 2007.

59. Ashby, M.F.; Evans, A.; Fleck, N.A.; Gibson, L.J.; Hutchinson, J.W.; Wadley, H.N.G. *Metal Foams: A Design Guide*; Elsevier Science: Burlington, MA, USA, 2000.

60. Ramamurty, U.; Paul, A. Variability in mechanical properties of a metal foam. *Acta Mater.* **2004**, *52*, 869–876.

61. Tekoğlu, C.; Gibson, L.J.; Pardoen, T.; Onck, P.R. Size effects in foams: Experiments and modeling. *Prog. Mater. Sci.* **2011**, *56*, 109–138. [CrossRef]

62. Xu, F.; Zhang, X.; Zhang, H. A review on functionally graded structures and materials for energy absorption. *Eng. Struct.* **2018**, *171*, 309–325.

63. The International Organization for Standardization. *ISO 7438:2005 Metallic Materials-Bend Test*; International Organization for Standardization: Geneva, Switzerland, 2005.

64. The International Organization for Standardization. *ISO 6892-1:2009 Metallic Materials—Tensile Testing—Part 1: Method of Test at Room Temperature*; International Organization for Standardization: Geneva, Switzerland, 2009.

65. Jung, A.; Diebels, S. Microstructural characterisation and experimental determination of a multiaxial yield surface for open-cell aluminium foams. *Mater. Des.* **2017**, *131*, 252–264. [CrossRef]

66. Taherishargh, M.; Katona, B.; Fiedler, T.; Orbulov, I.N. Fatigue properties of expanded perlite/aluminum syntactic foams. *J. Compos. Mater.* **2017**, *51*, 773–781. [CrossRef]

67. Nečemer, B.; Vesenjak, M.; Glodež, S. Fatigue of Cellular Structures—A Review. *J. Mech. Eng.* **2019**, *65*, 525–536. [CrossRef]

68. Tomažinčič, D.; Vesenjak, M.; Klemenc, J. Prediction of static and low-cycle durability of porous cellular structures with positive and negative Poisson's ratios. *Theor. Appl. Fract. Mech.* **2020**, *106*, 102479. [CrossRef]

69. Jacob, P.; Goulding, L. *An Explicit Finite Element Primer*; NAFEMS: Glasgow, UK, 2002.

70. Zheng, Z.; Yu, J.; Li, J. Dynamic crushing of 2D cellular structures: A finite element study. *Int. J. Impact Eng.* **2005**, *32*, 650–664. [CrossRef]

71. Taşdemirci, A.; Ergönenç, Ç.; Güden, M. Split Hopkinson pressure bar multiple reloading and modeling of a 316 L stainless steel metallic hollow sphere structure. *Int. J. Impact Eng.* **2010**, *37*, 250–259. [CrossRef]

72. GOM. *Material Testing/Material Properties—ARAMIS and the Detection of Flow Behaviour for Sheet Metal Material in High Speed Tensile Tests*; GOM: Leuven, Belgium, 2009.

73. Altenbach, H.; Oechsner, A. *Cellular and Porous Materials in Structures and Processes*; CISM International Centre for Mechanical Sciences/Springer: Wien, Austria, 2010.

74. Guo, H.; Takezawa, A.; Honda, M.; Kawamura, C.; Kitamura, M. Finite element simulation of the compressive response of additively manufactured lattice structures with large diameters. *Comput. Mater. Sci.* **2020**, *175*, 109610. [CrossRef]

75. Caty, O.; Maire, E.; Youssef, S.; Bouchet, R. Modeling the properties of closed-cell cellular materials from tomography images using finite shell elements. *Acta Mater.* **2008**, *56*, 5524–5534.

76. Kucewicz, M.; Baranowski, P.; Małachowski, J.; Popławski, A.; Płatek, P. Modelling, and characterization of 3D printed cellular structures. *Mater. Des.* **2018**, *142*, 177–189. [CrossRef]

77. Dassault Systémes SIMULIA, ABAQUS Documentation, 4.4.6 Models for Crushable Foams. Available online: https://classes.engineering.wustl.edu/2009/spring/mase5513/abaqus/docs/v6.6/books/stm/default.htm?startat=ch04s04ath118.html (accessed on 28 May 2020).

78. Fiedler, T.; Öchsner, A.; Gracio, J.; Kuhn, G. Structural modeling of the mechanical behavior of periodic cellular solids: Open-cell structures. *Mech. Compos. Mater.* **2005**, *41*, 277–290. [CrossRef]

79. Kovačič, A.; Vesenjak, M.; Borovinšek, M.; Ren, Z. Modelling large compression of advanced pore morphology foams with discrete element method. In Proceedings of the 11th World Congress on Computational Mechanics (WCCM XI) and 5th European Conference on Computational Mechanics (ECCM V) and 6th European Conference on Computational Fluid Dynamics (ECFD VI), Barcelona, Spain, 20–25 July 2014.

80. Zuo, Q.; He, K.; Mao, H.; Dang, X.; Du, R. Manufacturing process and mechanical properties of a novel periodic cellular metal with closed cubic structure. *Mater. Des.* **2018**, *153*, 242–258.

81. Xu, Z.; Meng, K.; Yang, C.; Zhang, W.; Fan, X.; Sun, Y. Elasto-Plastic Behaviour of Transversely Isotropic Cellular Materials with Inner Gas Pressure. *Metals* **2019**, *9*, 901.

82. Lee, M.-G.; Ko, G.-D.; Song, J.; Kang, K.-J. Compressive characteristics of a wire-woven cellular metal. *Mater. Sci. Eng. A* **2012**, *539*, 185–193.

83. Al-Ketan, O.; Rowshan, R.; Al-Rub, R.K.A. Topology-mechanical property relationship of 3D printed strut, skeletal, and sheet based periodic metallic cellular materials. *Addit. Manuf.* **2018**, *19*, 167–183. [CrossRef]

84. Kang, K.-J. Wire-woven cellular metals: The present and future. *Prog. Mater. Sci.* **2015**, *69*, 213–307. [CrossRef]

85. Harris, J.A.; McShane, G.J. Metallic stacked origami cellular materials: Additive manufacturing, properties, and modelling. *Int. J. Solids Struct.* **2020**, *185–186*, 448–466. [CrossRef]

86. Oliveira, J.P.; Lalonde, A.; Ma, J. Processing parameters in laser powder bed fusion metal additive manufacturing. *Mater. Des.* **2020**, *193*, 108762. [CrossRef]

87. Murr, L.E. Strategies for creating living, additively manufactured, open-cellular metal and alloy implants by promoting osseointegration, osteoinduction and vascularization: An overview. *J. Mater. Sci. Technol.* **2019**, *35*, 231–241. [CrossRef]

88. Attar, E. Simulation of Selective Electron Beam Melting Processes. Ph.D. Thesis, Universität Erlangen-Nürnberg, Erlangen, Germany, 2011.

89. Novak, N.; Borovinšek, M.; Vesenjak, M.; Wormser, M.; Körner, C.; Tanaka, S.; Hokamoto, K.; Ren, Z. Crushing Behavior of Graded Auxetic Structures Built from Inverted Tetrapods under Impact. *Phys. Status Solidi B* **2019**, *256*, 1800040. [CrossRef]

90. Yuan, L.; Ding, S.; Wen, C. Additive manufacturing technology for porous metal implant applications and triple minimal surface structures: A review. *Bioact. Mater.* **2019**, *4*, 56–70. [CrossRef] [PubMed]

91. SIS. *Non-Destructive Testing—Thermographic Testing—Part 1: General Principles (EN 16714-1:2016)*; Swedish Institute of Standards: Stockholm, Sweden, 2016.

92. SIS. *Non-Destructive Testing—Thermographic Testing—Part 2: Equipment (EN 16714-2:2016)*; Swedish Institute of Standards: Stockholm, Sweden, 2016.

93. SIS. *Non-Destructive Testing—Thermographic Testing—Part 3: Terms and Definitions (EN 16714-3:2016)*; Swedish Institute of Standards: Stockholm, Sweden, 2016.

94. Jung, A.; Majthoub, K.A.; Jochum, C.; Kirsch, S.-M.; Welsch, F.; Seelecke, S.; Diebels, S. Correlative digital image correlation and infrared thermography measurements for the investigation of the mesoscopic deformation behaviour of foams. *J. Mech. Phys. Solids* **2019**, *130*, 165–180. [CrossRef]

95. Krstulović-Opara, L.; Surijak, M.; Vesenjak, M.; Tonković, Z.; Kodvanj, J.; Domazet, Ž. Comparison of Infrared and 3D Digital Image Correlation Techniques Applied for Mechanical Testing of Materials. In Proceedings of the 31th Danubia-Adria Symposium on Advances in Experimental Mechanics, Kempten, Germany, 24–27 September 2014; pp. 232–233.

96. Krstulović-Opara, L.; Surijak, M.; Vesenjak, M.; Tonković, Z.; Kodvanj, J.; Domazet, Ž. Comparison of infrared and 3D digital image correlation techniques applied for mechanical testing of materials. *Infrared Phys. Technol.* **2015**, *73*, 166–174. [CrossRef]

97. Pehilj, A.; Krstulović-Opara, L.; Bagavac, P.; Vesenjak, M.; Duarte, I.; Domazet, Ž. The detection of plastic flow propagation based on the temperature gradient. *Mater. Today Proc.* **2017**, *4*, 5925–5930. [CrossRef]

98. Jung, A.; Bronder, S.; Diebels, S.; Schmidt, M.; Seelecke, S. Thermographic investigation of strain rate effects in Al foams and Ni/Al hybrid foams. *Mater. Des.* **2018**, *160*, 363–370. [CrossRef]

99. Osornio-Rios, R.A.; Antonino-Daviu, J.A.; Romero-Troncoso, R.J. Recent Industrial Applications of Infrared Thermography: A Review. *IEEE Trans. Ind. Inform.* **2019**, *15*, 615–625. [CrossRef]

100. Broxtermann, S.; Vesenjak, M.; Krstulović-Opara, L.; Fiedler, T. Quasi static and dynamic compression of zinc syntactic foams. *J. Alloys Compd.* **2018**, *768*, 962–969. [CrossRef]

101. Wei, L.C.; Ehrlich, L.E.; Powell-Palm, M.J.; Montgomery, C.; Beuth, J.; Malen, J.A. Thermal conductivity of metal powders for powder bed additive manufacturing. *Addit. Manuf.* **2018**, *21*, 201–208. [CrossRef]

102. ASTM International. *ASTM D5930-17: Standard Test Method for Thermal Conductivity of Plastics by Means of a Transient Line-Source Technique*; ASTM International: West Conshohocken, PA, USA, 2017.

103. The International Organization for Standardization. *ISO 22007-2:2015 Plastics—Determination of Thermal Conductivity and Thermal Diffusivity—Part 2: Transient Plane Heat Source (Hot Disc) Method*; International Organization for Standardization: Geneva, Switzerland, 2015.

104. Skibina, V.; Wulf, R.; Gross, U. Temperature Influence on the Effective Thermal Conductivity of Cellular Metals Measured in Stagnant Air. *Procedia Mater. Sci.* **2014**, *4*, 347–351. [CrossRef]

105. Solórzano, E.; Rodríguez-Perez, M.A.; de Saja, J.A. Thermal conductivity of metallic hollow sphere structures: An experimental, analytical and comparative study. *Mater. Lett.* **2009**, *63*, 1128–1130. [CrossRef]

106. Liu, P.S.; Qing, H.B.; Hou, H.L.; Wang, Y.Q.; Zhang, Y.L. EMI shielding and thermal conductivity of a high porosity reticular titanium foam. *Mater. Des.* **2016**, *92*, 823–828. [CrossRef]

107. Solórzano, E.; Reglero, J.A.; Rodríguez-Pérez, M.A.; Lehmhus, D.; Wichmann, M.; de Saja, J.A. An experimental study on the thermal conductivity of aluminium foams by using the transient plane source method. *Int. J. Heat Mass Transf.* **2008**, *51*, 6259–6267. [CrossRef]

108. Solórzano, E.; Rodríguez-Pérez, M.A.; Reglero, J.A.; de Saja, J.A. Density gradients in aluminium foams: Characterisation by computed tomography and measurements of the effective thermal conductivity. *J. Mater. Sci.* **2007**, *42*, 2557–2564. [CrossRef]

109. Solórzano, E.; Hirschmann, M.; Rodríguez-Pérez, M.A.; Körner, C.; de Saja, J.A. Thermal conductivity of AZ91 magnesium integral foams measured by the Transient Plane Source method. *Mater. Lett.* **2008**, *62*, 3960–3962. [CrossRef]

110. Carslaw, H.S.; Jaeger, J.C. *Conduction of Heat in Solids*, 2nd ed.; Clarendon Press: Oxford, UK, 1959.

111. Wulf, R. Waermeleitfaehigkeit von Hitzebestaendigen und Feuerfesten Daemmstoffen—Untersuchungen zu Ursachen Fuer Unterschiedliche Messergebnisse bei Verwendung Verschiedener Messverfahren. Ph.D. Thesis, Technischen Universitaet Bergakademie Freiberg, Freiberg, Germany, 2009.

112. ASTM. *E1461-13: Standard Test Method Thermal Diffusivity by Flash Method*; ASTM International: West Conshohocken, PA, USA, 2013.

113. Amani, Y.; Takahashi, A.; Chantrenne, P.; Maruyama, S.; Dancette, S.; Maire, E. Thermal conductivity of highly porous metal foams: Experimental and image based finite element analysis. *Int. J. Heat Mass Transf.* **2018**, *122*, 1–10. [CrossRef]

114. DIN Deutsches Institut für Normung. *DIN 52612-1: Testing of Thermal Insulating Materials; Determination of Thermal Conductivity by The Guarded Hot Plate Apparatus; Test Procedure and Evaluation of Results*; DIN Deutsches Institut für Normung: Berlin, Germany, 1979.

115. DIN Deutsches Institut für Normung. *DIN EN 12664: Thermal Performance of Building Materials and Products—Determination of Thermal Resistance by Means of Guarded Hot Plate and Heat Flow Meter Methods—Dry and Moist Products with Medium and Low Thermal Resistance*; DIN Deutsches Institut für Normung: Berlin, Germany, 2001.

116. ASTM International. *ASTM C177-19: Standard Test Method for Steady-State Heat Flux Measurements and Thermal Transmission Properties by Means of the Guarded-Hot-Plate Apparatus*; ASTM International: West Conshohocken, PA, USA, 2019.

117. ASTM International. *ASTM C 518: Standard Test Method for Steady-State Thermal Transmission Properties by Means of the Heat Flow Meter Apparatus*; ASTM International: West Conshohocken, PA, USA, 2017.

118. DIN Deutsches Institut für Normung. *DIN EN 1094-7: Insulating Refractory Products—Part 7: Methods of Test for Ceramic Fibre Products*; DIN Deutsches Institut für Normung: Berlin, Germany, 2005.

119. ASTM International. *ASTM C 201: Standard Test Method for Thermal Conductivity of Refractories*; ASTM International: West Conshohocken, PA, USA, 2019.

120. Gross, U.; Barth, G.; Wulf, R.; Tran, L.T.S. Messung der Wärmeleitfähigkeit von Dämmstoffen mit unterschiedlichen Methoden. *High Temp. High Press.* **2001**, *33*, 141–150. [CrossRef]

121. ASTM International. *ASTM E1225-13: Standard Test Method for Thermal Conductivity of Solids Using the Guarded-Comparative-Longitudinal Heat Flow Technique*; ASTM International: West Conshohocken, PA, USA, 2013.

122. Catchpole-Smith, S.; Sélo, R.R.J.; Davis, A.W.; Ashcroft, I.A.; Tuck, C.J.; Clare, A. Thermal conductivity of TPMS lattice structures manufactured via laser powder bed fusion. *Addit. Manuf.* **2019**, *30*, 100846. [CrossRef]

123. Abuserwal, A.F.; Elizondo Luna, E.M.; Goodall, R.; Woolley, R. The effective thermal conductivity of open cell replicated aluminium metal sponges. *Int. J. Heat Mass Transf.* **2017**, *108*, 1439–1448. [CrossRef]

124. Dyga, R.; Witczak, S. Investigation of Effective Thermal Conductivity Aluminum Foams. *Procedia Eng.* **2012**, *42*, 1088–1099. [CrossRef]

125. Mirabolghasemi, A.; Akbarzadeh, A.H.; Rodrigue, D.; Therriault, D. Thermal conductivity of architected cellular metamaterials. *Acta Mater.* **2019**, *174*, 61–80. [CrossRef]

126. Hu, Y.; Fang, Q.-Z.; Yu, H.; Hu, Q. Numerical simulation on thermal properties of closed-cell metal foams with different cell size distributions and cell shapes. *Mater. Today Commun.* **2020**, *24*, 100968. [CrossRef]

127. Iasiello, M.; Bianco, N.; Chiu, W.K.S.; Naso, V. Thermal conduction in open-cell metal foams: Anisotropy and Representative Volume Element. *Int. J. Therm. Sci.* **2019**, *137*, 399–409. [CrossRef]

128. Veyhl, C.; Belova, I.V.; Murch, G.E.; Oechsner, A.; Fiedler, T. Thermal analysis of aluminium foam based on micro-computed tomography. *Materwiss. Werksttech.* **2011**, *42*, 350–355. [CrossRef]

129. Fiedler, T.; Belova, I.V.; Murch, G.E. Theoretical and Lattice Monte Carlo analyses on thermal conduction in cellular metals. *Comput. Mater. Sci.* **2010**, *50*, 503–509. [CrossRef]

130. Wulf, R.; Mendes, M.A.A.; Skibina, V.; Al-Zoubi, A.; Trimis, D.; Ray, S.; Gross, U. Experimental and numerical determination of effective thermal conductivity of open cell FeCrAl-alloy metal foams. *Int. J. Therm. Sci.* **2014**, *86*, 95–103. [CrossRef]

131. Cops, M.J.; McDaniel, J.G.; Magliula, V.; Bamford, D.J.; Bliefnick, J. Measurement and analysis of sound absorption by a composite foam. *Appl. Acoust.* **2020**, *160*, 107138. [CrossRef]

132. Byakova, A.V.; Gnyloskurenko, S.; Bezimyanniy, Y.; Nakamura, T. Closed-Cell Aluminum Foam of Improved Sound Absorption Ability: Manufacture and Properties. *Metals* **2014**, *3*, 445–454. [CrossRef]

133. Yang, F.; Shen, X.; Bai, P.; Zhang, X.; Li, Z.; Yin, Q. Optimization and Validation of Sound Absorption Performance of 10-Layer Gradient Compressed Porous Metal. *Metals* **2019**, *9*, 588. [CrossRef]

134. Vaidya, U.A.; Pillay, S.; Bartus, S.; Ulven, C.A.; Grow, D.T.; Mathew, B. Impact and post-impact vibration response of protective metal foam composite sandwich plates. *Mater. Sci. Eng. A* **2006**, *428*, 59–66. [CrossRef]

135. Pang, X.; Du, H. Dynamic characteristics of aluminium foams under impact crushing. *Compos. Part B* **2017**, *112*, 265–277. [CrossRef]

136. Liu, H.; Wei, J.; Qu, Z. Prediction of aerodynamic noise reduction by using open-cell metal foam. *J. Sound Vib.* **2012**, *331*, 1483–1497. [CrossRef]

137. Otaru, A.J.; Morvan, H.P.; Kennedy, A.R. Modelling and optimisation of sound absorption in replicated microcellular metals. *Scr. Mater.* **2018**, *150*, 152–155. [CrossRef]

138. Yang, X.; Peng, K.; Shen, X.; Zhang, X.; Bai, P.; Xu, P. Geometrical and Dimensional Optimization of Sound Absorbing Porous Copper with Cavity. *Mater. Des.* **2017**, *131*, 297–306. [CrossRef]

139. Jiejun, W.; Chenggong, L.; Dianbin, W.; Manchang, G. Damping and sound absorption properties of particle reinforced Al matrix composite foams. *Compos. Sci. Technol.* **2003**, *63*, 569–574. [CrossRef]

140. Carneiro, V.H.; Puga, H.; Meireles, J. Positive, zero and negative Poisson's ratio non-stochastic metallic cellular solids: Dependence between static and dynamic mechanical properties. *Compos. Struct.* **2019**, *226*, 111239. [CrossRef]

141. ASTM International. *ASTM E 1050 Standard. Standard Test Method for Impedance and Absorption of Acoustical Materials Using a Tube, Two Microphones and a Digital Frequency Analysis System*; ASTM International: West Conshohocken, PA, USA, 2019.

MDPI

St. Alban-Anlage 66

4052 Basel

Switzerland

Tel. +41 61 683 77 34

Fax +41 61 302 89 18

www.mdpi.com

Metals Editorial Office

E-mail: metals@mdpi.com

www.mdpi.com/journal/metals